Springer Tracts in Advanced Robotics

Volume 33

Editors: Bruno Siciliano · Oussama Khatib · Frans Groen

Xianwen Kong and Clément Gosselin

Type Synthesis of Parallel Mechanisms

Springer

Professor Bruno Siciliano, Dipartimento di Informatica e Sistemistica, Universitá di Napoli Federico II, Via Claudio 21, 80125 Napoli, Italy, E-mail: siciliano@unina.it

Professor Oussama Khatib, Robotics Laboratory, Department of Computer Science, Stanford University, Stanford, CA 94305-9010, USA, E-mail: khatib@cs.stanford.edu

Professor Frans Groen, Department of Computer Science, Universiteit vanAmsterdam, Kruislaan 403, 1098 SJ Amsterdam, The Netherlands, E-mail: groen@science.uva.nl

Authors

Xianwen Kong
Département de Génie Mécanique
Université Laval, Québec
Québec G1K 7P4
Canada

Clément Gosselin
Département de Génie Mécanique
Université Laval, Québec
Québec G1K 7P4
Canada

Library of Congress Control Number: 2007925210

ISSN print edition: 1610-7438
ISSN electronic edition: 1610-742X
ISBN-10 3-540-71989-X Springer Berlin Heidelberg New York
ISBN-13 978-3-540-71989-2 Springer Berlin Heidelberg New York

Springer is a part of Springer Science+Business Media
springer.com

Printed in Germany

Typesetting: Digital data supplied by editor.
Data-conversion and production: SPS, Chennai, India
Printed on acid-free paper SPIN: 11820680 89/SPS 5 4 3 2 1 0

STAR (Springer Tracts in Advanced Robotics) has been promoted under the auspices of EURON (European Robotics Research Network)

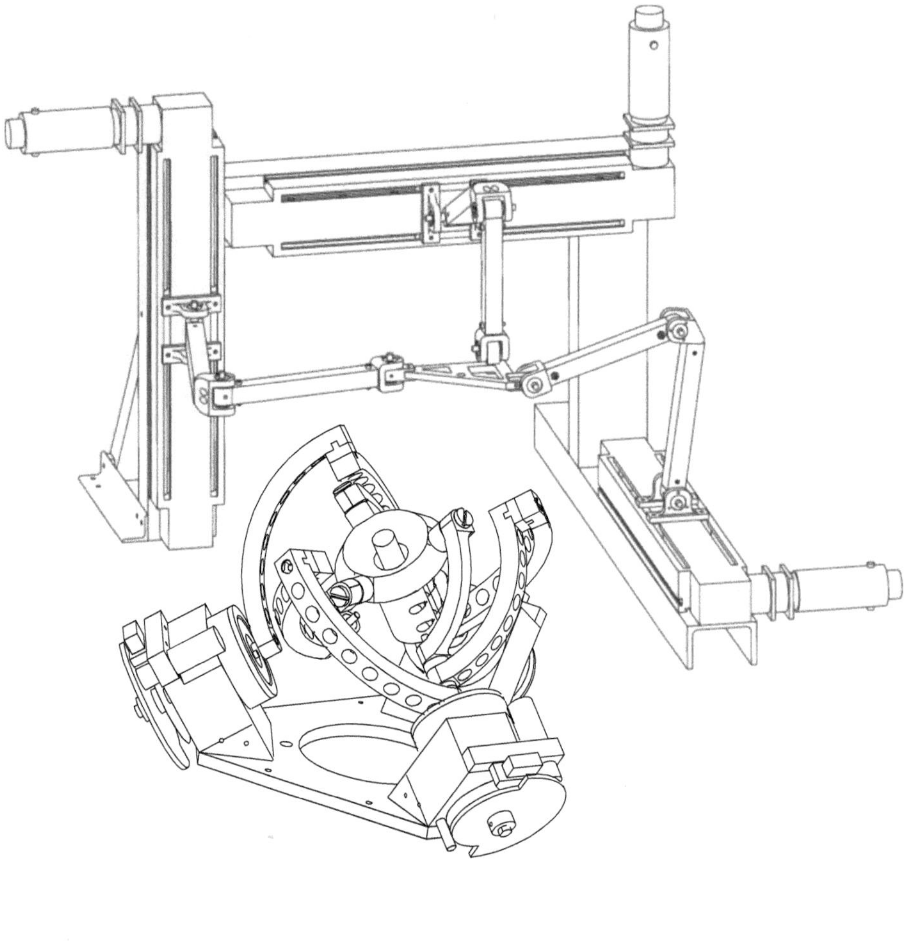

To Hao Ma, Qingmiao Kong and Mary Kong

To Annette Schwerdtfeger, Marie-Joëlle Gosselin and
Alexandrine Gosselin

Foreword

At the dawn of the new millennium, robotics is undergoing a major transformation in scope and dimension. From a largely dominant industrial focus, robotics is rapidly expanding into the challenges of unstructured environments. Interacting with, assisting, serving, and exploring with humans, the emerging robots will increasingly touch people and their lives.

The goal of the new series of Springer Tracts in Advanced Robotics (STAR) is to bring, in a timely fashion, the latest advances and developments in robotics on the basis of their significance and quality. It is our hope that the wider dissemination of research developments will stimulate more exchanges and collaborations among the research community and contribute to further advancement of this rapidly growing field.

The monograph written by Xianwen Kong and Clément Gosselin is the outcome of a decade of work accomplished by the authors on parallel mechanisms. Research in this area has a great potential for several applications, as more and more designs are adopting this type of kinematic arrangement, from motion simulators to haptic devices, from parallel manipulators to micro- and even nano-manipulators. The strength of the book is in the systematic approach used for the type synthesis of parallel mechanisms, which is then applied in a variety of case studies from three to five degrees of freedom. The result is a rigorous treatment of the subject matter, with several insights into creative design of mechanisms.

As the first focused STAR volume in the area of parallel mechanisms, this title constitutes a very fine addition to the series!

Naples, Italy
February 2007

Bruno Siciliano
STAR Editor

Preface

学而不思则罔，思而不学则殆。
——孔子(公元前 551-479)

Learning without thought is labor lost; thought without learning is perilous.
— Confucius (551–479 BC)

Parallel mechanisms (PMs) have been and are being used in a wide variety of applications such as motion simulators and parallel manipulators and even nano-manipulators and micro-manipulators. From the well-known Gough-Stewart platform to the Delta robot to the Agile eye and many other designs, PMs have been largely synthesized using intuition and ingenuity. As opposed to serial kinematic chains, in which the number of kinematic arrangements (types) is somewhat limited, PMs can lead to a very large number of kinematic arrangements for a given motion pattern. Therefore, a systematic approach is needed in order to reveal all types of PMs thereby allowing the development of the most promising designs. This fundamental issue, namely type synthesis, is the focus of this book.

This book is a summary and an extension of the work accomplished by the authors on the type synthesis of PMs over the last decade.[1] It includes two parts namely, a first part (Chaps. 2–5) in which the virtual-chain approach for the type synthesis of PMs is presented systematically and a second part (Chaps. 6–14) that presents the application of the method to a variety of motion patterns. Motion patterns have been chosen for their relevance to robotic applications and include, among others, Cartesian translations and SCARA motions. Additional concepts are also discussed in appendices, namely, the application of the method presented in the book to practical design problems, the determination of the DOF of PMs and the type synthesis of PMs using the displacement group theory.

[1] More details on publications and prototypes developed by the authors can be found on the web site of the Laval University Robotics Laboratory (www.robot.gmc.ulaval.ca).

While this book is primarily intended for researchers and developers working on parallel manipulators, parallel kinematic machines and haptic devices, we hope that it will also be of great interest to a broader class of readers: (a) graduate students and senior undergraduates working in the above areas since the methods proposed are mainly based on linear algebra and basic skills in kinematics, which they are familiar with; (b) researchers and graduate students in the area of nanotechnology and microelectromechanical systems (MEMS) since this book provides a solid starting point for the design of nano-manipulators and micro-manipulators; (c) researchers and graduate students in the area of creative mechanism design since this book is also an illustrative example of the creative design of mechanisms; and (d) researchers in screw theory since this book acts as a successful application of screw theory.

This book would not have been possible without the help and involvement of many people. In particular we would like to thank Mr. Pierre-Luc Richard and Mr. Mathieu Goulet for creating many CAD models, Thierry Laliberté for building several plastic models, and Simon Foucault for designing the Tripteron, which is one of the practical applications of the results of this book. Thanks also go to several former members of the Laval University Robotics Laboratory, especially Mr. Jonathan Levesque, for most of the figures of this book. The help of Mr. Boris Mayer-St-Onge in using LaTeX is also acknowledged. Last but not least, we would also like to acknowledge the financial support of the Natural Sciences and Engineering Research Council of Canada (NSERC) and of the Canada Research Chairs Program.

Quebec City, Canada
February 2007

Xianwen Kong
Clément Gosselin

Contents

Part I Synthesis Approach

List of Figures

List of Tables

Part I

Synthesis Approach

工欲善其事，必先利其器。

——孔子(公元前 551-479)

He that would perfect his work must first sharpen his tools.

— Confucius (551–479 BC)

1. Introduction

Over the last two decades, parallel mechanisms (PMs) evolved from rather marginal machines to widely used mechanical architectures. Current application of PMs include motion simulators, industrial robots, nano-manipulators and micro-manipulators, to name only a few. However, the existing architectures of PMs have been largely synthesized using intuition and ingenuity. As opposed to serial kinematic chains (KCs), in which the number of kinematic arrangements (types) is somewhat limited, PMs can lead to a very large number of kinematic arrangements for a given motion pattern. Therefore, a systematic approach is needed in order to determine all types of PMs thereby allowing the development of the most promising designs. This fundamental issue, namely type synthesis, is the focus of this book. In this chapter, the background of the type synthesis of PMs is presented. The state of the art of the research is also reviewed. Finally, the outline of the book is proposed.

1.1 Parallel Mechanisms and Their Applications

1.1.1 Parallel Mechanism

Although several definitions of PMs have been proposed (see for instance [99]), it is important, in a context of type synthesis, to adopt a definition that will formally circumscribe the types of KCs to be contemplated.

The following definition is used throughout this book: A *parallel mechanism* is a multi-degree-of-freedom (multi-DOF) mechanism composed of one moving platform and one base connected by at least two serial KCs in parallel (Fig. 1.1). These serial KCs are called *legs* (or *limbs*). PMs in which the legs include closed KCs are out of the scope of this book. Finally, it should be noted that in a PM, the actuated joints are usually located on or close to the base.

1.1.2 Characteristics and Applications

In comparison with serial mechanisms, properly designed PMs generally have higher stiffness and higher accuracy, although their workspace is usually smaller.

X. Kong and C. Gosselin: Type Syn. of Parallel Mech., STAR 33, pp. 5–17, 2007.
springerlink.com

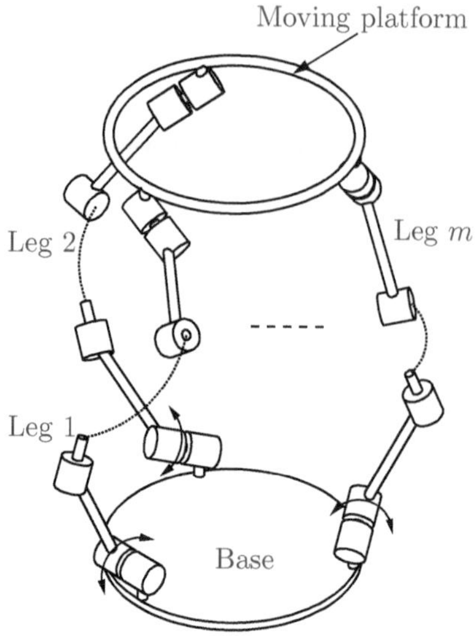

Fig. 1.1. Schematic representation of a PM

The variety of applications in which PMs are used is constantly expanding [5, 6, 7, 11, 98, 109, 117]. For an updated comprehensive list of PMs, see [8, 99]. The first application of a six-legged PM dates back to the 1950's when a tire testing machine based on a PM was developed by Gough (Fig. 1.2a[109]). In the 1970's, flight simulators [Fig. 1.2b] based on PMs were put into practice. Since the 1980's, the research on parallel manipulators has attracted the interest of many researchers and is still the focus of several important research projects. Parallel manipulators alone also cover a wide range of applications in assembly, inspection and others. Some parallel manipulators, such as the Gough-Stewart platform and the Delta robot [Fig. 1.2c], have become state of the art in the commercial world. During the past decade, PMs have also been used in machine tools, also referred to as parallel kinematic machines [Figs. 1.2d–1.2e], camera orienting devices (Fig. 1.2f[38]), haptic devices (Fig. 1.2g[5]), medical robots (Fig. 1.2h[117]), alignment devices (Fig. 1.2i[117]), coordinate measuring machines and force sensors.

In order to meet the needs of the development of nanotechnology and microsystem or microelectromechanical systems (MEMS), nano-manipulators [87, 93, 103, 120, 128] and micro-manipulators [60] were recently developed based on PMs. Figure 1.3a shows a non-magnetic Hexapod developed by PI (Physik Instrumente). Figure 1.3b shows a parallel nano-manipulator [103] developed by Prof. I-Ming Chen's group from Nanyang Technological University. This parallel manipulator is based on flexural joints. In addition to its high accuracy, this parallel manipulator can be selectively actuated and thus is easy to control. Figure 1.3c shows a parallel micro-manipulator [60] developed by Prof. Larry L. Howell's group from Brigham Young University. This micro-manipulator can be used for positioning microcomponents such as mirrors, lenses and gratings in

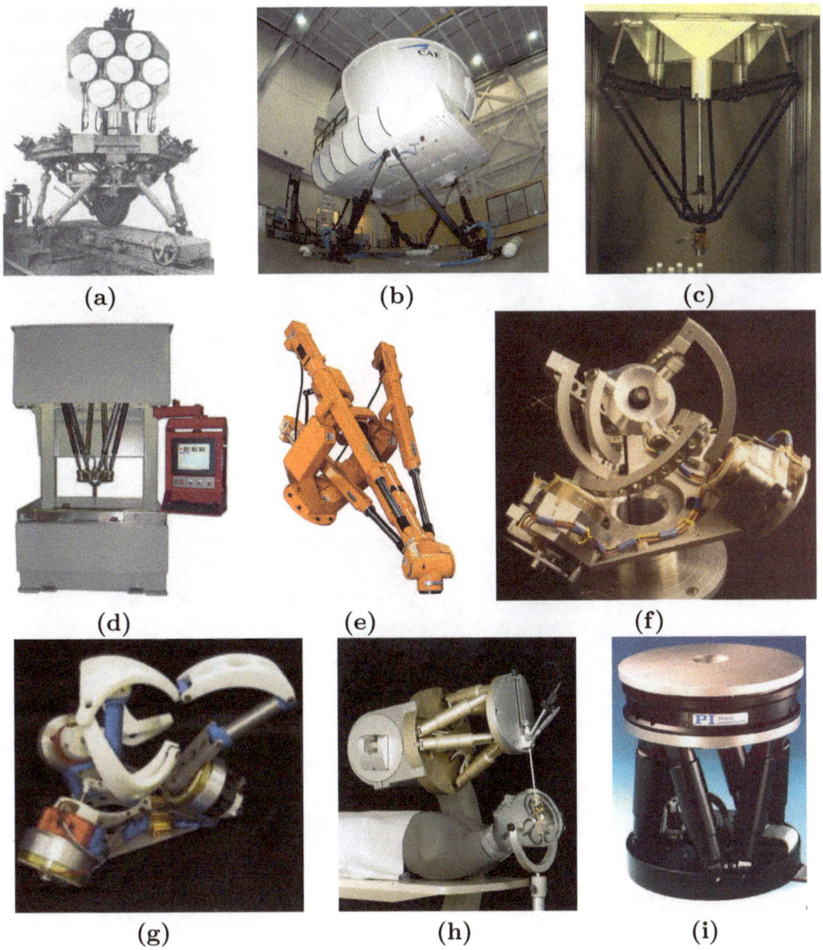

Fig. 1.2. Applications of PMs: (**a**) Gough's original tire testing machine (Courtesy of Proceedings of the IMechE), (**b**) CAE full-flight simulator (Courtesy of CAE), (**c**) Delta robot (Courtesy of Demaurex SA), (**d**) P 3000 Hexapod parallel kinematic machine (Courtesy of Parallel Robotic Systems Corporation), (**e**) Tricept Robot (Courtesy of ABB Ltd), (**f**) Laval University Agile Eye, (**g**) Laval University haptic device, (**h**) Medical robot (Courtesy of PI (Physik Instrumente)) and (**i**) Alignment device (Courtesy of PI)

three directions by three independent linear inputs. The design of this parallel micro-manipulator is based on a translational PM proposed in [115].

1.2 Type Synthesis of Parallel Mechanisms

For a given PM in which the moving platform has $\mathcal{F}$ DOF, the characteristics of the motion of the moving platform may be of several types. For example, a 3-DOF motion may be a 3-DOF translational motion, a 3-DOF spherical motion,

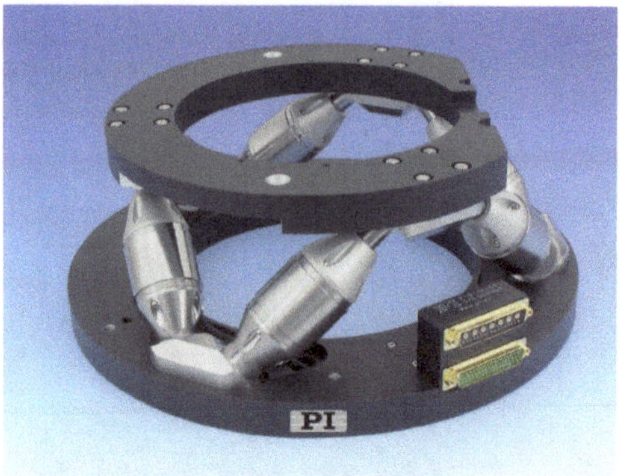

(a) Parallel nano-manipulator using piezo linear drives (Courtesy of PI (Physik Instrumente), www.pi.ws).

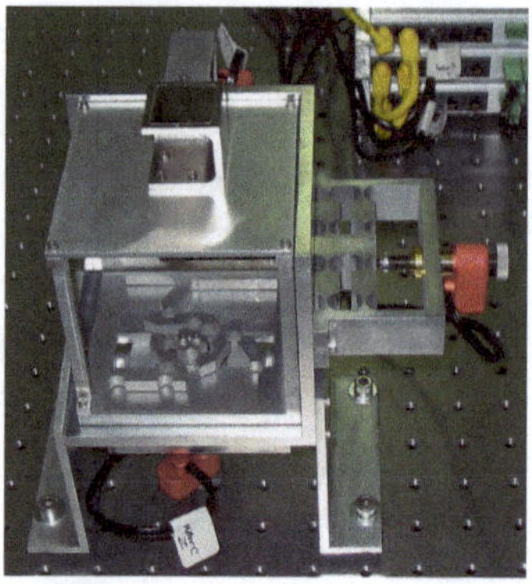

(b) Parallel nano-manipulator (Courtesy of Dr. Huy Hoang Pham and Prof. I-Ming Chen from Nanyang Technological University).

(c) Parallel micro-manipulator (Courtesy of Prof. Larry L. Howell from Brigham Young University).

Fig. 1.3. Parallel nano-manipulators and parallel micro-manipulator

a 3-DOF planar motion or any other 3-DOF motion. Each type of motion of the moving platform is called a *motion pattern*. A detailed classification of motion patterns is given in Chap. 4. Clearly, the specification of the DOF of a mechanism is not sufficient to determine its motion pattern.

The *type* synthesis of PMs consists in finding all the possible types of PMs generating a specified motion pattern of the moving platform. In most of the previous works, the type synthesis of PMs was defined as finding all the possible types of PMs generating a motion of the moving platform with a specified DOF. The former definition is used by default in this book, for it is much more useful than the latter in practice. Indeed, the selected motion pattern is a natural starting point in the type synthesis.

Type synthesis is a fundamental and important issue in the study of PMs. It is also the first logical step in the development of new motion simulators, parallel manipulators and other devices based on PMs. The reader is referred to Appendix A for the overall design process of devices based on PMs.

1.3 Representation of Parallel Mechanisms

A PM usually needs to satisfy certain geometric conditions within the same leg and/or among different legs. In a PM, the legs that should satisfy certain

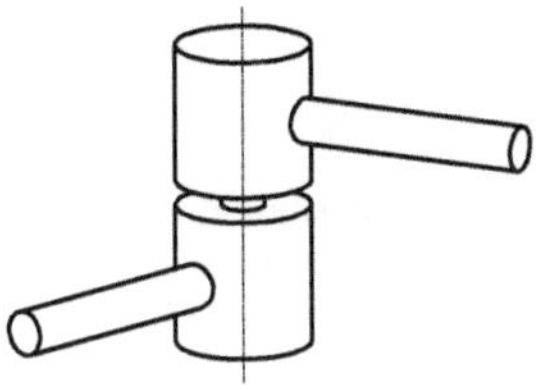

(a) R joint.

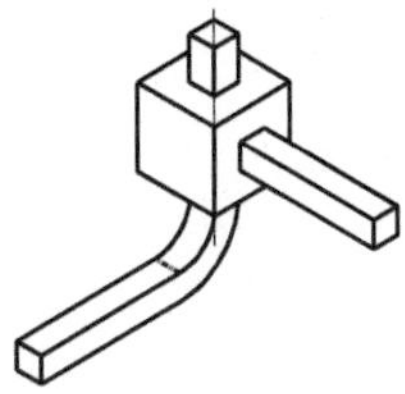

(b) P joint.

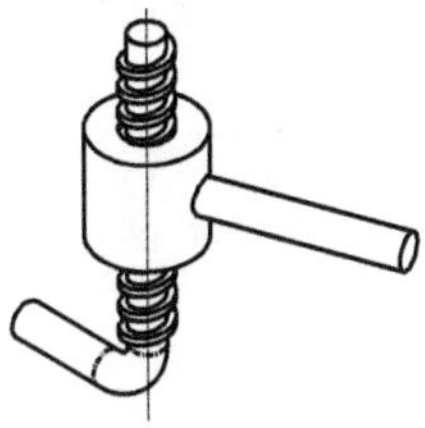

(c) H joint.

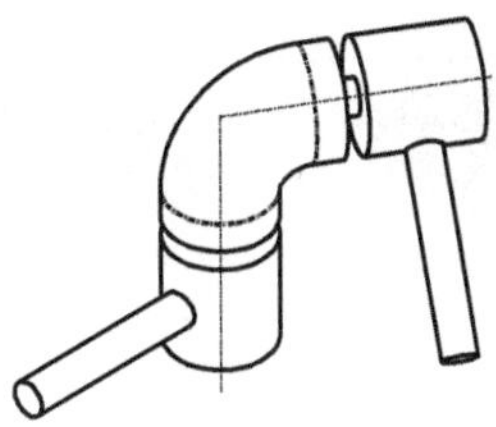

(d) U joint.

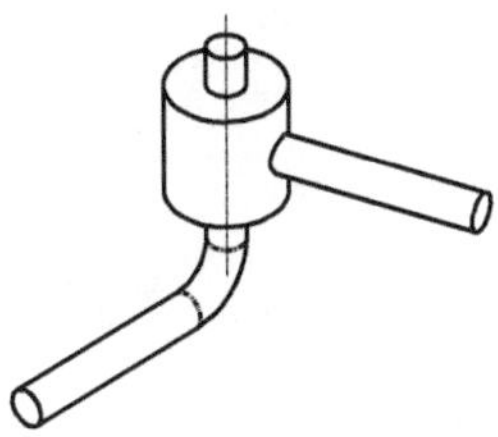

(e) C joint.

(f) S joint.

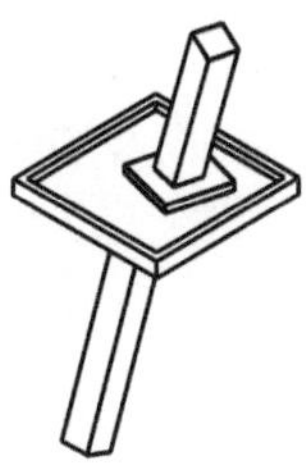

(g) E joint.

Fig. 1.4. Kinematic joints

Table 1.1. Representation of joints

C	Cylindrical joint
H	Helical joint
P	Prismatic joint
R	Revolute joint
Ŕ	R joints with parallel axes within a same leg
R̀	R joints with parallel axes within a same leg
Ṙ	R joints with intersecting axes within a same leg
R̂	R joints with concurrent axes within a leg-group
Ř	R joints with concurrent axes within a leg-group
R̈	R joints with coaxial axes, passing through the intersections of the axes of the Ṙ joints if any, within a leg-group
R̋	R joints with parallel axes, being parallel to the axes of the R̈ joints or a line passing through at least two intersections of the axes of the Ṙ joints, if any, within a leg-group
R̄	R joints with parallel axes within a leg-group
S	Spherical joint
U	Universal joint
$()_A$	Successive joints that are arranged such that the axes of all the R joints are parallel and the direction of any P joint, if any is not perpendicular to the axes of the R joints.
$()_E$	Successive joints within a leg that are arranged such that all the links move along parallel planes
$()_E^{\vert}$	Each of $()_E$ within a leg-group for which all the associated planes of motion are parallel to the same line
$()_E^{\Vert}$	Each of $()_E$ within a leg-group for which all the associated planes of motion are parallel
$()_L$	At least one coaxial R joint or at least one codirectional P joint
$()_S$	Successive joints that are arranged such that all the links move along concurrent spherical surfaces
$()_T$	Successive P joints within a leg that are arranged such that all the links move along parallel planes

geometric conditions can be divided into one (Chaps. 6–13) or more leg-groups (Chap. 14). A *leg-group* is defined as the maximum set of legs in which all the legs must satisfy certain geometric conditions both within the same leg and among all the legs within the set.

1.3.1 Representation of Joints

Kinematic joints used in PMs include (Fig. 1.4): a) revolute (R) joints, b) prismatic (P) joints, c) helical (H) joints, d) universal (U) joints, e) cylindrical (C) joints, f) spherical (S) joints and g) planar (E) joints.

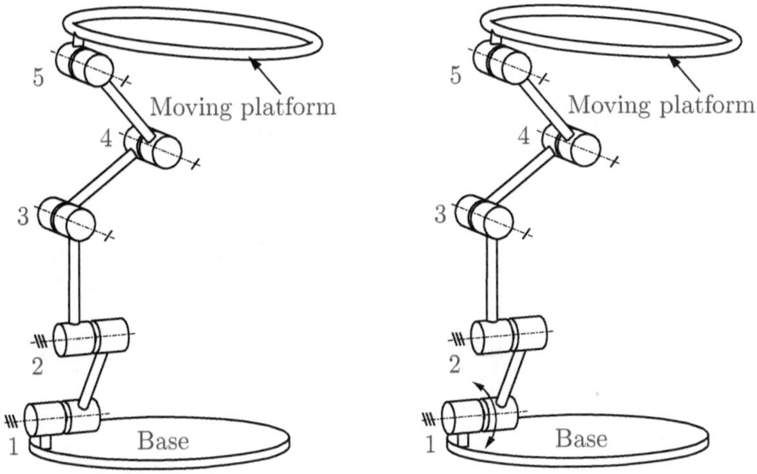

(a) $\acute{\mathrm{R}}\acute{\mathrm{R}}\grave{\mathrm{R}}\grave{\mathrm{R}}\grave{\mathrm{R}}$ leg. (b) $\underline{\acute{\mathrm{R}}}\acute{\mathrm{R}}\grave{\mathrm{R}}\grave{\mathrm{R}}\grave{\mathrm{R}}$ leg.

Fig. 1.5. Representation of legs

Since the U and S joints can be respectively regarded as a combination of two R joints with intersecting and orthogonal axes and a combination of three R joints with concurrent axes, in the type synthesis of PMs, we consider mainly PMs that are composed of R and P joints.

In order to obtain a compact representation of the geometric relations between the axes of the joints in a KC, the notations listed in Table 1.1 are introduced. In addition, $\underline{\mathrm{R}}$ and $\underline{\mathrm{P}}$ are used to represent the actuated R and P joints.

1.3.2 Representation of Legs

A *leg* of PM is represented by a chain of characters representing the types of joints starting from the base to the moving platform (Fig. 1.5). For example, the $\acute{\mathrm{R}}\acute{\mathrm{R}}\grave{\mathrm{R}}\grave{\mathrm{R}}\grave{\mathrm{R}}$ leg [Fig. 1.5a] is composed of five R joints. The axes of the first two R joints are parallel to each other, and the axes of the last three R joints are also parallel. Similarly, the $\underline{\acute{\mathrm{R}}}\acute{\mathrm{R}}\grave{\mathrm{R}}\grave{\mathrm{R}}\grave{\mathrm{R}}$ leg [Fig. 1.5b] is composed of five R joints, the first of which is actuated. The axes of the first two R joints are parallel to each other, and the axes of the last three R joints are also parallel.

1.3.3 Representation of PMs

A parallel kinematic chain (PKC) or PM is represented by the types of its legs connected by “-” and “_” . The types of legs within the same leg-group are successively connected by “-”. The types of legs of different leg-groups are

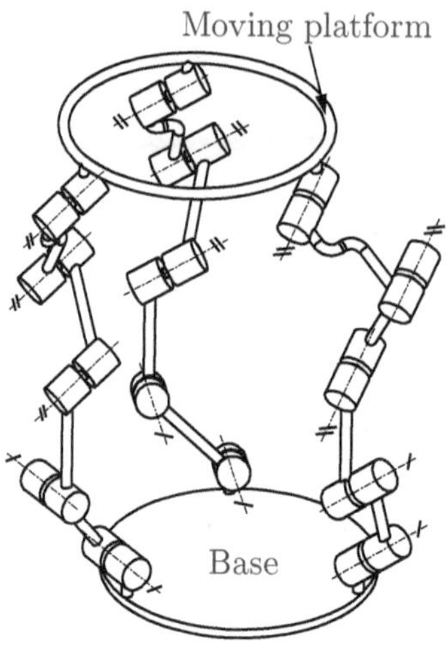

(a) 3-ŔŔR̀R̀R̀ PKC.

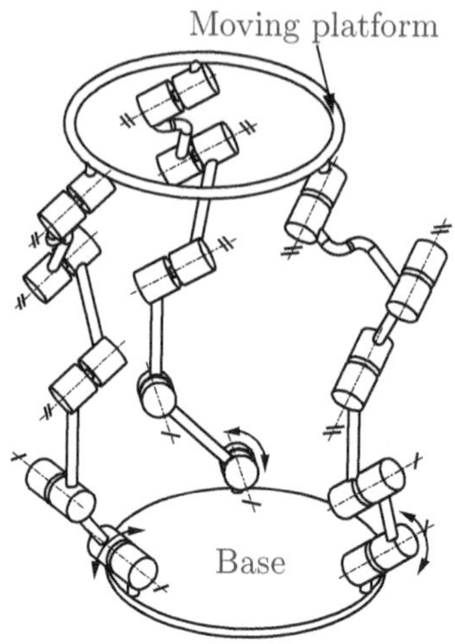

(b) 3-Ŕ̲ŔR̀R̀R̀ PM.

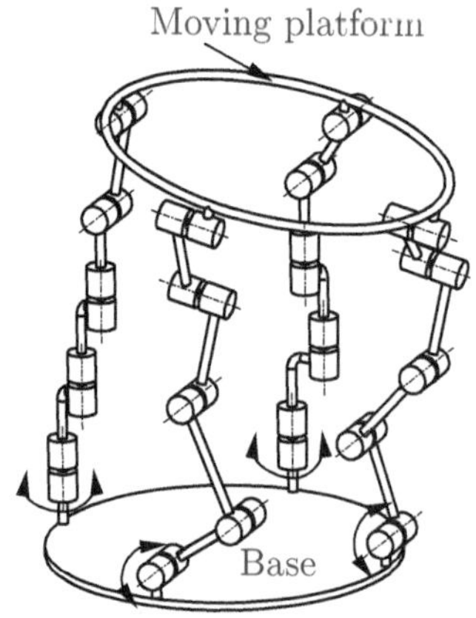

(c) 2-R̲̋R̋R̋R̄R̄_2-R̲̋R̋R̋R̄R̄ PM.

Fig. 1.6. Representation of PKCs and PMs

connected by "_". Legs that are not subject to any geometric constraint are also connected successively to one another by "-" and connected to a leg within a leg-group, if any, by "-". When l legs within the same leg-group are of the same type, the l legs are represented by the type of the legs preceded by "l-" (Fig. 1.6).

For example, the 3-ŔŔR̀R̀R̀ PKC [Fig. 1.6a] is composed of three ŔŔR̀R̀R̀ legs, while the 3-Ŕ̲ŔR̀R̀R̀ PM [Fig. 1.6b] is composed of three Ŕ̲ŔR̀R̀R̀ legs. Figure 1.6c shows a 2-R̲̋R̋R̋R̄R̄_2-R̲̋R̋R̋R̄R̄ PM, which is composed of two leg-groups. Each leg-group is composed of two R̲̋R̋R̋R̄R̄ legs. Within each leg-group, the axes of all the R̋ joints are parallel, and the axes of all the R̄ are also parallel.

1.4 The State of the Art

PMs are a class of multi-loop spatial mechanisms. In the review of the type synthesis of PMs, the literature on the type synthesis of multi-loop spatial mechanisms and the previous work on the type synthesis of PMs generating a motion of the moving platform having a specified DOF should both be taken into consideration. The next subsections will cover each of these topics in order to clearly establish the context in which the type synthesis of PMs should be placed. Furthermore, it should be specified, from the outset, that the type synthesis of $\mathcal{F}$-DOF mechanisms can be essentially divided into two stages. The first is to perform the type synthesis of $\mathcal{F}$-DOF KCs and the second is to select $\mathcal{F}$ actuated joints in an $\mathcal{F}$-DOF KC to obtain $\mathcal{F}$-DOF mechanisms.

1.4.1 Type Synthesis of Multi-loop Spatial Mechanisms

The type synthesis of multi-loop spatial mechanisms deals with the generation of all types (architectures) of multi-loop spatial mechanisms for a specified DOF. The type synthesis of multi-loop spatial mechanisms began in the 1960's and was perhaps the least explored area of research in the science of mechanisms for a few decades [30]. However, since the beginning of 1990's, some progress has been made on this topic. The type synthesis of multi-loop spatial mechanisms is usually based on the mobility criterion of a mechanism which takes one of the following forms [29, 59, 114, 122], namely:

$$\mathcal{F} = d(n - g - 1) + \sum_{j=1}^{g} f_j \tag{1.1}$$

where $\mathcal{F}$ is the mobility or relative DOF of a KC, n is the number of links including the base, g is the number of joints, f_j is the freedom of the j-th joint, d is the number of independent constraint equations within a loop, or

$$f = \sum_{j=1}^{g} f_j - dv \tag{1.2}$$

where v is the number of independent loops in the mechanism, or

$$f = \sum_{j=1}^{g} f_j - \min \sum_{i=1}^{v} d_i \tag{1.3}$$

where d_i is the number of independent constraint equations within loop i, $\sum_{i=1}^{v} d_i$ is the sum of d_i in a set of v independent loops, $min\sum_{i=1}^{v} d_i$ is the minimum of $\sum_{i=1}^{v} d_i$ of all the sets of v independent loops.

Equation (1.1) is usually referred to as the Chebychev-Grübler-Kutzbach criterion or the general mobility criterion.

The type synthesis of multi-loop spatial mechanisms in which all the loops have the same number of independent constraint equations has been dealt with by several authors (such as [125]) based on the mobility criterion [(1.2)]. In 1994, the type synthesis of spatial mechanisms involving R and P joints in which all the loops have 6 independent constraint equations was dealt with in [68, 85]. Spatial KCs with inactive joints due to P joints were identified and discarded. In 1998, the type synthesis of spatial mechanisms in which not all the loops have the same number of independent constraint equations [(1.3)] was dealt with in [96].

As mentioned above, the second step in the type synthesis of PMs is the selection of the actuated joints. However, the selection of actuated joints has been overlooked for a long time. One reason is that most of the work on the type synthesis of spatial linkages focusses on 1-DOF KCs. For a 1-DOF KC, any one of the joints can be actuated. There is no invalid actuated joint appearing in 1-DOF mechanisms. The other reason is that the validity condition of actuated joints has been proposed (see for example [122]) and stated in the following fashion:

"For an $\mathcal{F}$-DOF mechanism, a set of $\mathcal{F}$ actuated joints is valid if the DOF of the KC obtained from the mechanism by blocking all the actuated joints is 0. "

However, in the selection of actuated joints using the above condition, the calculation of the DOF encountered is in fact very difficult. As pointed out in [78], the sets of actuated joints for two of the PMs proposed in [126] do not work.

In 1999, a validity condition of actuated joints was proposed in [67] for spatial mechanisms involving R and P joints in which each loop has six independent constraint equations.

At present, both the type synthesis of spatial KCs and the selection of actuated joints of spatial KCs are not yet fully solved.

1.4.2 Type Synthesis of PMs with a Specified Number of DOF

While most published works proposed new PMs on a case by case basis, a few [29, 47, 59] presented systematic approaches for the type synthesis of PMs with a specified number of DOF.

The type synthesis of PMs with a specified number of DOF is performed based on the mobility criterion [(1.1) and (1.3)]. In [59], the type synthesis of PMs was solved for d=2, 3, 4 and 6. Some PMs generating 2-DOF planar translation, 3-DOF spatial translation, planar motion, spherical motion and 3T1R (three DOF spatial translation and one DOF rotation) motion were obtained. In [29, 114], the type synthesis of PMs for d=6 are dealt with. This approach is most appropriate for the type synthesis of PMs with full-DOF (six for spatial PM and three for spherical and planar PMs). PMs shown in Fig. 1.2 were indeed obtained using this approach. However, PMs that do not satisfy the general mobility criterion (see Fig. 1.7 for example) could not be obtained using this approach. The PMs shown in Figs. 1.7a[39, 71, 72] and 1.7b[78] belong to the classes of PMs to be dealt with in Chaps. 6 and 9 respectively.

1.4.3 Type Synthesis of PMs with a Specified Motion Pattern

Due to the large variety of applications of PMs, the motion patterns of the moving platform required vary to a great extent. There is still a great need

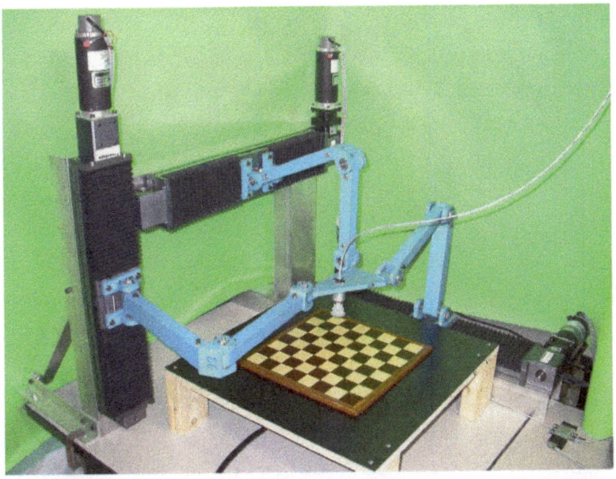

(a) Tripteron robot.

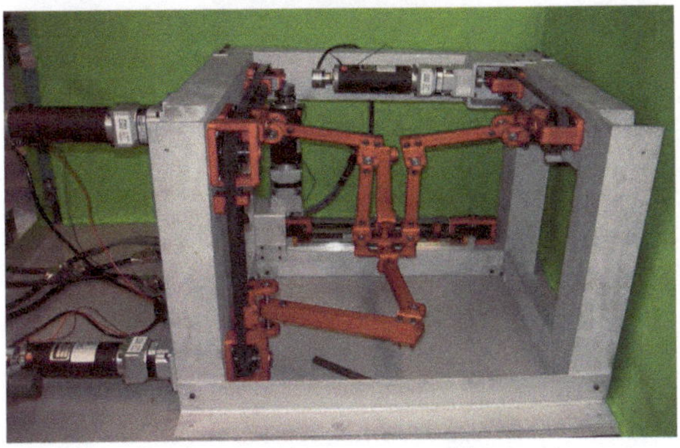

(b) Quadrupteron robot.

Fig. 1.7. Two prototypes of PMs developed at Laval University

to find new PMs [10, 100] generating desired motion patterns. Identifying new types of PMs also facilitates the development of hybrid kinematic machine tools in which two PMs are used cooperatively.

Since the motion pattern of the moving platform contains more information than the mobility of the moving platform and because of the need for PMs with less than 6-DOF for a broad variety of applications, the type synthesis of PMs (with a specified motion pattern) has received much attention since the early 1990s [3, 13, 33, 42, 45, 47, 55, 62, 70, 75, 126, 134]. It may be argued that 6-DOF PMs could be used in all applications and the need to develop PMs with fewer than 6 DOFs may be questioned. One of the main reasons for using PMs with fewer than 6-DOFs is the reduction of the cost. Another reason is that, in general, reducing the number of DOF will increase the maximum motion range of the remaining DOF.

It is worth noticing that the type synthesis of PKCs with a specified motion pattern was addressed by Hunt as early as 1973 [58]. In [58], PKCs are used as constant-velocity couplings. Unfortunately, this work remained virtually unknown to the robotics research community for a long time, since Hunt himself did not mention his pioneer work on the type synthesis of PMs in [59]. Several authors worked independently on the type synthesis of translational PKCs with 5-DOF legs [13, 33, 70] and the same results were re-obtained. The contribution of [13, 33, 70] lies in that the full-cycle mobility conditions, which are given in [58] without detailed explanations, were derived algebraically. It was thus implicitly proved that there are no 5-DOF legs composed of R and P joints that can be used to build translational PKCs, except those identified in [58].

The approaches to the type synthesis of PKCs generating a specified motion pattern can be divided into four classes: (1) The approach based on screw theory (see for instance [13, 33, 55, 58, 70]), (2) The approach based on displacement group theory (see for example [3, 42, 45, 47, 64]), (3) The approach based on single-opened-chain units (see for instance [62, 126]), and (4) The virtual-chain approach (see for example [75, 76, 78, 80]).

Each approach has its own characteristics. One way to assess the merits of each of the different approaches is to consider the type synthesis of translational PMs, which was dealt with using each of the four classes of approaches mentioned above. In [80], the virtual-chain approach — which is based on the concept of virtual chain and screw theory — is used to synthesize translational PMs and is compared to the other approaches found in the literature. It is observed that:

- The virtual-chain approach requires fewer derivations than the approaches proposed in [13, 33, 65, 79].
- The virtual-chain approach is conceptually simpler and therefore easier to understand than the method proposed in [58].
- The virtual-chain approach is more general than the approaches presented in [47, 121]. When the latter approaches are used for the type synthesis, some 3-DOF and 4-DOF PMs with 5-DOF legs may not be obtained.

In most of the literature on the type synthesis of PMs, very little work has been devoted to the selection of actuated joints for PMs. By contrast, in the virtual-chain approach, the selection of actuated joints is considered an important step of the type synthesis of PMs. The scarcity of the literature on the selection of actuated joints for PMs may be due to the fact that for most of the proposed $\mathcal{F}$-DOF PMs, any set of $\mathcal{F}$ joints can be actuated. In [1], a validity condition of actuated joints was proposed based on screw theory and the selection of actuated joints for a 2-DOF planar PM was discussed in detail. In [56], an alternative validity condition of actuated joints was proposed based on screw theory, and the selection of actuated joints for a 3-DOF PM and a 4-DOF PM was discussed in detail. For a PM, the validity detection of the actuated joints requires the calculation of a 6×6 determinant when the above approaches are applied. In fact, for most $\mathcal{F}$-DOF PMs, the validity detection of the actuated joints requires the calculation of an $\mathcal{F} \times \mathcal{F}$ determinant [75, 76, 78, 80].

It is also noted that the inactive joints in a PKC make no contribution to the movement of the moving platform. This may explain why PKCs with inactive joints have been discarded in previous work on the type synthesis of PMs [33,

42, 70]. In this book, PKCs with inactive joints are considered valid and are kept since for a PKC with inactive joints and its kinematic equivalent PKC without inactive joints, the number of overconstraints as well as the reaction forces in the joints are different.

1.5 Objective and Organization of This Book

As a major objective, this book aims at presenting systematically the virtual-chain approach to the type synthesis of PMs. In addition, the type synthesis of PMs generating the most commonly used motion patterns is performed. Moreover, as a by-product of the work on the type synthesis, we also present a novel approach for the mobility analysis of PMs. This is also a classical but still not well solved basic problem in the theory of mechanisms.

This book includes two main parts. Part 1, including Chaps. 2–5, presents systematically the virtual-chain approach for the type synthesis of PMs. Chap. 2 deals with the structural analysis of PMs based on screw theory. The type synthesis of overconstrained single-loop KCs is presented in Chap. 3. The classification of PMs is discussed in Chap. 4 where the concept of virtual-chain is introduced to represent the motion pattern of the moving platform. In Chap. 5, a procedure is proposed for the type synthesis of PMs based on the concept of virtual-chain and on screw theory. Part 2, including nine chapters, is the application of the virtual-chain approach to the type synthesis of PMs. It deals with the type synthesis of PMs generating specified motion patterns that are deemed to have the broadest application potential. Examples of PMs are those generating the same motion patterns as the popular Cartesian serial robots and the SCARA serial robots. In the last chapter, conclusions are drawn and future work is suggested. In addition, three appendices are included. In Appendix A, we illustrate how to apply the results presented in this book to the design of innovative parallel manipulators, haptic devices, medical robots, nano-manipulators and micro-manipulators. In Appendix B, we present a novel approach for the mobility analysis of PMs in order to help the reader understand the principle of a PM that is not included in this book. This proposed procedure for the mobility analysis can be thought of as the reverse procedure of the type synthesis based on the virtual-chain approach. In Appendix C, we present briefly the method based on the displacement group theory for the type synthesis of PMs. The inclusion of this appendix is justified mainly by the following reasons: (a) this method is efficient at generating PMs having simple kinematic characteristics [43, 81]; and (b) this method needs to be systematically developed.

2. Structural Analysis

In the type synthesis of PMs, one needs to deal with the constraints induced by kinematic joints and KCs. Screw theory provides an efficient tool to address this issue and it will therefore be used throughout this book. In this chapter, a review of important results from screw theory — such as the principle of reciprocity of screws — is first given. Subsequently, an instantaneous mobility criterion for PMs is proposed that is different from existing ones, and which will facilitate the type synthesis of PKCs. Finally, a validity condition is proposed for the selection of the actuated joints of a PKC that allow the synthesis of valid PMs.

2.1 Screw Theory

In this section, some relevant results of screw theory[57, 115] are given in order to allow a better understanding of this book.

2.1.1 Screws

A (normalized) *screw* is defined by (See Fig. 2.1)

$$\$ = \begin{bmatrix} \$_F \\ \$_S \end{bmatrix} = \begin{cases} \begin{bmatrix} \mathbf{s} \\ \mathbf{s} \times \mathbf{r} + h\mathbf{s} \end{bmatrix} & \text{if } h \text{ is finite} \\ \begin{bmatrix} \mathbf{0} \\ \mathbf{s} \end{bmatrix} & \text{if } h \to \infty \end{cases} \tag{2.1}$$

where $\mathbf{s}$ is a unit vector along the axis of the screw $\$$, $\mathbf{r}$ is a vector directed from any point on the axis of the screw to the origin of the reference frame O-XYZ, and h is called the *pitch*. It is noted that there are two vector components or six scalar components in the above presentation of the screw.

For convenience, $\$_0$ and $\$_\infty$ are used to represent a screw of 0-pitch and a screw of ∞-pitch respectively. In diagrams, the arrowheads shown in Fig. 2.2 are used to differentiate these types of screws.

X. Kong and C. Gosselin: Type Syn. of Parallel Mech., STAR 33, pp. 19–42, 2007.
springerlink.com

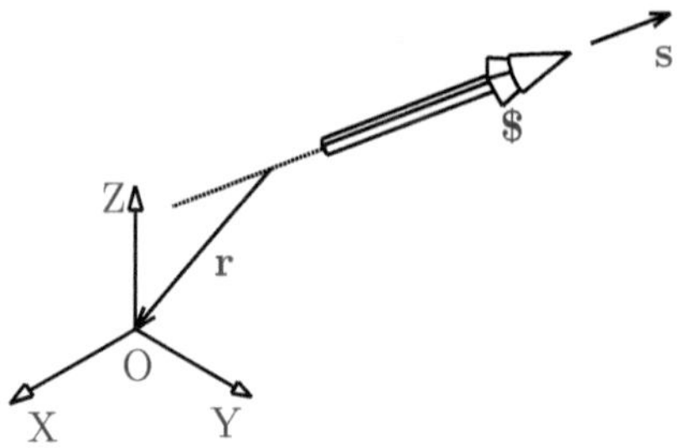

Fig. 2.1. A screw

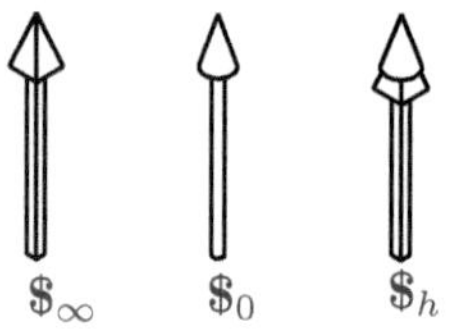

Fig. 2.2. Representation of screws

2.1.2 Screw Systems

A *screw system* of *order* n ($0 \leq n \leq 6$) comprises all the screws that are linearly dependent on n given linearly independent screws. Examples of screw systems of order 1, 2, 3 and 4 are shown in Figs. 2.3–2.6, respectively. A screw system of order n is also called an n-system. Any set of n linearly independent screws within an n-system forms a *basis* of the n-system. Usually, the basis of an n-system can be chosen in different ways. Figure 2.7 shows two bases of a 1-$\$_\infty$-1-$\$_0$ system: (a) $\$_{\infty 1}$ and $\$_{02}$ and (b) $\$_{01}$ and $\$_{02}$.

There are many types of screw systems, see for instance [57]. Here, only the screw systems that are the most relevant for the analysis and synthesis of PKCs will be given. These screw systems are illustrated in Figs. 2.3–2.6. They are described in more detail in the following in order to facilitate the understanding of the upcoming sections and chapters.

- 1-systems
  - 1-$\$_\infty$-system: A 1-$\$_\infty$-system is composed of all the $\$_\infty$ along a same direction [Fig. 2.1.2].
  - 1-$\$_0$-system: A 1-$\$_0$-system is composed of all the $\$_0$ along a same line [Fig. 2.1.2].
- 2-systems
  - 2-$\$_\infty$-system: A 2-$\$_\infty$-system is composed of all the $\$_\infty$ whose directions are parallel to a same plane [Fig. 2.4a].
  - 1-$\$_\infty$-1-$\$_0$-system: A 1-$\$_\infty$-1-$\$_0$-system is composed of all the $\$_0$ whose axes are coplanar and parallel as well as the $\$_\infty$ whose direction is perpendicular to the axes of all the $\$_0$ [Fig. 2.4b].

(a) 1-$\$_{\infty}$-system. (b) 1-$\$_0$-system.

Fig. 2.3. Some 1-systems

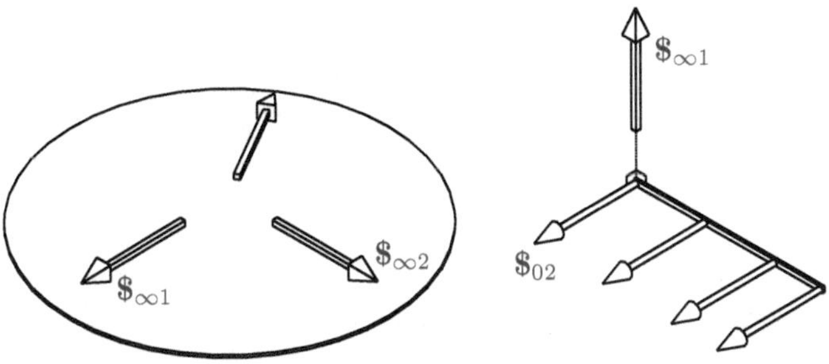

(a) 2-$\$_{\infty}$-system. **(b)** 1-$\$_{\infty}$-1-$\$_0$-system.

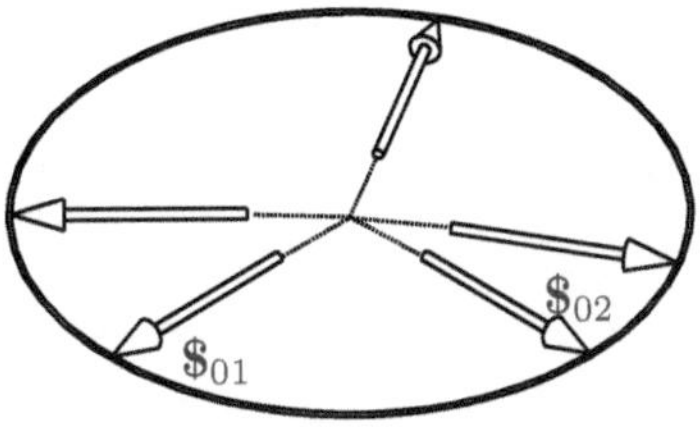

(c) 2-$\$_0$-system.

Fig. 2.4. Some 2-systems

- 2-$\$_0$-system: A 2-$\$_0$-system is composed of all the $\$_0$ whose axes intersect at a common point and are coplanar [Fig. 2.4c]. The common point is called the centre of the 2-$\$_0$-system.

- 3-systems
  - 3-$\$_{\infty}$-system: The 3-$\$_{\infty}$-system is composed of all the $\$_{\infty}$ [Fig. 2.5a].
  - 2-$\$_{\infty}$-1-$\$_0$-system (Perpendicular case): A 2-$\$_{\infty}$-1-$\$_0$-system is composed of all the $\$_0$ whose axes are parallel as well as all the $\$_{\infty}$ whose directions are perpendicular to the axes of all the $\$_0$ [Fig. 2.5b].
  - 2-$\$_{\infty}$-1-$\$_0$-system (General case): A 2-$\$_{\infty}$-1-$\$_0$-system is composed of a $\$_0$ and all the $\$_{\infty}$ whose directions are parallel to a plane that is not perpendicular to the axis of the $\$_0$ as well as any linear combination of the $\$_0$ and the above $\$_{\infty}$. In such a system, there also exist screws of non-zero finite pitch [Fig. 2.5c].
  - 1-$\$_{\infty}$-2-$\$_0$-system: A 1-$\$_{\infty}$-2-$\$_0$-system is composed of a $\$_{\infty}$ as well as all the $\$_0$ whose axes are located on a plane which is perpendicular to the direction of the $\$_{\infty}$ [Fig. 2.5d].

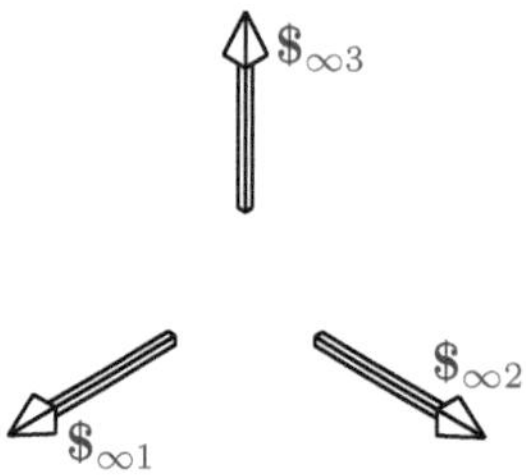

(a) 3-$\$_\infty$-system.

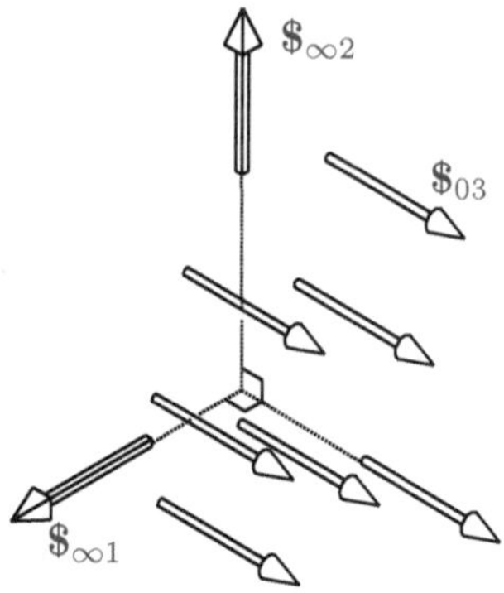

(b) 2-$\$_\infty$-1-$\$_0$-system (Perpendicular case).

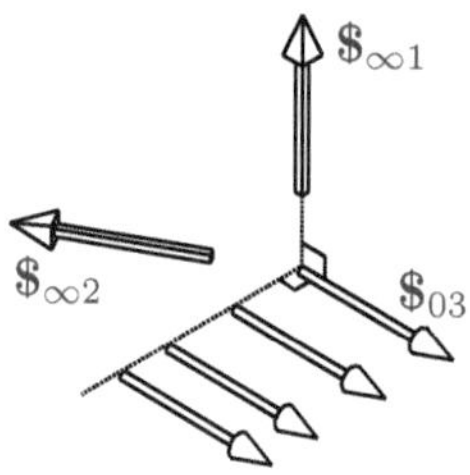

(c) 2-$\$_\infty$-1-$\$_0$-system (General case).

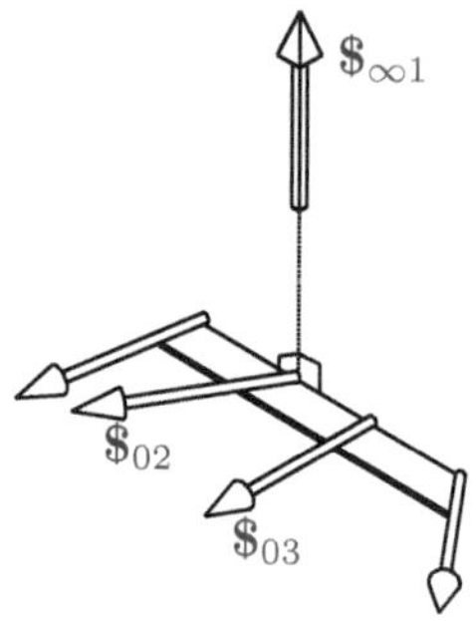

(d) 1-$\$_\infty$-2-$\$_0$-system.

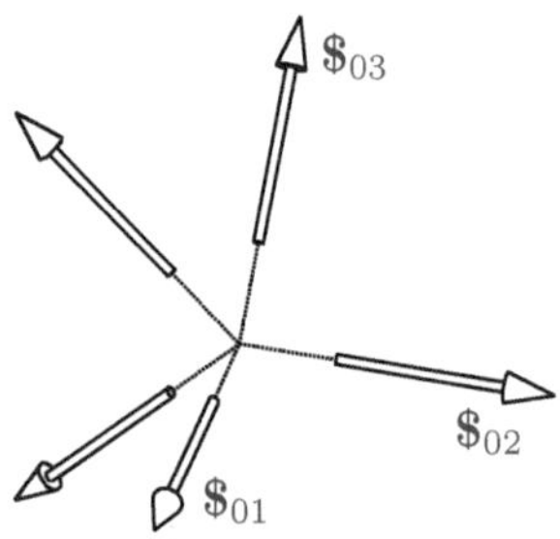

(e) 3-$\$_0$-system.

Fig. 2.5. Some 3-systems

 - 3-$\$_0$-system: A 3-$\$_0$-system is composed of all the $\$_0$ whose axes intersect at a common point [Fig. 2.5e]. The common point is called the centre of the 3-$\$_0$-system.

- 4-systems
  3-$\$_\infty$-1-$\$_0$-system: A 3-$\$_\infty$-1-$\$_0$-system is composed of all the $\$_\infty$ and all the $\$_0$ whose axes are all parallel to one line (Fig. 2.6).

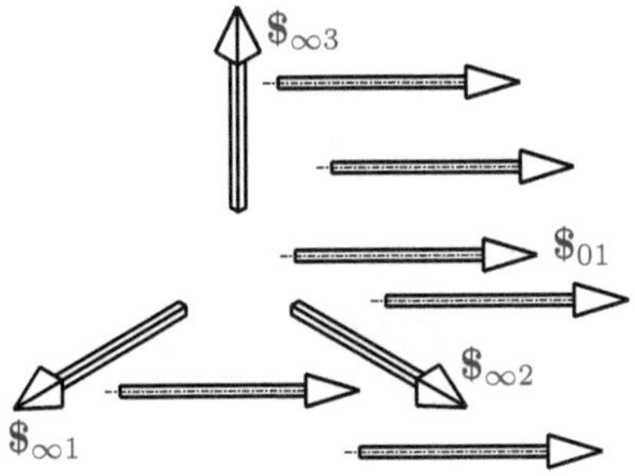

Fig. 2.6. A 4-system: 3-$\$_\infty$-1-$\$_0$-system

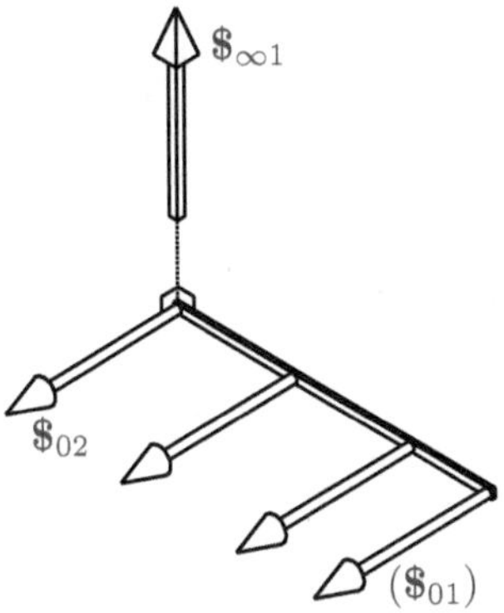

Fig. 2.7. Two bases of a 1-$\$_\infty$-1-$\$_0$-system

2.1.3 Operations on Screw Systems

Subsystem of a Screw System. A *sub-system* of an n-system is composed of all the screws which are linear combinations of $n_s(0 \leq n_s \leq n)$ basis screws of the n-system.

For example, the sub-systems of the 3-$\$_\infty$-system include (Fig. 2.8): (a) a 0-system, (b) a 1-$\$_\infty$-system, (c) a 2-$\$_\infty$-system, and (d) the 3-$\$_\infty$-system.

In the type synthesis of PMs, the sub-systems whose bases are each composed of screws of infinite and/or zero pitches are of particular interest.

Linear Combination of Screw Systems. The linear combination of two screw systems is composed of all the linear combinations of all the basis screws of the two screw systems. The order of the new screw system is less than or equal to the sum of the orders of the two screw systems.

For example, the linear combination of a 1-$\$_\infty$-system and a 2-$\$_\infty$-system is (a) a 2-$\$_\infty$-system if the direction of the $\$_\infty$ within the 1-$\$_\infty$-system is parallel to a plane which is parallel to the directions of all the $\$_\infty$ within the 2-$\$_\infty$-system [Fig. 2.9a] or (b) the 3-$\$_\infty$-system if the direction of the $\$_\infty$ within the

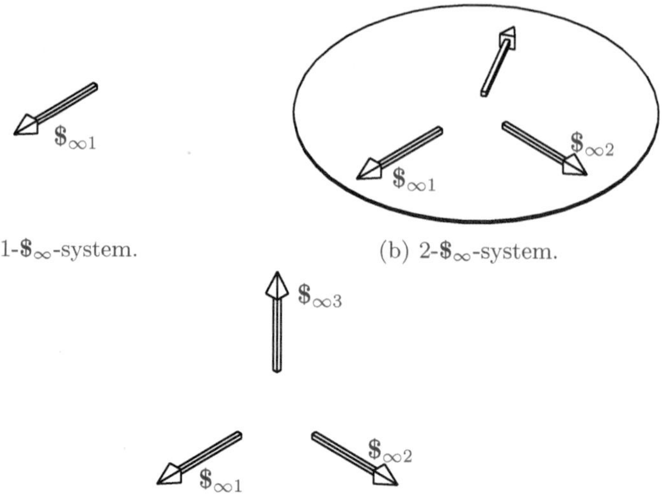

(a) 1-$\$_\infty$-system.

(b) 2-$\$_\infty$-system.

(c) 3-$\$_\infty$-system.

Fig. 2.8. Sub-systems of the 3-$\$_\infty$-system

1-$\$_\infty$-system is not parallel to any plane which is parallel to the directions of all the $\$_\infty$ within the 2-$\$_\infty$-system [Fig. 2.9b].

Reciprocal Screws and Reciprocal Screw Systems

Reciprocal Screws

Two screws, $\$_1$ and $\$_2$, are said to be *reciprocal* if they satisfy the following condition:

$$\$_1 \circ \$_2 = [\boldsymbol{\Pi}\$_1]^T \$_2 = 0 \tag{2.2}$$

where

$$\boldsymbol{\Pi} = \begin{bmatrix} \mathbf{0} & \mathbf{I}_3 \\ \mathbf{I}_3 & \mathbf{0} \end{bmatrix} \tag{2.3}$$

where $\mathbf{I}_3$ is the 3×3 identity matrix and $\mathbf{0}$ is the 3×3 zero matrix. The operator $\circ$ is defined as the reciprocal product of two screws.

This reciprocity condition can be derived as (Fig. 2.10)

$$\begin{cases} \text{no constraint} & \text{if } h_1 \text{ and } h_2 \text{ are both } \infty \\ \cos\lambda = 0 & \text{if } h_1 \text{ or } h_2 \text{ is } \infty \\ (h_1 + h_2)\cos\lambda - r_{12}\sin\lambda = 0 & \text{if } h_1 \text{ and } h_2 \text{ are both finite} \end{cases} \tag{2.4}$$

where r_{12} is the offset distance along the common perpendicular leading from screw $\$_1$ to screw $\$_2$ and λ is the angle between the axes of $\$_1$ and $\$_2$, measured

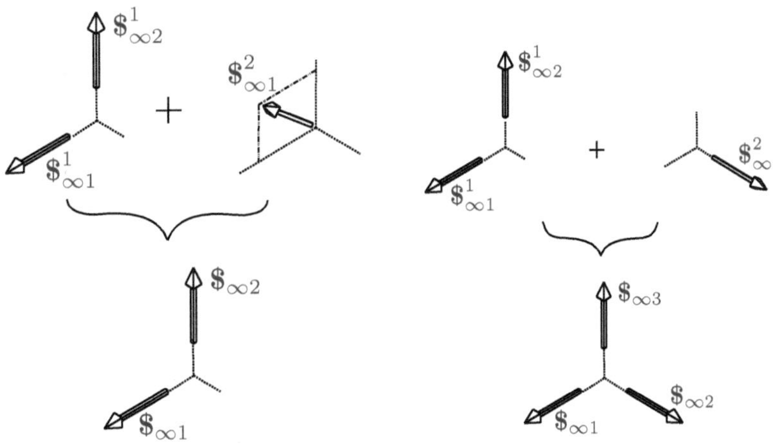

(a) Case a. (b) Case b.

Fig. 2.9. Linear combination of a 1-$\$_\infty$-system and a 2-$\$_\infty$-system

from $\$_1$ to $\$_2$ about the common perpendicular according to the right-hand rule as shown in Fig. 2.10.

It can be concluded from (2.4) that

(1) Two $\$_\infty$ are always reciprocal to each other.
(2) A $\$_\infty$ is reciprocal to a $\$_0$ if and only if their axes are perpendicular to each other.
(3) Two $\$_0$ are reciprocal to each other if and only if their axes are coplanar.

Reciprocal Screw Systems

Given an n-system, there is a unique *reciprocal screw system* of order $(6 - n)$ which comprises all the screws reciprocal to the original screw system. Let $\mathcal{T}$ and $\mathcal{T}^\perp$ denote a screw system and its reciprocal screw system. We have

$$\mathcal{T} = (\mathcal{T}^\perp)^\perp \tag{2.5}$$

where $()^\perp$ denotes the reciprocal screw system of the screw system within the parentheses.

The reciprocal screw system of a given screw system can be obtained using the reciprocity condition of screws presented at the beginning of this subsection. For example, the reciprocal system of the 3-$\$_\infty$-system is still the 3-$\$_\infty$-system. This is illustrated in Fig. 2.11a where the 3-$\$_\infty$-system is represented by a basis $\$_{\infty 1}$, $\$_{\infty 2}$ and $\$_{\infty 3}$ and its reciprocal system — which is also the 3-$\$_\infty$-system — is represented by a basis $\$_{r\infty 1}$, $\$_{r\infty 2}$ and $\$_{r\infty 3}$. Similarly, the reciprocal system of a 3-$\$_0$-system is still a 3-$\$_0$-system. This is illustrated in Fig. 2.11b where the 3-$\$_0$-system is represented by a basis $\$_{\infty 1}$, $\$_{02}$ and $\$_{03}$ and its reciprocal system — which is also a 3-$\$_0$-system — is represented by a basis $\$_{r01}$, $\$_{r02}$ and $\$_{r03}$. Here, the reciprocal screw systems are represented using dashed arrows. Alternatively, one can also calculate the reciprocal screw system of a given screw system nu-

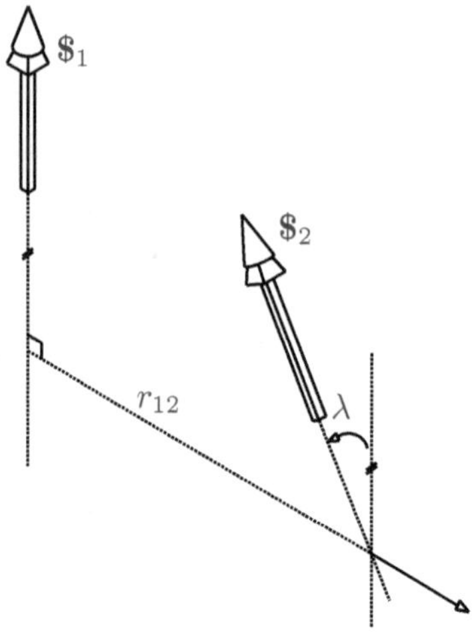

Fig. 2.10. Reciprocal screws

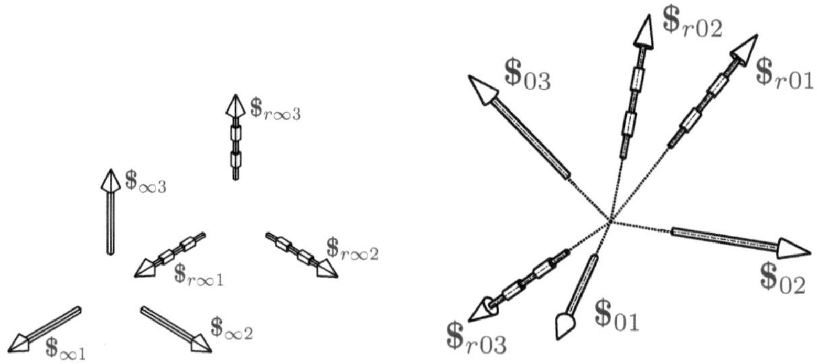

(a) Reciprocal screw system of the 3-$\$_\infty$-system.

(b) Reciprocal screw system of a 3-$\$_0$-system.

Fig. 2.11. Reciprocal screw systems

merically [15] or symbolically [20]. For a detailed discussion on interrelationship between screw systems and corresponding reciprocal systems, see [21].

2.1.4 Twist Systems and Wrench Systems of Kinematic Chains

Consider two links connected by a KC in the form of kinematic joint, serial KC or PKC shown schematically in Fig. 2.12. The instantaneous relative motion between the two links is represented by a screw system which is usually called the *twist system* of the KC. The constraint on one link by another link through the KC is represented by the reciprocal screw system of the twist system which is usually called the *wrench system* of the KC.

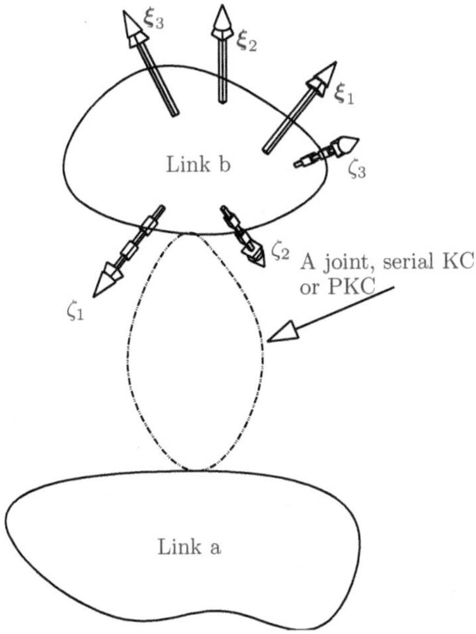

Fig. 2.12. Two links connected by a joint, serial KC or a PKC

The twist system of a KC is a $\mathcal{C}$-system where $\mathcal{C} \leq \mathcal{F}$ and $\mathcal{F}$ denotes the DOF of the KC. The wrench system of a KC is a c-system, where $c = 6 - \mathcal{C}$. The twist system of a KC is the reciprocal screw system of its wrench system, and vice versa.

Let $\mathcal{T}$ and $\mathcal{W}$ represent, respectively, a twist system and its wrench system. From (2.5), we have

$$\begin{cases} \mathcal{W} = \mathcal{T}^{\perp} \\ \mathcal{T} = \mathcal{W}^{\perp} \end{cases} \tag{2.6}$$

Let $\boldsymbol{\xi}$ denote a twist in a twist system and $\boldsymbol{\zeta}$ denote a wrench in the wrench system corresponding to the twist system. Throughout this book, $\boldsymbol{\xi}$ and $\boldsymbol{\zeta}$ are respectively represented by arrows in solid line and arrows in dashed line. For convenience, $\boldsymbol{\xi}_0$, $\boldsymbol{\xi}_\infty$, $\boldsymbol{\zeta}_0$ and $\boldsymbol{\zeta}_\infty$ are used to represent a twist of 0-pitch, a twist of ∞-pitch, a wrench of 0-pitch and a wrench of ∞-pitch respectively. Based on the reciprocity condition of screws, we obtain the geometric relation between a twist system and its wrench system as follows:

(a1) The axis of a $\boldsymbol{\xi}_0$ is coplanar with the axis of any $\boldsymbol{\zeta}_0$.
(a2) The direction of a $\boldsymbol{\xi}_\infty$ is perpendicular to the axis of any $\boldsymbol{\zeta}_0$.
(a3) The axis of a $\boldsymbol{\xi}_0$ is perpendicular to the direction of any $\boldsymbol{\zeta}_\infty$.

The relation between a twist system and its wrench system can also be stated in the following way: the virtual power developed by any $\boldsymbol{\zeta}$ along any $\boldsymbol{\xi}$ is equal to zero.

Based on the relation between a twist system and its wrench system, one can identify the wrench systems of the twist systems that are encountered in many PKCs.

Kinematic Joints

The most commonly used kinematic joints are R, P, C, U and S joints. It can easily be shown that all these joints can be decomposed into a KC including only P and R joints. Therefore, only the twist systems and wrench systems associated with the R and P kinematic joints (Fig. 2.13) are presented below.

- R joint
  The twist system of an R joint is a 1-system. The twist in the 1-system is a $\boldsymbol{\xi}_0$ directed along the joint axis. The wrench system is a 5-system which includes all the $\boldsymbol{\zeta}_0$ whose axes intersect the joint axis, all the $\boldsymbol{\zeta}_\infty$ whose axes are perpendicular to the axis of the R joint and all the linear combinations of the above wrenches.
- P joint
  The twist system of a P joint is a 1-system. The twist in the 1-system is a $\boldsymbol{\xi}_\infty$ in the direction of the joint axis. The wrench system is a 5-system which includes all the $\boldsymbol{\zeta}_0$ whose axes are perpendicular to the direction of the joint, all the $\boldsymbol{\zeta}_\infty$ and all the linear combinations of the above wrenches.

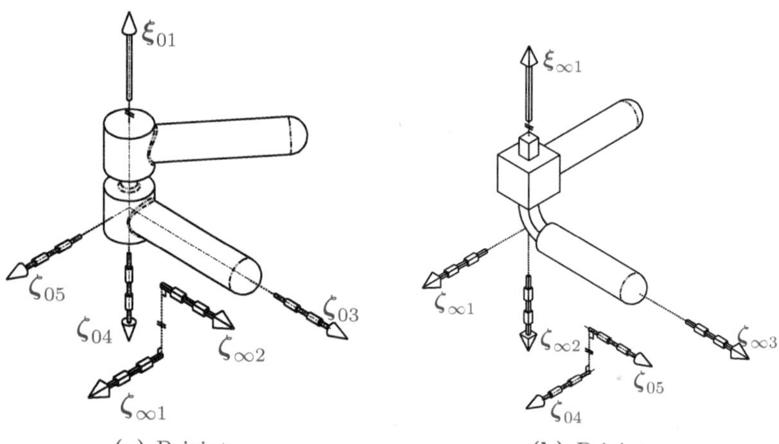

(a) R joint. (b) P joint.

Fig. 2.13. Twist system and wrench system of the R and P joints

Serial Kinematic Chains

For simplicity and without loss of generality, we make the assumption that a serial KC is composed of 1-DOF joints since an l-DOF joint can be treated as a serial KC of l 1-DOF joints. A serial KC is illustrated schematically in Fig. 2.14.

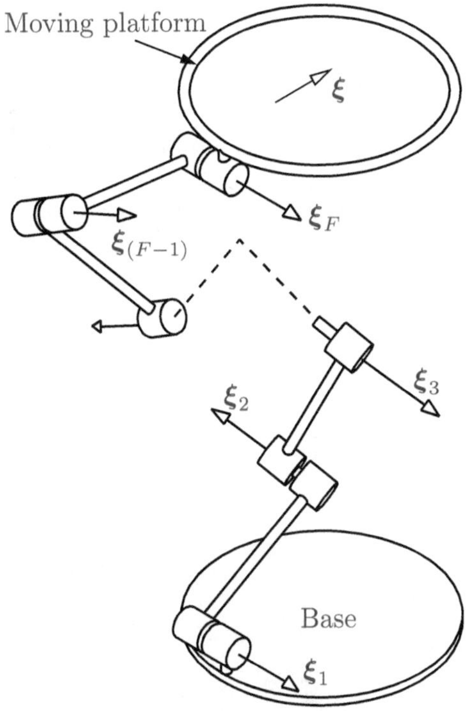

Fig. 2.14. Serial kinematic chain

The output twist of the moving platform — or end-effector — in the serial KC is given by

$$\rho\boldsymbol{\xi} = \sum_{j=1}^{f} \boldsymbol{\xi}_j \dot{\theta}_j \tag{2.7}$$

where $\boldsymbol{\xi}_j$ and $\dot{\theta}_j$ are respectively the twist and the velocity of the j-th joint. Variables f and ρ denote respectively the DOF of the serial KC and the amplitude of the output twist.

From (2.7), we can conclude that the twist system $\mathcal{T}$ of a serial KC is the linear combination of the twist systems $\mathcal{T}_j$ of all the joints in the serial KC, i.e.,

$$\mathcal{T} = \sum_{j=1}^{f} \mathcal{T}_j \tag{2.8}$$

where the subscript, j, denotes the j-th kinematic joint. From (2.8) and (2.6), we obtain

$$\mathcal{W} = \bigcap_{j=1}^{f} \mathcal{W}_j. \tag{2.9}$$

Equation (2.9) states that the wrench system $\mathcal{W}$ of a serial KC is the intersection of the wrench systems $\mathcal{W}_j$ of all the joints in the serial KC.

Consider the PR serial KC shown in Fig. 2.15. The twist system of the PR serial KC is the linear combination of the twist systems of the P and R joints, which is a 2-system. One basis for this system is composed of a $\boldsymbol{\xi}_0$ along the axis of the R joint and a $\boldsymbol{\xi}_\infty$ along the direction of the P joint. The wrench system of the PR serial KC is the intersection of the wrench system of the R joint with that of the P joint. This is a 4-system which includes all the $\boldsymbol{\zeta}_\infty$ whose axes are perpendicular to the axis of the R joint, all the $\boldsymbol{\zeta}_0$ whose axes are perpendicular to the axis of the P joint and intersect the axis of the R joint and the linear combinations of the above wrenches.

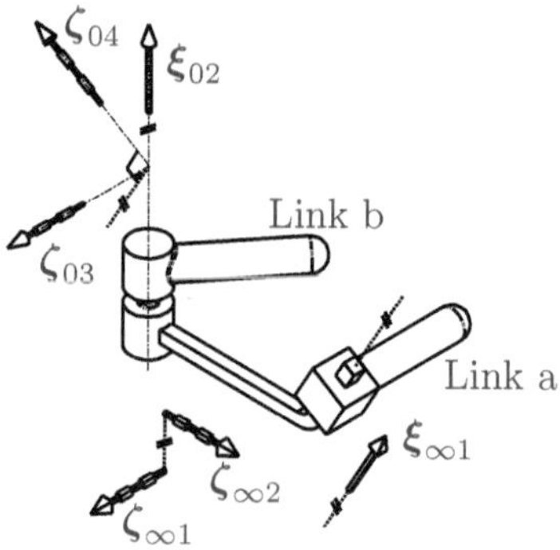

Fig. 2.15. Twist system and wrench system of a PR serial KC

There exist several classes of serial KCs with the following special kinematic characteristic, namely: *for any configuration of these KCs, the wrench system of each of these KCs always includes a specified number of independent wrenches of zero-pitch or infinity-pitch.* These serial KCs, each exhibiting certain specific constraint characteristics, will be used in the construction of single-loop KCs (Chap. 3) and PKCs (Chaps. 5–14) and are thus called *compositional units.*

The basic types of compositional units are illustrated in Fig. 2.16 and are described below:

- Class 1 (Parallelaxis compositional units). Serial KCs composed of at least one R joint and at least one P joint in which the axes of all the R joints are parallel and not all the directions of the P joints are perpendicular to the axes of the R joints. The characteristic of a compositional unit of this class is that the axes of all the R joints are always parallel. The wrench system of this compositional unit always includes a 2-$\boldsymbol{\zeta}_\infty$-system. The 2-$\boldsymbol{\zeta}_\infty$-system is composed of all the $\boldsymbol{\zeta}_\infty$ whose directions are perpendicular to all the axes of the R joints [Fig. 2.16.a]. A parallelaxis compositional unit is denoted by $()_A$.

- Class 2 (Spatial translational compositional units). Serial KCs composed of three or more P joints whose directions are not parallel to a plane. The wrench system of a compositional unit of this class is always the 3-$\boldsymbol{\zeta}_\infty$-system [Fig. 2.16b]. A spatial translational compositional unit is denoted by $()_T$.
- Class 3 (Planar compositional units). Serial KCs composed of at least two R and/or P joints which include at least one R joint and in which all the links are moving along parallel planes. In a compositional unit of this class, the axes of all the R joints are parallel, and the directions of the P joints are perpendicular to all the axes of the R joints. The wrench system of this compositional unit always includes a 2-$\boldsymbol{\zeta}_\infty$-1-$\boldsymbol{\zeta}_0$-system. The 2-$\boldsymbol{\zeta}_\infty$-1-$\boldsymbol{\zeta}_0$-system is composed of all the $\boldsymbol{\zeta}_0$ whose axes are parallel to the axes of the R joints as well as all the $\boldsymbol{\zeta}_\infty$ whose directions are perpendicular to the axes of all the R joints [Fig. 2.16c]. A planar compositional unit is denoted by $()_E$.
- Class 4 (Planar translational compositional units). Serial KCs composed of two or more P joints whose directions are all parallel to a plane. The characteristic of a compositional unit of this class is that all the directions of the P joints are always parallel to a plane. The wrench system of this compositional unit always includes a 3-$\boldsymbol{\zeta}_\infty$-1-$\boldsymbol{\zeta}_0$-system. The 3-$\boldsymbol{\zeta}_\infty$-1-$\boldsymbol{\zeta}_0$-system is composed of all the $\boldsymbol{\zeta}_\infty$ as well as all the $\boldsymbol{\zeta}_0$ whose axes are perpendicular to the directions of all the P joints [Fig. 2.16d]. A planar translational compositional unit is a special case of a planar compositional unit and is therefore denoted by $()_E$.
- Class 5 (Spherical compositional units). Serial KCs composed of two or more concurrent R joints. The characteristic of a KC of this class is that the axes of all the R joints are always concurrent. The wrench system of this serial KC always includes a 3-$\boldsymbol{\zeta}_0$-system with its centre at the intersection of the axes of the R joints [Fig. 2.16e]. A spherical compositional unit is denoted by $()_S$.
- Class 6 (Coaxial compositional units). Serial KCs composed of one or more coaxial R joints. The characteristic of a compositional unit of this class is that all the axes of the R joints are always coaxial. The wrench system of this compositional unit is always a 2-$\boldsymbol{\zeta}_\infty$-3-$\boldsymbol{\zeta}_0$-system [Fig. 2.16f], which includes all the $\boldsymbol{\zeta}_0$ whose axes intersect the axes of the R joints, all the $\boldsymbol{\zeta}_\infty$ whose axes are perpendicular to the axes of the R joints and all the linear combinations of the above wrenches. A coaxial compositional unit is denoted by $()_L$.
- Class 7 (Codirectional compositional units). Serial KCs composed of one or more P joints whose directions are parallel. The characteristic of a compositional unit of this class is that all the directions of the P joints are always parallel. The wrench system of this compositional unit is always a 3-$\boldsymbol{\zeta}_\infty$-2-$\boldsymbol{\zeta}_0$-system [Fig. 2.16g], which includes all the $\boldsymbol{\zeta}_0$ whose axes are perpendicular to the direction of the P joints, all the $\boldsymbol{\zeta}_\infty$ and all the linear combinations of the above wrenches. A codirectional compositional unit is denoted by $()_L$.

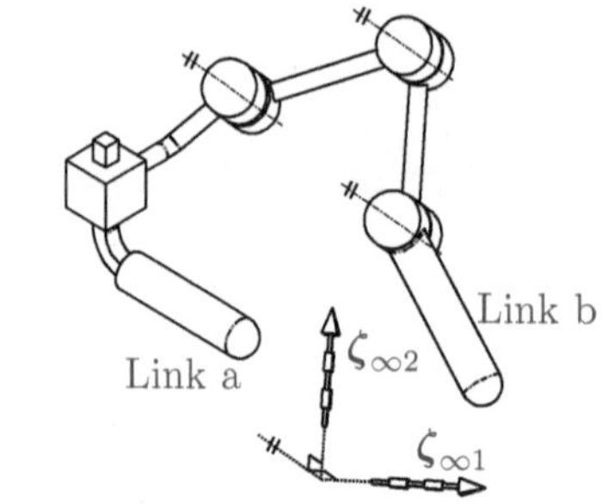

(a) (PRRR)$_A$ parallelaxis unit.

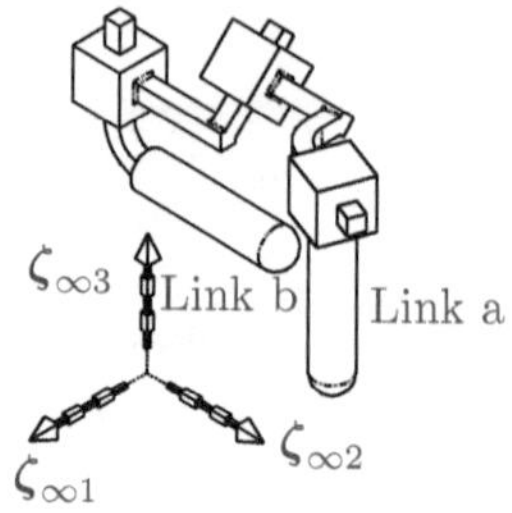

(b) (PPP)$_T$ spatial translational unit.

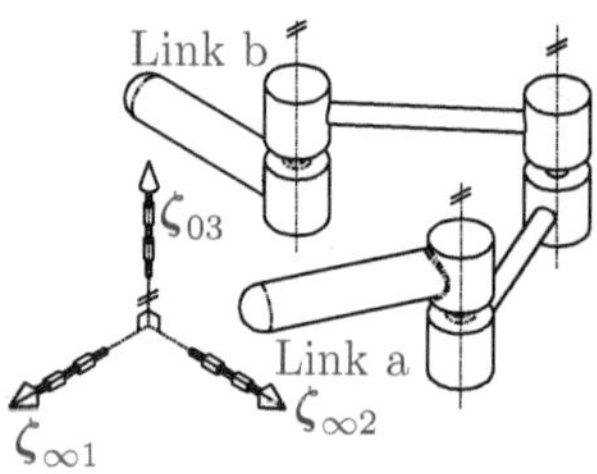

(c) (RRR)$_E$ planar unit.

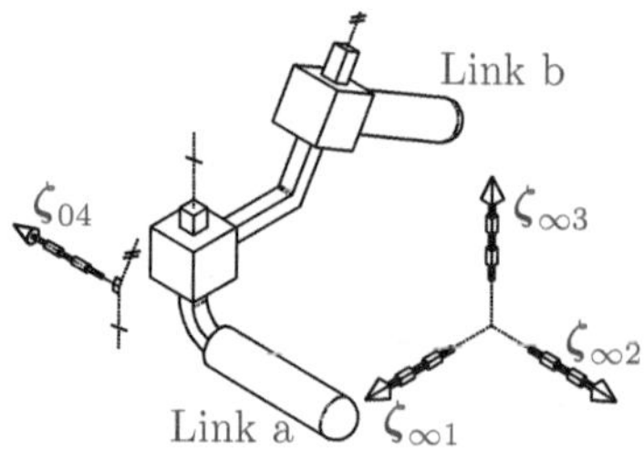

(d) (PP)$_E$ planar translational unit.

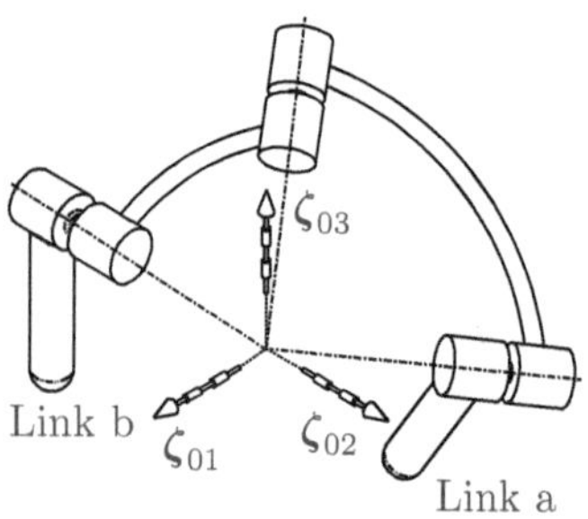

(e) (RRR)$_S$ spherical unit.

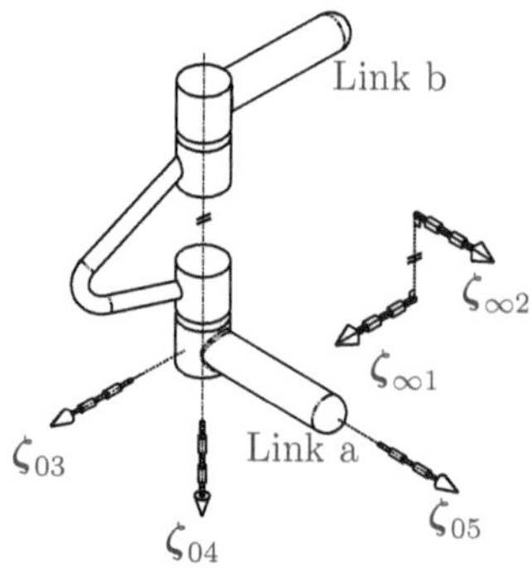

(f) (RR)$_L$ coaxial unit.

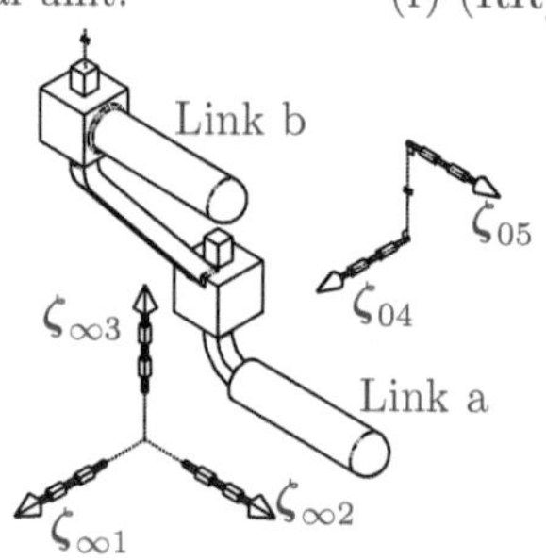

(g) (PP)$_L$ codirectional unit.

Fig. 2.16. Compositional units

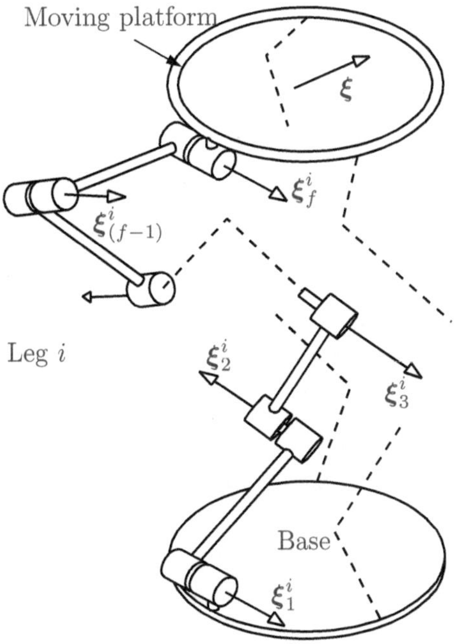

Fig. 2.17. Parallel kinematic chain

Parallel Kinematic Chains [88]
A PKC is illustrated schematically in Fig. 2.17. It is composed of a set of m serial KCs mounted on a common base and attached to a common moving platform. The output twist of the moving platform can be written as

$$\rho \boldsymbol{\xi} = \sum_{j=1}^{f^i} \boldsymbol{\xi}_j^i \dot{\theta}_j^i, \qquad i = 1, 2, \cdots, m \tag{2.10}$$

where the subscript and superscript, ^i_j, denote the j-th joint in the i-th leg, while m and f^i respectively denote the number of legs in the PKC and the DOF of the i-th leg.

From (2.10), we can conclude that the twist system $\mathcal{T}$ of a PKC is the intersection of the twist systems $\mathcal{T}^i$ of all its legs, i.e.,

$$\mathcal{T} = \bigcap_{i=1}^{m} \mathcal{T}^i \tag{2.11}$$

where

$$\mathcal{T}^i = \sum_{j=1}^{f^i} \mathcal{T}_j^i$$

and $\mathcal{T}_j^i$ denotes the twist system of joint j in leg i.

From (2.11) and (2.6), we obtain

$$\mathcal{W} = \sum_{i=1}^{m} \mathcal{W}^i \tag{2.12}$$

where

$$\mathcal{W}^i = \bigcap_{j=1}^{f^i} \mathcal{W}^i_j$$

and $\mathcal{W}^i_j$ denotes the wrench system of joint j in leg i.

Equation (2.12) states that the wrench system $\mathcal{W}$ of a PKC is the linear combination of the wrench systems $\mathcal{W}^i$ of all its legs.

2.2 Instantaneous Mobility Analysis of Kinematic Chains

The *mobility* or *degree of freedom (DOF)* of a KC is the number of independent parameters required to determine the relative configuration of all its links. It is well known that the classical Chebychev-Grübler-Kutzbach mobility criterion, which is based solely on topology, fails to provide the correct mobility in many instances. On the other hand, *instantaneous mobility*, not the *full-cycle mobility*, of a KC is used in the type synthesis of PMs if the virtual-chain approach is applied. Therefore, based on the geometric analysis completed above, alternative instantaneous mobility criteria are now proposed. As shown in [74] or Appendix B, the full-cycle mobility analysis of PMs can indeed be obtained by reversing the process of the type synthesis of PMs proposed in this book. Readers interested in the full-cycle mobility analysis of PMs are referred to [74] or Appendix B.

2.2.1 Serial Kinematic Chains

The mobility of a serial KC is equal to the sum of the DOF of all its joints.

For example, the PRRR serial KC shown in Fig. 2.18 is composed of one P joint and three R joints. Its mobility is thus equal to four.

2.2.2 Single-Loop Kinematic Chain

A single-loop KC can be regarded as a chain formed by closing the two end-links of a serial KC rigidly. An example is shown in Fig. 2.19, where the $(\text{PRRRP})_A$ single-loop KC illustrated in Fig. 2.19a can be formed by rigidly attaching the two end-links of the serial KC shown in Fig. 2.19b. Let $\mathcal{C}$ and c denote the order of the twist system and the order of the wrench system of the serial KC, and f denote the sum of DOF of all the joints.

Since the mobility of the serial KC is f, f independent parameters should be used to describe the motion of the serial KC. When the two end-links are closed in the construction of the single-loop KC, $\mathcal{C}$ independent parameters, which

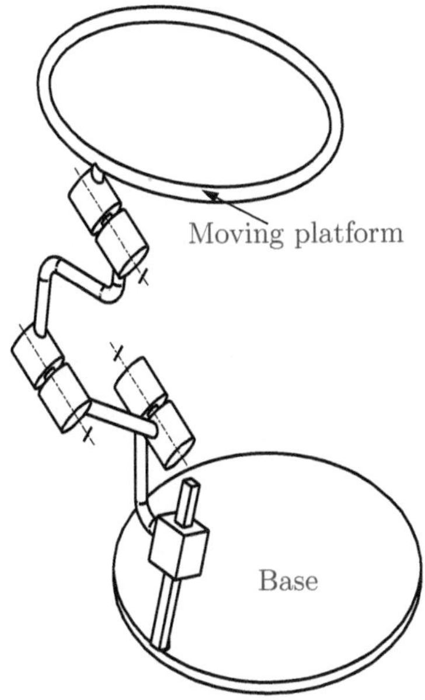

Fig. 2.18. PRRR serial KC

describe the relative motion between the two end-links, are specified. Thus, the mobility, $\mathcal{F}$, of a single-loop KC is

$$\mathcal{F} = f - \mathcal{C}. \tag{2.13}$$

Since we have

$$c = 6 - \mathcal{C} \tag{2.14}$$

Equation (2.13) can be written in the following form

$$\mathcal{F} = f - \mathcal{C} = f - (6 - c) = f - 6 + c. \tag{2.15}$$

For example, the $(\text{PRRRP})_A$ single-loop KC shown in Fig. 2.19a is formed by closing the two end-links of a parallelaxis serial KC [Fig. 2.16a] rigidly. The wrench system of the parallelaxis serial KC is a 2-$\boldsymbol{\zeta}_\infty$ system, which is composed of all the $\boldsymbol{\zeta}_\infty$ whose directions are perpendicular to the axes of the R joints within the parallelaxis serial KC. We have $c = 2$. Using (2.15), we obtain the mobility of the single-loop KC as

$$\mathcal{F} = f - 6 + c = 5 - 6 + 2 = 1.$$

It is noted that there may exist inactive joints in a KC. An *inactive joint* is a joint in a KC which cannot be moved due to the constraints induced by other joints in the KC. When an inactive joint is removed (blocked) from a KC, the relative motion within the KC remains unchanged.

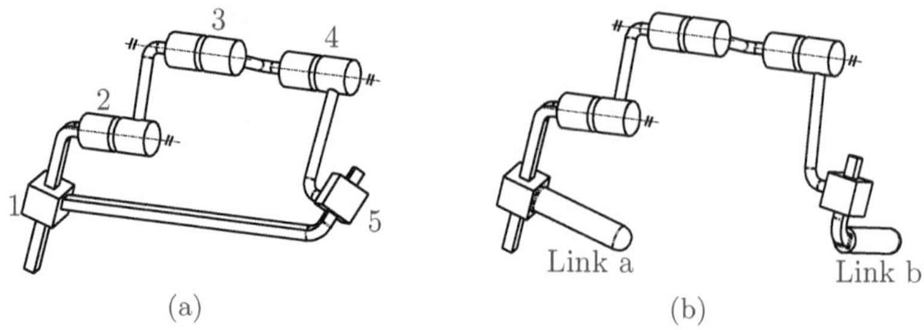

Fig. 2.19. (**a**) $(\text{PRRRP})_A$ single-loop KC and (**b**) one of its corresponding serial kinematic chains

An inactive joint in a single-loop KC can be detected in the following way[1]: A joint in a single-loop KC is inactive if and only if

$$\mathcal{C}' = \mathcal{C} - 1 \tag{2.16}$$

or

$$c' = c + 1 \tag{2.17}$$

where $\mathcal{C}$ and c denote respectively the order of the twist system and the wrench system of the serial KC corresponding to the single-loop KC while $\mathcal{C}'$ and c' denote respectively the order of the twist system and the wrench system of the serial KC corresponding to the single-loop KC with the considered joint blocked.

As an example, consider the $(\text{RRR(R)}_L\text{R})_E$ single-loop KC shown in Fig. 2.20a. In this KC, the axes of all the R joints except joint 4 are all parallel. The order c of the wrench system of its corresponding serial KC is 2. When we block joint 4, we obtain an $(\text{RRRR})_E$ single-loop KC as illustrated in Fig. 2.20b, and the order c' of the wrench system of the corresponding serial KC is 3. According to (2.17), joint 4 is inactive.

2.2.3 Parallel Kinematic Chains

Consider an m-legged PKC (Fig. 2.17). Let c and $\mathcal{F}$ denote the order of the wrench system $\mathcal{W}$ and mobility (DOF) of the PKC, and c^i and f^i denote the order of the wrench system $\mathcal{W}^i$ and DOF of leg i.

The mobility $\mathcal{F}$ of a PKC is the sum of (1) the number of independent parameters required to determine the relative configuration of the moving platform and (2) the number of independent parameters required to determine the configuration of all the links in all the legs with the relative configuration of the moving platform specified.

The number of independent parameters required to determine the relative configuration of the moving platform is equal to the order $\mathcal{C}$ of the twist system

[1] This condition can be proven directly using (2.13) and (2.14). It is left to the reader to prove.

(a) $(\mathrm{RRR(R)}_L\mathrm{R})_E$. (b) $(\mathrm{RRRR})_E$.

Fig. 2.20. Single-loop KC with an inactive joint

of the PKC (also called the *connectivity* of the moving platform). $\mathcal{C}$ can be calculated using

$$\mathcal{C} = 6 - c. \tag{2.18}$$

Since the order of the twist system of leg i is $(6 - c^i)$, the number of independent parameters needed to determine the configuration of all the links in leg i with the relative configuration of the moving platform specified can be determined using

$$R^i = f^i - (6 - c^i) = f^i - 6 + c^i \tag{2.19}$$

where R^i is called the redundant DOF of leg i.

The number of independent parameters needed to determine the configuration of all the links in all legs with the relative configuration of the moving platform specified is

$$R = \sum_{i=1}^{m} R^i \tag{2.20}$$

where R is called the *redundant DOF* of the PKC. Then, we obtain the *mobility* (or DOF) $\mathcal{F}$ of the PKC as

$$\mathcal{F} = \mathcal{C} + R = 6 - c + \sum_{i=1}^{m} R^i. \tag{2.21}$$

The mobility obtained using (2.21) is usually instantaneous. When c, c^i and R^i are the same in different general configurations, the DOF is full-cycle.

In addition to the mobility, another important index of a PKC is defined as

$$\Delta = \sum_{i=1}^{m} c^i - c \tag{2.22}$$

where Δ is called the *number of overconstraints* (also *passive constraints* or *redundant constraints*) if $\Delta > 0$.

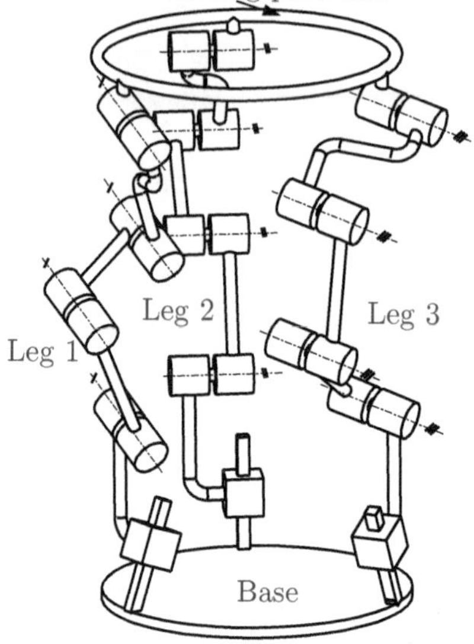

Fig. 2.21. 3-(PRRRR)$_A$ PKC

The mobility analysis proposed above will now be illustrated using two examples of PKCs.

Example 2.1. Consider the 3-(PRRRR)$_A$ PKC shown in Fig. 2.21. In this PKC, all the axes of the R joints within a same leg are parallel. The direction of a P joint is not perpendicular to the axes of the R joints within the same leg. Not all the axes of the R joints on the moving platform are parallel. The wrench system of each leg is a 2-ζ_∞-system. The wrench system of the PKC is the 3-ζ_∞-system. We have $c^i = 2$, $c = 3$, $R^i = 5 - (6 - 2) = 1$. Then

$$\mathcal{C} = 6 - c = 3$$

and

$$\mathcal{F} = \mathcal{C} + \sum_{i=1}^{3} R^i = 6.$$

It is noted that the axes of the successive R joints with parallel axes always remain parallel throughout any arbitrary motion. Hence, c, c^i and R^i do not change when the moving platform undergoes a small displacement from a general configuration and the mobility obtained is thus full-cycle. The number of overconstraints of this 3-legged PKC is

$$\Delta = \sum_{i=1}^{3} c^i - c = 6 - 3 = 3.$$

Example 2.2. Consider the 3-(PRRR)$_A$ PKC shown in Fig. 2.22. In this PKC, all the axes of the R joints within a same leg are parallel. The direction of a P joint is not perpendicular to the axes of the R joints within the same leg. The axes of the R joints on the moving platform are not all parallel. The wrench system of each leg is a 2-ζ_∞-system. The wrench system of the PKC is the 3-ζ_∞-system. We have $c^i = 2$, $c = 3$, $R^i = 4 - (6 - 2) = 0$. Then

$$\mathcal{C} = 6 - c = 3$$

and

$$\mathcal{F} = \mathcal{C} + \sum_{i=1}^{3} R^i = 3.$$

Similarly to the previous example, it is noted that the axes of the successive R joints with parallel axes always remain parallel throughout any arbitrary motion. Hence, c, c^i and R^i do not change when the moving platform undergoes a small displacement from a general configuration and the mobility obtained is thus full-cycle. The number of overconstraints of this 3-legged PKC is

$$\Delta = \sum_{i=1}^{3} c^i - c = 6 - 3 = 3.$$

To facilitate the type synthesis of PMs, we can substitute (2.22) into (2.21) and then obtain

$$\sum_{i=1}^{m} c^i = 6 - \mathcal{C} + \Delta = 6 - \mathcal{F} + \Delta + R. \tag{2.23}$$

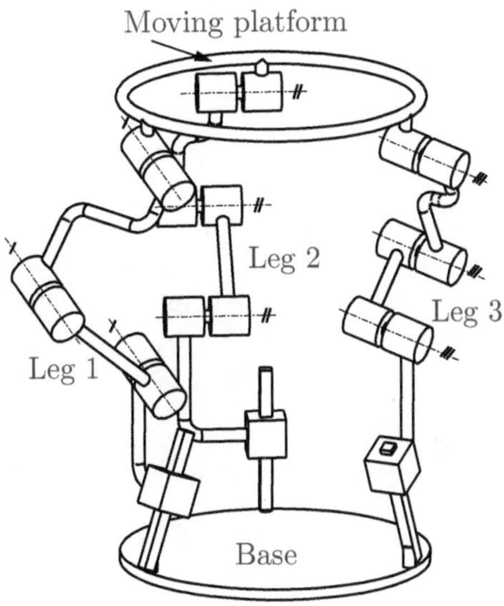

Fig. 2.22. 3-(PRRR)$_A$ PKC

Equations (2.12), (2.20) and (2.19) can then be respectively rewritten as

$$\sum_{i=1}^{m} \mathcal{W}^i = \mathcal{W} \tag{2.24}$$

$$\sum_{i=1}^{m} R^i = R \tag{2.25}$$

$$f^i = 6 + R^i - c^i. \tag{2.26}$$

Equations (2.23)–(2.26) will be used in the type synthesis of PMs.

In this book, we focus on non-redundant PMs (see for instance the PM of Fig. 2.22) for which

$$R = R^i = 0, \qquad i = 1, 2, \cdots, m. \tag{2.27}$$

In this case, (2.23) and (2.26) can be reduced to

$$\sum_{i=1}^{m} c^i = 6 - \mathcal{C} + \Delta = 6 - \mathcal{F} + \Delta \tag{2.28}$$

and

$$f^i = 6 - c^i. \tag{2.29}$$

Therefore, three equations, i.e., (2.28), (2.24) and (2.29), will be used in the type synthesis of PMs in this book. The general methodology based on these equations will be elaborated and applied in Chaps. 5–13.

2.3 Validity Condition of Actuated Joints in Parallel Mechanisms

For an $\mathcal{F}$-DOF mechanism, $\mathcal{F}$ actuated joints should be selected. There are many different ways of selecting the actuated joints of a mechanism. Usually, the actuated joints cannot be selected arbitrarily. The selection of actuated joints should ensure that, in a general configuration, the DOF of the mechanism with the $\mathcal{F}$ actuated joints blocked will be zero.

For general PMs, the validity condition of actuated joints can be obtained based on the mobility criterion derived in Sect. 2.2. As mentioned in the preceding section, this book focuses on non-redundant PMs. Therefore, non-redundant PMs are assumed in the following derivation.

2.3.1 Actuation Wrenches

Let $\mathcal{W}^i_{\not{D}j}$ (j=1,2,...,c^i) be the set of all the wrenches which are not reciprocal to the twist of joint j and reciprocal to all the twists of the other joints within leg i.

Physically speaking, $\mathcal{W}^i_{\not{\mathcal{D}}j}$ is the set of wrenches that can be exerted on the moving platform through the actuation of joint j in leg i. This set of wrenches has been previously defined by several authors (such as in [2, 12]).

Let $\boldsymbol{\zeta}^i_j$ denote a basis of the wrench system $\mathcal{W}^i$ of leg i and $\boldsymbol{\zeta}^i_{\not{\mathcal{D}}j}$ denote any one arbitrary wrench which belongs to $\mathcal{W}^i_{\not{\mathcal{D}}j}$. Then, any wrench in $\mathcal{W}^i_{\not{\mathcal{D}}j}$ can be expressed as

$$\boldsymbol{\zeta}'^i_{\not{\mathcal{D}}j} = \alpha\boldsymbol{\zeta}^i_{\not{\mathcal{D}}j} + \sum_{k=1}^{c^i}(\beta^i_k\boldsymbol{\zeta}^i_k), \quad \alpha \neq 0. \tag{2.30}$$

For convenience, $\boldsymbol{\zeta}^i_{\not{\mathcal{D}}j}$ is called the *actuation wrench* of the joint j in leg i.

For example, in each $(\underline{\mathrm{P}}\mathrm{RRR})_A$ leg represented in Fig. 2.23b of the 3-$(\underline{\mathrm{P}}\mathrm{RRR})_A$ PM of Fig. 2.23a, the first P joint is actuated. The actuation wrench of a P joint is any $\boldsymbol{\zeta}_0$ whose axis is parallel to the axes of the three R joints within the same leg.

2.3.2 Validity Condition of Actuated Joints

For a non-redundant $\mathcal{F}$-DOF PM, a given set of actuated joints is valid if and only if $\mathcal{F}'$=0, where $\mathcal{F}'$ denotes the mobility of the PM with all its actuated joints blocked. From (2.21), we have

$$c' = 6 - \mathcal{F}' = 6 \tag{2.31}$$

where c' denotes the order of the wrench system of the PM with all its actuated joints blocked.

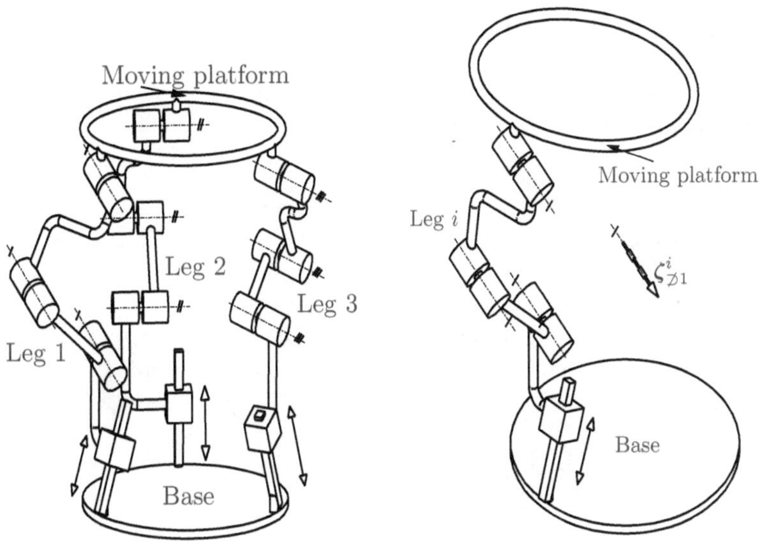

(a) 3-$(\underline{\mathrm{P}}\mathrm{RRR})_A$ PM. **(b)** Actuation wrench of $(\underline{\mathrm{P}}\mathrm{RRR})_A$ leg.

Fig. 2.23. Validity detection of actuated joints for 3-$(\underline{\mathrm{P}}\mathrm{RRR})_A$ PM

In fact, the wrench system of the non-redundant $\mathcal{F}$-DOF PM with its $\mathcal{F}$ actuated joints blocked is a linear combination of the wrench system $\mathcal{W}$ of the PKC and all the actuation wrenches, $\zeta^i_{\mathcal{D}j}$, of the $\mathcal{F}$ actuated joints. Thus, the validity condition of actuated joints for PMs can be stated as follows:

For an $\mathcal{F}$-DOF PM in which all the twists within the same leg are linearly independent in a general configuration, a set of $\mathcal{F}$ actuated joints is valid if and only if, in a general configuration, all the actuation wrenches, $\zeta^i_{\mathcal{D}j}$, of the $\mathcal{F}$ actuated joints together with a set of basis wrenches of the wrench system $\mathcal{W}$ of the PKC constitute a 6-system.

Example 2.3. Let us now determine whether or not the set of actuated joints is valid for the 3-($\underline{\text{P}}$RRR)$_A$ PM shown in Fig. 2.23a [39, 71, 72]. In this PM, we assume that the axes of the R joints on the moving platform are not parallel to a plane. Let $[\mathbf{0} \quad \mathbf{i}^T]^T$, $[\mathbf{0} \quad \mathbf{j}^T]^T$, $[\mathbf{0} \quad \mathbf{k}^T]^T$ denote a basis of $\mathcal{W}$, and let $[\zeta^i_{F\mathcal{D}j} \quad \zeta^i_{S\mathcal{D}j}]^T$ represent $\zeta^i_{\mathcal{D}j}$. Here, $\mathbf{i}$, $\mathbf{j}$ and $\mathbf{k}$ denote respectively the unit vectors along the X-, Y- and Z-axes. The validity condition of actuated joints of the PM can be expressed as

$$\begin{vmatrix} \zeta^1_{F\mathcal{D}j} & \zeta^2_{F\mathcal{D}j} & \zeta^3_{F\mathcal{D}j} & \mathbf{0} & \mathbf{0} & \mathbf{0} \\ \zeta^1_{S\mathcal{D}j} & \zeta^2_{S\mathcal{D}j} & \zeta^3_{S\mathcal{D}j} & \mathbf{i} & \mathbf{j} & \mathbf{k} \end{vmatrix} = \begin{vmatrix} \mathbf{i} & \mathbf{j} & \mathbf{k} \end{vmatrix} \begin{vmatrix} \zeta^1_{F\mathcal{D}j} & \zeta^2_{F\mathcal{D}j} & \zeta^3_{F\mathcal{D}j} \end{vmatrix} \neq 0 \tag{2.32}$$

where $\mathbf{0}$ stands for the three-dimensional zero vector. As $\begin{vmatrix} \mathbf{i} & \mathbf{j} & \mathbf{k} \end{vmatrix} = 1$, (2.32) can be reduced to

$$\begin{vmatrix} \zeta^1_{F\mathcal{D}j} & \zeta^2_{F\mathcal{D}j} & \zeta^3_{F\mathcal{D}j} \end{vmatrix} \neq 0. \tag{2.33}$$

Since the axes of the R joints on the moving platform are not parallel to a plane, (2.33) is satisfied. Thus, the set of actuated joints is valid for the 3-($\underline{\text{P}}$RRR)$_A$ PM shown in Fig. 2.23a.

2.4 Summary

In this chapter, basic concepts of screw theory were first recalled. More specifically, screw systems and the operations than can be performed on them were reviewed. Also, the important concept of reciprocity was recalled. Using this concept, the wrench system of a KC can be obtained from the twist system of the KC, and vice visa. Then, classes of KCs whose wrench systems always includes a specified number of independent wrenches of zero-pitch or infinity-pitch were given. These classes of KCs are referred to as compositional units because of the role that they play in the type synthesis of PMs. The instantaneous mobility analysis of KCs was also performed in this chapter. Important fundamental relationships that will be used in the type synthesis were established. Finally a condition was given for the determination of the validity of a given set of actuated joints in a PM.

3. Type Synthesis of Single-Loop Kinematic Chains

This chapter deals with the type synthesis of (full-cycle mobility) multi-DOF single-loop KCs with a specified c-ζ-system. The rationale for addressing this problem is that it constitutes one of the important foundations for the type synthesis of PMs. Based on the results presented in Chap. 2 — especially the reciprocal screw systems and the compositional units with specific constraint characteristics — multi-DOF single-loop kinematic chains with a specified c-ζ-system are constructed in a straightforward way. Although the results presented in this chapter may not seem to have direct practical relevance, they will be used in the type synthesis of PMs in the following chapters. In fact, it will be shown later that the type synthesis approach proposed here for single-loop kinematic chains is a key component of the type synthesis methodology developed in this book for PMs. Therefore, this chapter is devoted solely to the type synthesis of single-loop kinematic chains.

3.1 Introduction

Previous research on the type synthesis of overconstrained single-loop KCs mainly focused on single-DOF single-loop KCs [118, 127]. Although several multi-DOF overconstrained single-loop KCs have been proposed [118], the work on the type synthesis of multi-DOF overconstrained single-loop KCs is not systematic. In this chapter, we will discuss the type synthesis of single-loop KCs with a specified c-ζ-system.

3.2 Procedure for the Type Synthesis

Based on the results of Chap. 2, it is possible to derive a systematic synthesis procedure. The type synthesis of single-loop KCs with a c-ζ-system can be performed using the following three steps:

X. Kong and C. Gosselin: Type Syn. of Parallel Mech., STAR 33, pp. 43–53, 2007.
springerlink.com

Step 1. Determine the number of joints.

From the mobility criterion of the single-loop KC (2.15), we obtain

$$f = \mathcal{F} + (6 - c) \tag{3.1}$$

where it is recalled that f is the total number of 1-DOF joints, $\mathcal{F}$ is the mobility of the single-loop KC, and c is the order of the wrench system of the serial KC associated with the single-loop KC.

Step 2. Find the geometric conditions on joint axes based on the twist-wrench relations.

Based on the reciprocity condition of screws (see Sect. 2.1.3), we have:

(a1) The axis of an R joint is coplanar with the axis of any ζ_0 within the c-ζ-system.

(a2) The direction of a P joint is perpendicular to the axis of any ζ_0 within the c-ζ-system.

(a3) The axis of an R joint is perpendicular to the axis of any ζ_∞ within the c-ζ-system.

The geometric conditions on joint axes can be further revealed without difficulty for single-loop KCs with a c-ζ-system in a general configuration.

Step 3. Identify the types of single-loop KCs with a c-ζ-system.

Based on the specific geometric conditions on joint axes obtained in Step 2, a number of types of single-loop KCs with a c-ζ-system can be constructed from one or two compositional units with specific characteristics except for the coaxial compositional units and the codirectional compositional units.

In addition, single-loop KCs with a c-ζ-system can be obtained from the single-loop KCs with a $(c+p)$-ζ-system, by inserting p coaxial compositional units and/or codirectional compositional units. The DOF $\mathcal{F}'$ and the number of joints f' of a single-loop KCs with a $(c+p)$-ζ-system are respectively

$$\mathcal{F}' = \mathcal{F} - (f_p - p) \tag{3.2}$$

where f_p denotes the total number of 1-DOF joints within the p coaxial or codirectional compositional units, and

$$f' = f - f_p. \tag{3.3}$$

Example 3.1. Consider the type synthesis of single-loop KCs with the 3-ζ_∞-system. The procedure proposed above can be applied as follows:

Step 1. Determine the number of joints. Since the wrench system is a 3-ζ-system, we have $c = 3$. Therefore, according to (3.1), the number of joints, f, is equal to $(\mathcal{F} + 3)$.

Step 2. Find the geometric conditions on joint axes based on the twist-wrench relations.

A basis of the 3-ζ_∞-system can be selected as three ζ_∞ [Fig. 2.5a] that are not parallel to a plane. Therefore, based on the twist-wrench relations, we can obtain that the specific geometric condition on joint axes for a single-loop KCs with the 3-ζ_∞-system is that the number of R joints is zero since

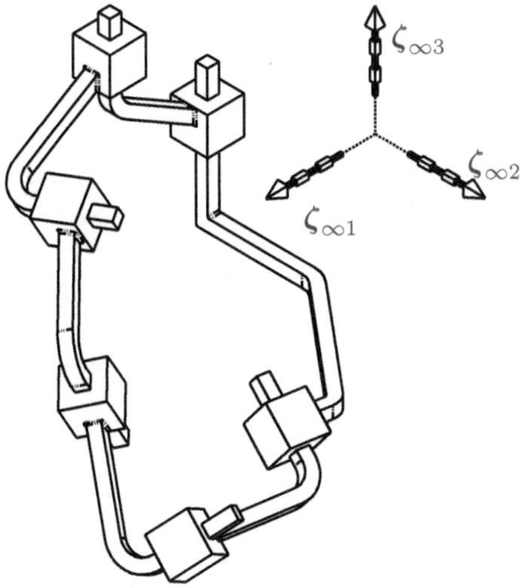

Fig. 3.1. $(\text{PPPPPP})_T$ 3-DOF single-loop KC with the 3-$\boldsymbol{\zeta}_\infty$-system

there exists no R joint whose axis is perpendicular to three lines that are not parallel to one plane.

Step 3. Identify the types of single-loop KCs with the 3-$\boldsymbol{\zeta}_\infty$-system.

The types of single-loop KCs with the 3-$\boldsymbol{\zeta}_\infty$-system are the single-loop KCs that are composed of one spatial translational compositional unit which is composed of $f = (F+3)$ P joints. The 3-$\boldsymbol{\zeta}_\infty$-system is composed of all the $\boldsymbol{\zeta}_\infty$. For instance, if the desired mobility is 3, we have $F = 3$ and therefore $f = 6$. The result is illustrated schematically in Fig. 3.1 where a 3-DOF single-loop KC with the 3-$\boldsymbol{\zeta}_\infty$-system is shown.

Example 3.2. Consider the type synthesis of single-loop KCs with a 2-$\boldsymbol{\zeta}_\infty$-system.

Step 1. Determine the number of joints. Since the wrench system is a 2-$\boldsymbol{\zeta}$-system, we have $c = 2$. According to (3.1), the number of joints, f, is equal to $(\mathcal{F} + 4)$.

Step 2. Find the geometric conditions on joint axes based on the twist-wrench relations.

In a 2-$\boldsymbol{\zeta}_\infty$-system [Fig. 2.4a], all the $\boldsymbol{\zeta}_\infty$ are always parallel to a plane. Therefore, based on the twist-wrench relations, we can obtain that all the axes of the R joints should be parallel to a line that is perpendicular to the directions of the $\boldsymbol{\zeta}_\infty$ within the 2-$\boldsymbol{\zeta}_\infty$-system.

Step 3. Identify the types of single-loop KCs with a 2-$\boldsymbol{\zeta}_\infty$-system.

Since all the axes of the R joints must be parallel, it follows that all single-loop KCs with a 2-$\boldsymbol{\zeta}_\infty$-system are composed of one Parallelaxis compositional unit. The 2-$\boldsymbol{\zeta}_\infty$-system is composed of all the $\boldsymbol{\zeta}_\infty$ whose axes are

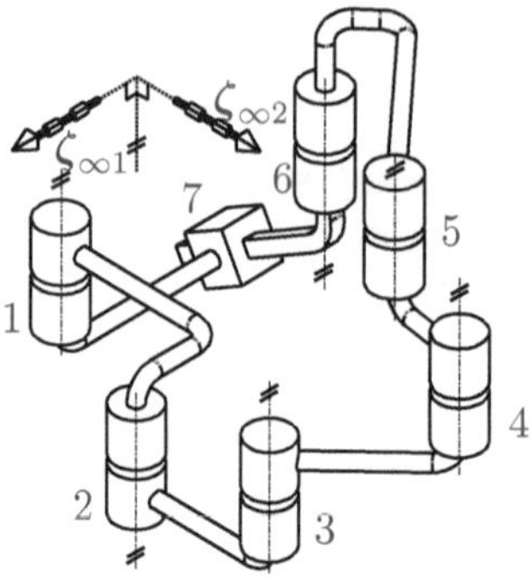

(a) $(\text{RRRRRRP})_A$.

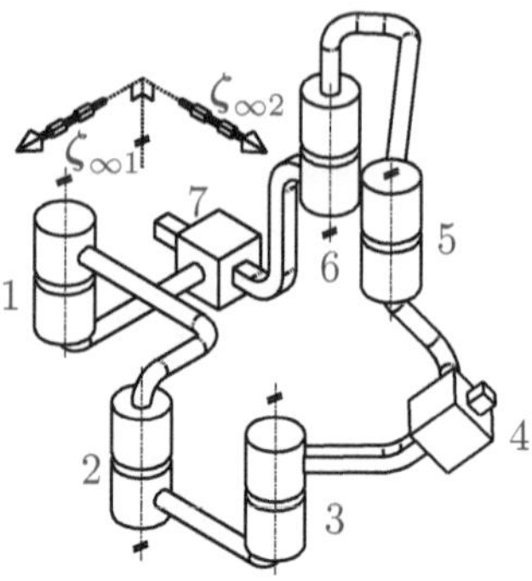

(b) $(\text{RRRPRRP})_A$.

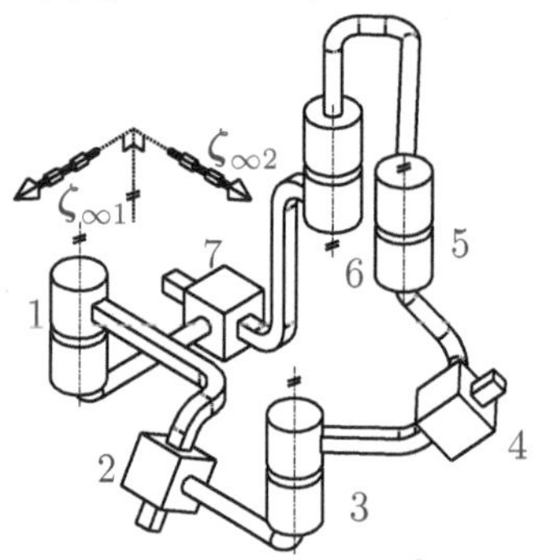

(c) $(\text{RPRPRRP})_A$.

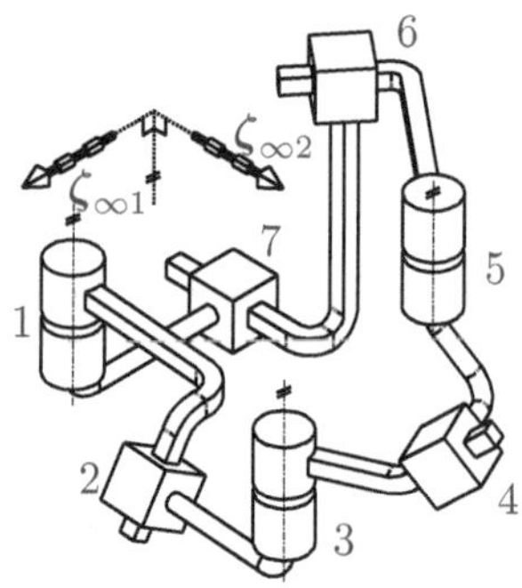

(d) $(\text{RPRPRPP})_A$.

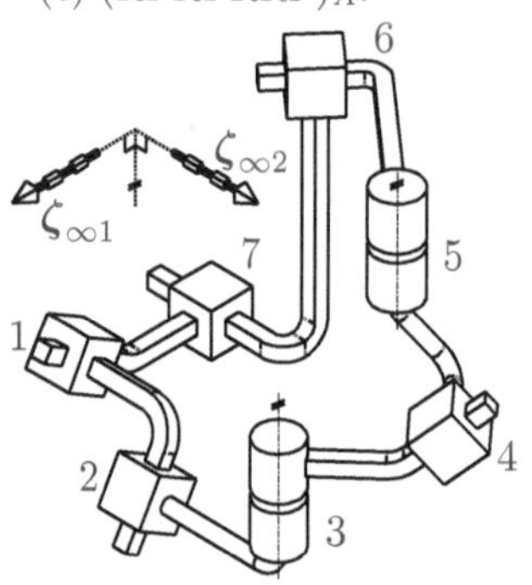

(e) $(\text{PPRPRPP})_A$.

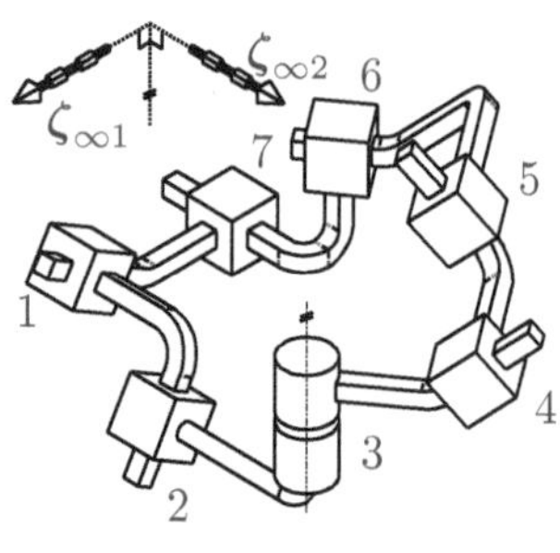

(f) $(\text{PPRPPPP})_A$.

Fig. 3.2. 3-DOF single-loop KCs with a 2-ζ_∞-system

perpendicular to all the axes of the R joints. Figure 3.2 shows several 3-DOF single-loop KCs with a 2-ζ_∞-system.

Example 3.3. Consider now the type synthesis of single-loop KCs with a 3-ζ_0-system.

Step 1. Determine the number of joints. Since the wrench system is a 3-ζ-system, we have $c = 3$. According to (3.1), the number of joints is $f = \mathcal{F} + 3$.

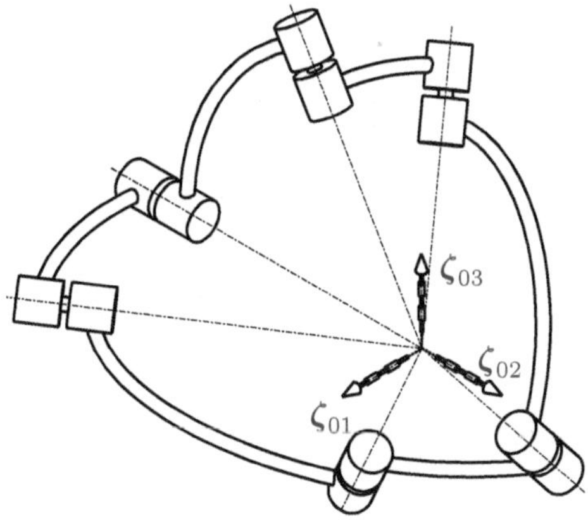

Fig. 3.3. $(\mathrm{RRRRRR})_S$ 3-DOF single-loop KC with a 3-ζ_0-system

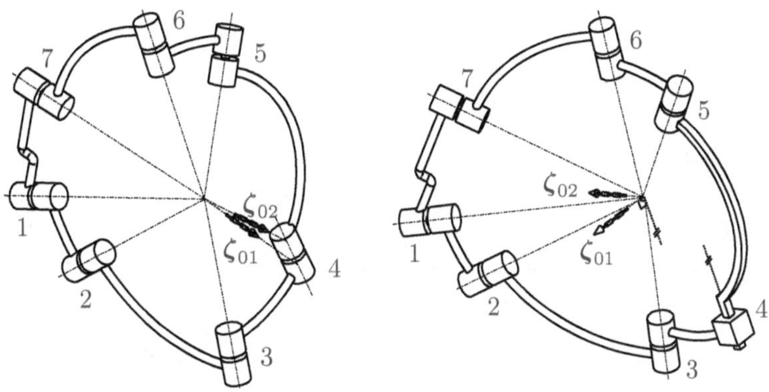

(a) $(\mathrm{RRR(R)}_L\mathrm{RRR})_S$. **(b)** $(\mathrm{RRR(P)}_L\mathrm{RRR})_S$.

Fig. 3.4. 3-DOF single-loop KCs with a 2-ζ_0-system

Step 2. Find the geometric conditions on joint axes based on the twist-wrench relations.

A basis for the 3-ζ_0-system [Fig. 2.5e] can be any three non-coplanar ζ_0 intersecting at one common point. Therefore, based on the twist-wrench relations, we can obtain that the specific geometric conditions on joint axes for a single-loop KCs with a 3-ζ_0-system are that (a) all the axes of the R joints pass through the centre of the 3-ζ_0-system, and (b) the number of P joints is zero since there exists no P joint whose direction is perpendicular to the axes of three ζ_0 that are not parallel to one plane.

Step 3. Identify the types of single-loop KCs with a 3-ζ_0-system.

Since the axes of all the R joints must intersect in one point, the types of single-loop KCs with a 3-ζ_0-system are the single-loop KC that are composed of one spherical compositional unit which is composed of $f = (F+3)$ R joints. The 3-ζ_0-system is composed of all the ζ_0 whose axes pass through the centre of the spherical serial compositional unit. This is illustrated in Figure 3.3 where a 3-DOF single-loop KC with a 3-ζ_0-system is shown.

Example 3.4. Consider the type synthesis of single-loop KCs with a 2-ζ_0-system.

Step 1. Determine the number of joints. Since in this case we have $c = 2$, the number of joints given by (3.1) is $\mathcal{F} + 4$.

Step 2. Find the geometric conditions on joint axes based on the twist-wrench relations.

A basis of the 2-ζ_0-system [Fig. 2.4c] can be obtained by considering two intersecting ζ_0 which define the plane and the centre of the 2-ζ_0-system. Based on the twist-wrench relations, we can obtain that (a) all the axes of the R joints either pass through the centre of the 2-ζ_0-system or are located on the plane of the 2-ζ_0-system, and (b) all the P joints are in a direction perpendicular to the plane of the 2-ζ_0-system.

Step 3. Identify the types of single-loop KCs with a 2-ζ_0-system.

From the above geometric conditions, it is easily determined that single-loop KCs with a 2-ζ_0-system can be obtained by (a) inserting one coaxial compositional unit into one single-loop KC with a 3-ζ_0-system, or (b) inserting one codirectional compositional unit into one single-loop KC with a 3-ζ_0-system. The 2-ζ_0-system is composed of all the ζ_0 whose axes pass through the centre of the spherical compositional unit, are perpendicular to all the axes of the P joints within the codirectional compositional unit and intersect the axes of the R joints within the coaxial compositional unit. Examples are given in Fig. 3.4 where two 3-DOF single-loop KCs with a 2-ζ_0-system are represented schematically.

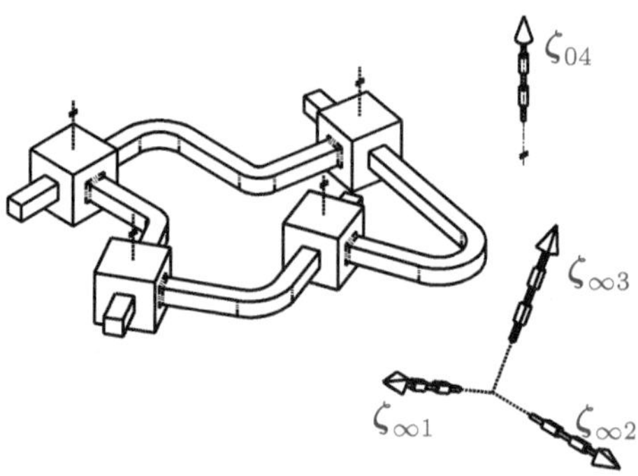

Fig. 3.5. $(\mathrm{PPPP})_E$ single-loop KCs with a 3-ζ_∞-1-ζ_0-system

3.3 Types of Single-Loop Kinematic Chains with a c-ζ-System

The examples of the preceding section illustrate how the synthesis methodology can be applied to generate types of single-loop KCs with a given wrench system. The procedure is systematic and rather straightforward, given the classes of serial KCs provided in Fig. 2.16. This procedure can be repeated systematically for the type synthesis of single-loop KCs with a c-ζ-system, thereby leading to

Table 3.1. Types of single-loop KCs with a 4-ζ-system

Wrench system	Geometric conditions on joint axes	Composition
3-ζ_∞-1-ζ_0-system	The number of R joints is zero since there exists no R joint whose axis is perpendicular to three directions that are not parallel to one plane. All the directions of the P joints are parallel to a plane that is perpendicular to the axis of the ζ_0.	One planar translational compositional unit (Fig. 3.5).

Table 3.2. Types of single-loop KCs with a 3-ζ-system

Wrench system	Geometric conditions on joint axes	Composition
3-ζ_∞-system	The number of R joints is zero since there exists no R joint whose axis is perpendicular to three directions that are not parallel to one plane.	One spatial translational compositional unit [Fig. 3.6a].
2-ζ_∞-1-ζ_0-system (Perpendicular case)	All the axes of the R joints are parallel to the axis of the ζ_0, and all the directions of the P joints are parallel to a plane that is perpendicular to the axis of the ζ_0.	One planar compositional unit [Fig. 3.6b].
2-ζ_∞-1-ζ_0-system (General case)	All the axes of the R joints are perpendicular to the directions of the ζ_∞ and coplanar with the axis of the ζ_0, and all the directions of the P joints are parallel to a plane that is perpendicular to the axis of the ζ_0.	One planar translational compositional unit + one coaxial compositional unit [Fig. 3.6c].
3-ζ_0-system	All the axes of the R joints pass through the centre of the 3-ζ_0-system, and the number of P joints is zero since there exists no P joint whose direction is perpendicular to axes of three ζ_0 that are not parallel to one plane.	One spherical compositional unit [Fig. 3.6d].

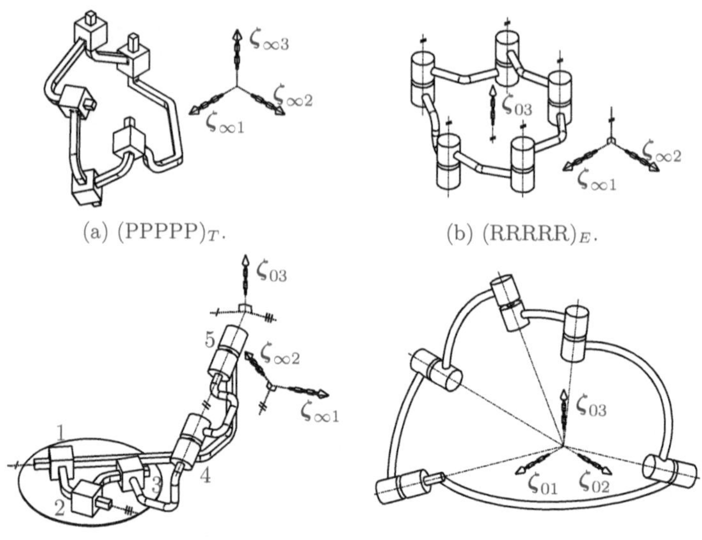

(a) $(PPPPP)_T$. (b) $(RRRRR)_E$.

(c) $(PPP(RR)_L)_E$. (d) $(RRRRR)_S$.

Fig. 3.6. Some 2-DOF single-loop KCs with a 3-ζ-system: (**a**) KC with the 3-ζ_∞-system, (**b**) KC with a 2-ζ_∞-1-ζ_0-system (perpendicular case), (**c**) KC with a 2-ζ_∞-1-ζ_0-system (general case) and (**d**) KC with a 3-ζ_0-system

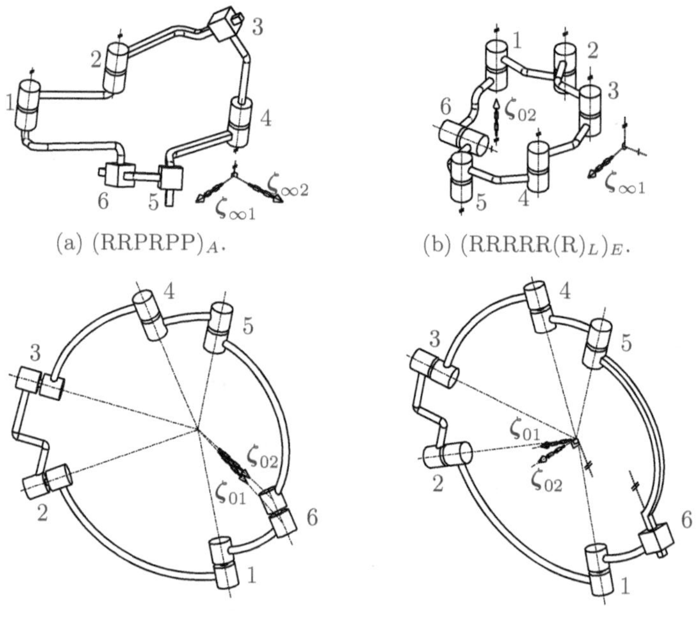

(a) $(RRPRPP)_A$. (b) $(RRRRR(R)_L)_E$.

(c) $(RRRRR(R)_L)_S$. (d) $(RRRRR(P)_L)_S$.

Fig. 3.7. Some 2-DOF single-loop KCs with a 2-ζ-system: (**a**) KC with a 2-ζ_∞-system, (**b**) KC with a 1-ζ_∞-1-ζ_0-system and (**c**)–(**d**) KCs with a 2-ζ_0-system

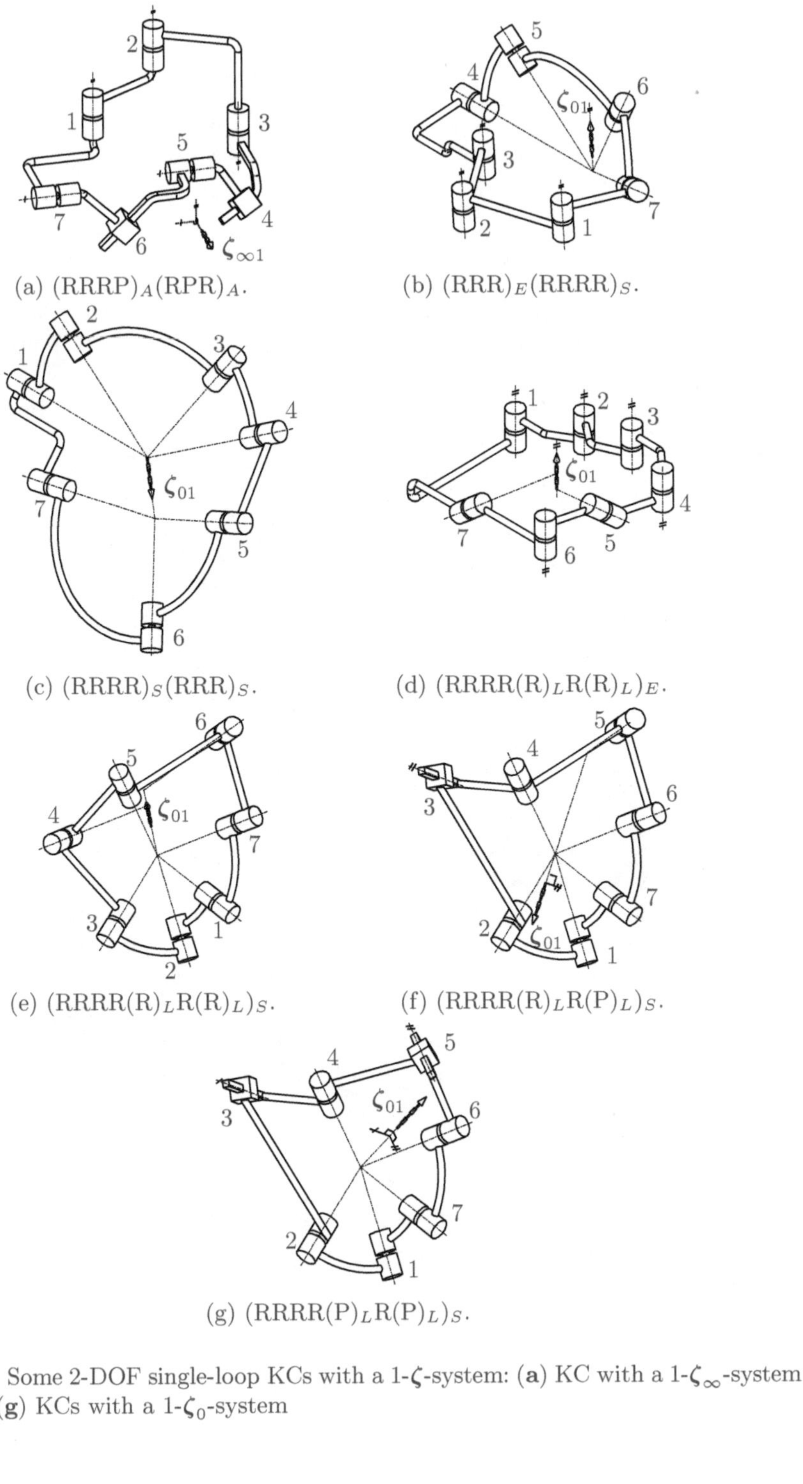

(a) $(\mathrm{RRRP})_A(\mathrm{RPR})_A$.
(b) $(\mathrm{RRR})_E(\mathrm{RRRR})_S$.
(c) $(\mathrm{RRRR})_S(\mathrm{RRR})_S$.
(d) $(\mathrm{RRRR(R)}_L\mathrm{R(R)}_L)_E$.
(e) $(\mathrm{RRRR(R)}_L\mathrm{R(R)}_L)_S$.
(f) $(\mathrm{RRRR(R)}_L\mathrm{R(P)}_L)_S$.
(g) $(\mathrm{RRRR(P)}_L\mathrm{R(P)}_L)_S$.

Fig. 3.8. Some 2-DOF single-loop KCs with a 1-$\boldsymbol{\zeta}$-system: (**a**) KC with a 1-$\boldsymbol{\zeta}_\infty$-system and (**b**)–(**g**) KCs with a 1-$\boldsymbol{\zeta}_0$-system

Table 3.3. Types of single-loop KCs with a 2-ζ-system

Wrench system	Geometric conditions on joint axes	Composition
2-ζ_∞-system	All the axes of the R joints are parallel to a line which is perpendicular to the directions of the ζ_∞.	One Parallelaxis compositional unit [Fig. 3.7a].
1-ζ_∞-1-ζ_0-system	All the axes of the R joints are coplanar with the axis of the ζ_0 and parallel to a plane which is perpendicular to the direction of the ζ_∞. The directions of the P joints are parallel to a plane that is perpendicular to the axis of the ζ_0.	Inserting one coaxial compositional unit into one single-loop KC composed of one planar compositional unit [Fig. 3.7b].
2-ζ_0-system	All the axes of the R joints pass through the centre of the 2-ζ_0-system or are located on the plane of the 2-ζ_0-system, and all the directions of the P joints are perpendicular to the plane of the 2-ζ_0-system.	(a) Inserting one coaxial compositional unit into one single-loop KC composed of one spherical compositional unit [Fig. 3.7c], or (b) Inserting one codirectional compositional unit into one single-loop KC composed of one spherical compositional unit [Fig. 3.7d].

a number of single-loop KCs with a c-ζ-system. The single-loop KCs obtained are listed in Tables 3.1–3.4, according to their wrench systems.

As it will be seen in Chap. 5 and in subsequent chapters, the single-loop KCs listed in Tables 3.1–3.4 constitute a set of building blocks that will be used in the type synthesis of PKCs. Therefore, the contents of these tables and the companion figures should be reviewed carefully before proceeding further.

Finally, it is noted that the list provided in Tables 3.1–3.4 is not exhaustive and there may exist other types of multi-DOF single-loop KCs with a specified

Table 3.4. Types of single-loop KCs with a 1-ζ-system

Wrench system	Geometric conditions on joint axes	Composition
1-ζ_∞-system	All the axes of the R joints are parallel to a plane which is perpendicular to the direction of the ζ_∞.	(a) Two Parallelaxis compositional units or planar compositional units [Fig. 3.8a], or (b) Inserting one coaxial compositional unit into a single-loop KC composed of one Parallelaxis compositional unit.

Table 3.4. (*continued*)

Wrench system	Geometric conditions on joint axes	Composition
1-ζ_0-system	All the axes of the R joints are coplanar with the axis of the ζ_0, all the directions of the P joints are parallel to a plane which is perpendicular to the axis of the ζ_0.	(a) One planar compositional unit + one spherical compositional unit [Fig. 3.8b], (b) Two spherical compositional units with distinct centres [Fig. 3.8c], (c) Inserting two coaxial compositional units into one single-loop KC composed of one planar compositional unit [Fig. 3.8d], (d) Inserting two coaxial compositional units into one single-loop KC composed of one spherical compositional unit [Fig. 3.8e], (e) Inserting one coaxial compositional unit and one codirectional compositional unit into one single-loop KC composed of one spherical compositional unit [Fig. 3.8f], or (f) Inserting two codirectional compositional units into one single-loop KC composed of one spherical compositional unit [Fig. 3.8g].

c-ζ-system. However, these other types of single-loop KCs, such as those encountered in some of the PMs proposed in [84], would require additional geometric constraints on the linkage parameters (link dimensions). Therefore the PMs related to these single-loop KCs may not be of practical use and these types of single-loop KCs are omitted in this book.

3.4 Summary

This chapter introduced a systematic methodology for the type synthesis of multi-DOF single-loop KCs with a specified c-ζ-system. Based on the proposed approach, types of single-loop KCs with a specified c-ζ-system were constructed using the compositional units given in Chap. 2. These single-loop KCs will be used to construct the types of PMs in Chaps. 6–14 and to determine whether a PM has full-cycle mobility in Appendix B.

4. Classification of Parallel Mechanisms

One of the first challenges to be addressed in the type synthesis of PMs with less than 6 DOF (or constrained mobility) is the accurate and unambiguous description of the desired motion. In some instances, this issue is not trivial, as evidenced by the plurality of notations and conventions used in the literature to describe the motion of 4-DOF and 5-DOF PMs. This chapter aims at proposing a classification of the motion spaces of PMs in order to make the synthesis more systematic. To this end, the concepts of motion pattern and virtual chain are introduced to represent the motion of the moving platform of PMs. A preliminary classification of the motion patterns and PMs is then proposed. This chapter, together with Chaps. 2 and 3, lays the foundation for the type synthesis of PMs generating a specified motion pattern presented in the following chapters.

4.1 Motion Patterns of Parallel Mechanisms

Before performing the type synthesis of PMs, the desired motion of the moving platform should be specified. In some cases, this issue is trivial such as for planar or translational PMs. However, for other types of motions (e.g. for 4-DOF and 5-DOF PMs), the specification of the desired motion of the moving platform is more complex. Indeed, there are many types of possible motion of the moving platform and the sole DOF is not sufficient to describe a motion. For instance, a 3-DOF motion may be a 3-DOF translational motion, a 3-DOF spherical motion, a 3-DOF planar motion or any other 3-DOF motion.

A *motion pattern* is defined as a (possibly unbounded) continuous set of poses which describes the type of motion desired at the moving platform. For example, for translational PMs, the motion pattern is described by the set of rigid body translations. Another example of motion pattern is the SCARA motion for 4-DOF PMs, which consists of all translations as well as rotations about any axis in a given fixed direction. This motion pattern can be described by the Schönflies motion [47].

X. Kong and C. Gosselin: Type Syn. of Parallel Mech., STAR 33, pp. 55–61, 2007.
springerlink.com

It is pointed out that a motion pattern cannot always be represented by a displacement group. In fact, the concept of motion pattern is very useful because it is more general than the concept of displacement group.

4.2 The Concept of Virtual Chain

Since the existing methods of description of the motion patterns of moving platforms are inconvenient in certain cases, the concept of virtual chain will be introduced in this section to represent the motion pattern of a moving platform.

A *virtual chain* associated with a motion pattern is a serial or parallel KC whose moving platform has the given prescribed motion pattern. Clearly, the concept of virtual chain is not unique and several virtual chains may be defined for one given motion pattern. In general, for a given motion pattern, a virtual chain is proposed by a comprehensive wrench system analysis of a large number of serial chains and parallel chains and by considering the possible changes in the wrench system with a change of the configuration of the serial chain or the parallel chain. Also, as a general rule, the simplest possible virtual chain will be selected.

For example, the wrench system of the $(PPPR)_A$, $(PPRR)_A$, $(PRPR)_A$, and $(PRRR)_A$ serial chains are all a constant 2-$\boldsymbol{\zeta}_\infty$-system with the directions of all the $\boldsymbol{\zeta}_\infty$ perpendicular to the axes of the R joints. Any of these KCs can be used to represent the same motion pattern. However, the $(PPPR)_A$ serial KC is selected as the virtual chain, since it is the most intuitive chain that represents the motion pattern.

The virtual chains that are serial chains are called *serial virtual chains*. Examples of serial virtual chains are given in Figs. 4.1–4.3. Similarly, the virtual chains that are parallel chains are called *parallel virtual chains*. Examples of parallel virtual chains are shown in Fig. 4.4.

A serial virtual chain is denoted by the chain of characters that denote the type of joints in the virtual chain from the base to the moving platform. For example, the PPP virtual chain [Fig. 4.1a] is composed of three P joints connected serially. In addition to the PPP virtual chain, the serial virtual chains of practical or theoretical interest include: E [Fig. 4.1b], S [Fig. 4.1c], PPR [Fig. 4.1d], PPPR [Fig. 4.2a], PS [Fig. 4.2b], SP [Fig. 4.2c], PPPU [Fig. 4.3a], PPS [Fig. 4.3b], US [Fig. 4.3c] and UE [Fig. 4.3d] virtual chains.

A parallel virtual chain is denoted by the types of all its legs connected by '-'. For example, the 3-PPS virtual chain [Fig. 4.4a] is composed of three PPS legs, and the 2-PPPU virtual chain [Fig. 4.4b] is composed of 2 PPPU legs. Further classification of parallel virtual chains is still an open issue.

By inspection of Figs. 4.1–4.4, it can be observed that the concept of virtual chain is a very intuitive and convenient method of representing motion patterns without any ambiguity. In Chap. 5, it will be shown that the concept of virtual chain is also an effective tool in the type synthesis of PMs.

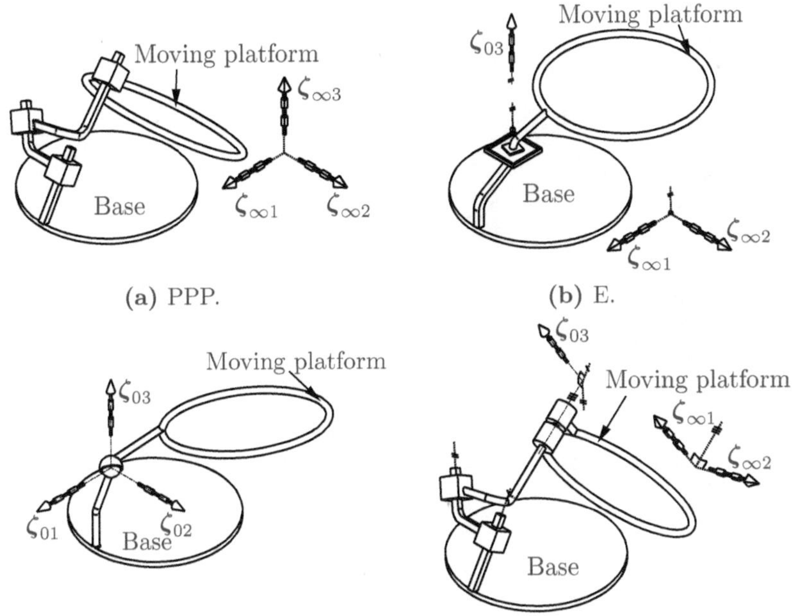

(a) PPP. (b) E. (c) S. (d) PPR.

Fig. 4.1. 3-DOF serial virtual chains

4.3 A Preliminary Classification of Motion Patterns and the Corresponding Parallel Mechanisms

The motion pattern corresponding to a virtual chain is called a V-motion. A PM generating a V-motion is called a V= PM, which stands for a V equivalent PM.

In order to make the synthesis process systematic and in order to maximize the practical relevance of the results, a preliminary classification of motion patterns and PMs is performed. The motion patterns and PMs of the greatest application potential are presented below:

- PPP motion: In this type of motion pattern [Fig. 4.1a], the moving platform can translate arbitrarily with respect to the base while its orientation must be constant. The corresponding class of PM is often called TPM (*translational PM*) in the literature. However, it will be referred to as PPP= PM in order to highlight the virtual chain to represent the motion. The wrench system of a PPP= PM is the 3-$\boldsymbol{\zeta}_\infty$-system.
- E motion: In this type of motion pattern [Fig. 4.1b], the moving platform can undergo planar motion with respect to the base. The corresponding class of PM is called *planar PM* in the literature. Here, it will be referred to as *E= PM*. The wrench system of an E= PM is a 2-$\boldsymbol{\zeta}_\infty$-1-$\boldsymbol{\zeta}_0$-system (Perpendicular case), which is composed of all the $\boldsymbol{\zeta}_\infty$ whose directions are parallel to the

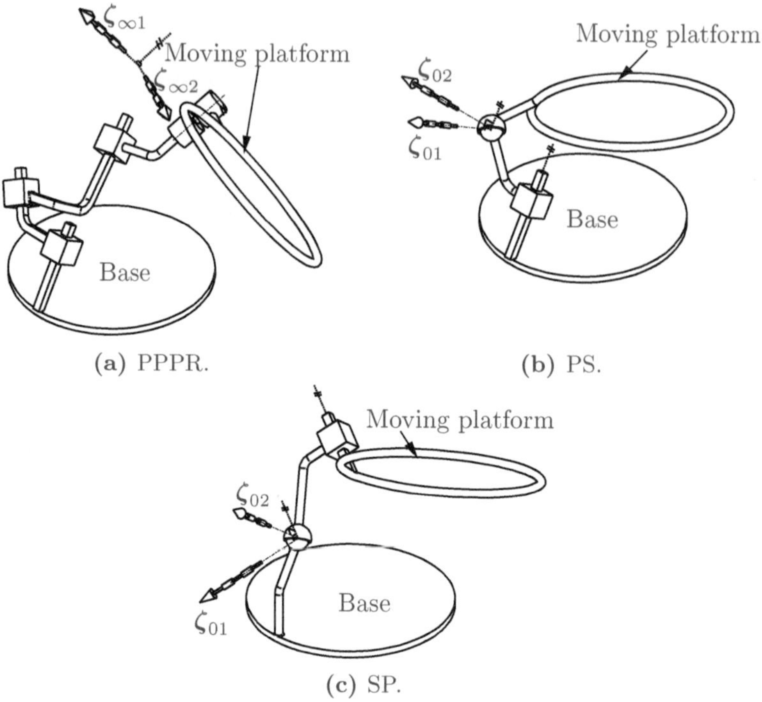

Fig. 4.2. 4-DOF serial virtual chains

plane of the E joint and all the $\boldsymbol{\zeta}_0$ whose axes are perpendicular to the plane of the E joint.

- S motion: In this type of motion pattern [Fig. 4.1c], there must be a fixed common point between the moving platform and the base while the moving platform can rotate arbitrarily with respect to the base. The corresponding class of PM is usually called *spherical PM*, *orientational PM* or *rotational PM* in the literature. It will be referred to as S= PM here. The wrench system of an S= PM is a 3-$\boldsymbol{\zeta}_0$-system in which all the axes of the $\boldsymbol{\zeta}_0$ pass through the centre of the S joint.
- PPR motion: In this type of motion pattern [Fig. 4.1d], the moving platform can rotate about an axis with a given constant direction which can translate along a plane. The corresponding class of PM is generally called cylindrical PM in the literature. It will be termed *PPR= PM* here. The wrench system of a PPR= PM is a 2-$\boldsymbol{\zeta}_\infty$-1-$\boldsymbol{\zeta}_0$-system (General case) which is composed of (a) all the $\boldsymbol{\zeta}_\infty$ whose axes are perpendicular to the axis of the R joint and (b) all the $\boldsymbol{\zeta}_0$ whose axes are perpendicular to the directions of the P joints and coplanar with the axis of the R joint, and (c) other $\boldsymbol{\zeta}$ which are linear combinations the above $\boldsymbol{\zeta}_\infty$ and $\boldsymbol{\zeta}_0$.
- PPPR motion: In this type of motion pattern [Fig. 4.2a], the moving platform can rotate about an axis which has a given constant direction and which can

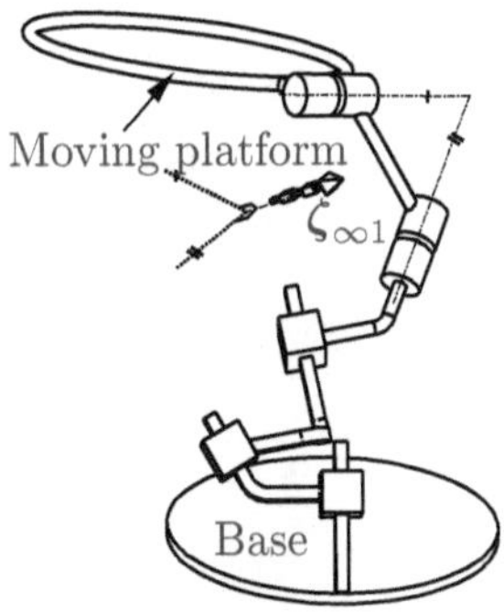

(a) PPPU.

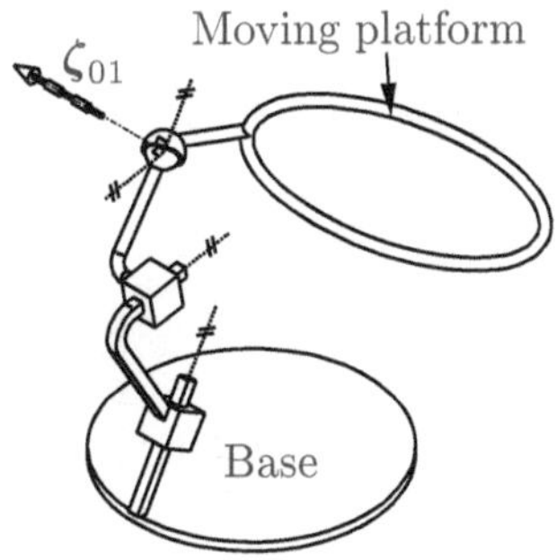

(b) PPS.

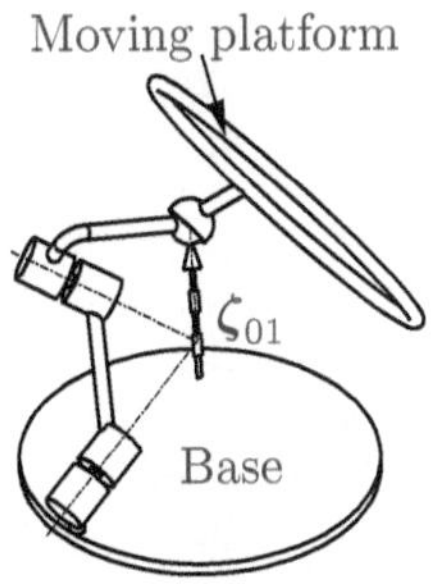

(c) US.

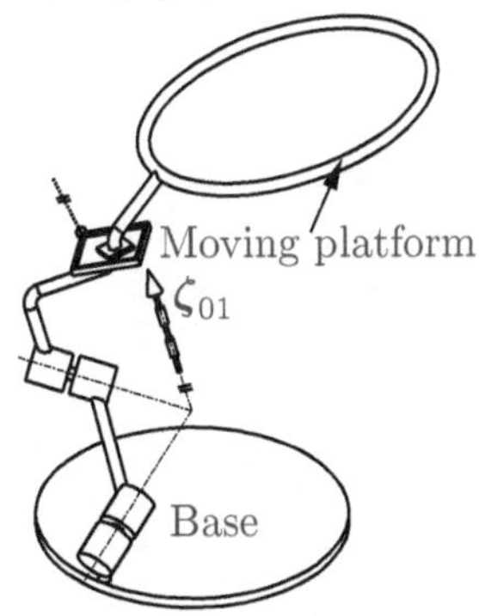

(d) UE.

Fig. 4.3. 5-DOF serial virtual chains

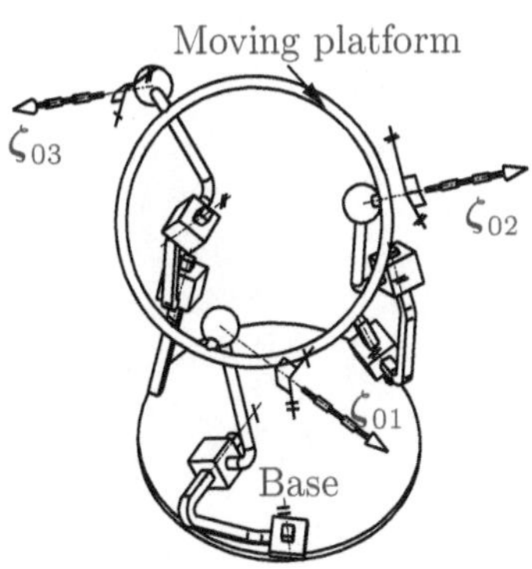

(a) 3-PPS.

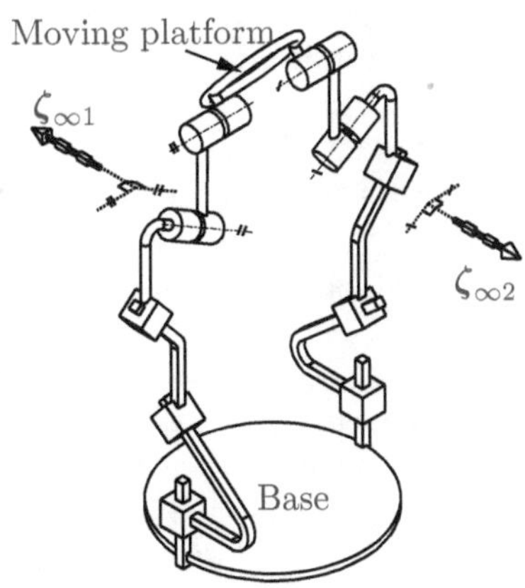

(b) 2-PPPU.

Fig. 4.4. Parallel virtual chains

translate arbitrarily. The PPPR motion is also called 3T1R motion, SCARA motion or Schönflies motion [3]. The corresponding class of PM is sometimes called *3T1R PM* in the literature. Here, it will be referred to as *PPPR= PM*. The wrench system of a PPPR= PM is a 2-ζ_∞-system in which the directions of the ζ_∞ are all perpendicular to the axis of the R joint within the PPPR virtual chain.

- PS motion: In this type of motion pattern [Fig. 4.2b], the moving platform can rotate about a point which translates along a given direction. The corresponding class of PM is termed *PS= PM*. The wrench system of a PS= PM is a 2-ζ_0-system in which all the axes of the ζ_0 pass through the centre of the S joint and are perpendicular to the direction of the P joint within the PS virtual chain.
- SP motion: In this type of motion pattern [Fig. 4.2c], the moving platform can undergo cylindrical motion along a line which rotates according to a U joint. The corresponding class of PM is termed *SP= PM*. The wrench system of an SP= PM is a 2-ζ_0-system in which all the axes of the ζ_0 pass through the centre of the S joint and are perpendicular to the direction of the P joint within the SP virtual chain.
- PPPU motion: In this type of motion pattern [Fig. 4.3a], the moving platform can rotate about a U joint whose centre can translate arbitrarily. The corresponding class of PM is often called *3T2R PM* in the literature [31, 35, 55, 62]. Here, it will be referred to as *PPPU= PM*. The wrench system of a PPPU= PM is a 1-ζ_∞-system in which the direction of the ζ_∞ is perpendicular to the axes of the U joints within the PPPU virtual chain.
- PPS motion: In this type of motion pattern [Fig. 4.3b], the moving platform can rotate about a point which translates along a plane. The corresponding class of PM is often called *2T3R PM* in the literature[31, 41, 51, 55]. Here, it will be termed *PPS= PM*. The wrench system of a PPS= PM is a 1-ζ_0-system in which the axis of the ζ_0 passes through the centre of the S joint and is perpendicular to the directions of the two P joints within the PPS virtual chain.
- US motion: In this type of motion pattern [Fig. 4.3c], the moving platform can rotate arbitrarily about the centre of the S joint which moves along a spherical surface with its centre located at the centre of the U joint. The corresponding class of PM is termed *US= PM*. The wrench system of a US= PM is a 1-ζ_0-system in which the axis of the ζ_0 passes through the centre of the U joint and the centre of the S joint.
- UE motion: In this type of motion pattern [Fig. 4.3d], the moving platform can undergo planar motion along a plane which rotates according to a U joint. The corresponding class of PM is termed *UE= PM*. The wrench system of a UE= PM is a 1-ζ_0-system in which the axis of the ζ_0 passes through the centre of the U joint and is perpendicular to the plane of the E joint within the UE virtual chain.
- 3-PPS motion: This type of motion pattern [Fig. 4.4a] is not easy to describe. The corresponding class of PM is called *zero-torsion PM* [9] in the literature.

It will be referred to as *3-PPS= PM*. The wrench system of a 3-PPS= PM is a special 3-ζ-system which is in fact a linear combination of the three 1-ζ_0-systems of the three PPS legs within the 3-PPS virtual chain.

- 2-PPPU motion: Like the 3-PPS motion, this type of motion pattern [Fig. 4.4b] is also not easy to describe. This corresponding class of PM is called *2-PPPU= PM*. One such PM was first proposed in [31]. The wrench system of a 2-PPPU= PM is a 2-ζ_∞-system which is in fact a linear combination of the two 1-ζ_∞-systems of the two PPPU legs within the 2-PPPU virtual chain.

Clearly, the classification of motion patterns and PMs proposed above is not exhaustive. However, it is thought that the proposed classes cover the most practically relevant cases. Furthermore, this classification being based on virtual chains, it alleviates any potential ambiguity in the description of the motion pattern and wrench system.

4.4 Summary

In this chapter, the concept of motion pattern was introduced in order to rigorously describe the motion to be produced by a PM. Based on this notion, virtual chains were introduced that describe, simply and without ambiguity, the motion patterns of interest in the type synthesis of PMs. As it will be seen in the next chapter, the concept of virtual chain is at the core of the type synthesis approach proposed in this book.

5. Virtual-Chain Approach for the Type Synthesis of Parallel Mechanisms

In this chapter, a method for the type synthesis of parallel mechanisms (PMs) with a specified motion pattern that can be represented by a serial virtual chain is developed based on screw theory and the concept of virtual chain (Chaps. 2–4). A general procedure is proposed which consists of four main steps, namely: (1) the decomposition of the wrench system of a parallel kinematic chain associated with a V virtual chain (V= PKC), (2) the type synthesis of legs, (3) the assembly of legs to generate V= PKCs, and (4) the selection of the actuated joints. These steps are discussed successively in separate sections. Using the procedure proposed here, the type synthesis of several classes of PMs will be dealt with in details in Chaps. 6–13. The results presented in this chapter also lay the foundations for the type synthesis of PMs with a parallel virtual chain (Chap. 14).

5.1 Introduction

Most of the work reported to date on the type synthesis of PMs uses, as a starting point, a specified DOF of a PM. Only few reported synthesis initiatives rely on a specified motion pattern. The type synthesis of PMs presented in this book is based on a specified motion pattern and a specified number of overconstraints (also referred to as redundant constraints or passive constraints), which is denoted by Δ. As explained in Chap. 4, the rationale behind this approach is twofold: (1) In many applications, PMs generating a specified motion pattern are required. A specified motion pattern contains more information than a specified DOF, and (2) The number of overconstraints Δ is also an important index characterizing the mechanical properties of PMs. Indeed, the complexity, the cost and the performance of PMs generating the same motion pattern usually vary consistently with a change of Δ.

Equations (2.28), (2.24) and (2.29) proposed in Chap. 2 will be used in the type synthesis of PMs generating a specified motion pattern. Additionally, for the sake of completeness of the results presented in Chaps. 6–13, Δ may take any one of the possible values during the type synthesis of PMs, thereby leading to a very general approach.

X. Kong and C. Gosselin: Type Syn. of Parallel Mech., STAR 33, pp. 63–83, 2007.
springerlink.com

5.2 Conditions for a Parallel Kinematic Chain to Be a V= Parallel Kinematic Chain

Let us consider an $\mathcal{F}$-DOF V= PKC (Fig. 5.1 where $\mathcal{F} = 3$ is used as an example). When we connect the base and the moving platform of a PKC by an appropriate virtual chain [Fig.5.1b] matching the motion pattern of the PKC, the function of the PKC is not affected. Any of its legs and the virtual chain thereby constitute an $\mathcal{F}$-DOF single loop KC [Fig.5.1c]. Meanwhile, the wrench system of the original PKC [Fig.5.1a] must be the same as that of the virtual chain in any general configuration. Otherwise, the DOF of the moving platform will be greater than that of the virtual chain. For clarity, a virtual chain is occasionally represented by a rectangle drawn with dashed lines.

Based on the concept of virtual chain, it becomes clear that a PKC is a V= PKC if it satisfies the following two conditions:

(1) Each leg of the PKC and a same virtual chain constitute an $\mathcal{F}$-DOF single-loop KC.
(2) The wrench system of the PKC is the same as that of the virtual chain in any general configuration.

Condition (1) for V= PKCs guarantees that the moving platform can undergo the V-motion, while Condition (2) for V= PKCs guarantees further that the DOF of the moving platform is the same as that of the virtual chain.

It is noted that when the orders of the leg-wrench system and the wrench system of the virtual chain are both greater than 0, the single-loop KC constructed must be an overconstrained KC. This is so because the virtual chain constrains the moving platform to the motion pattern and therefore already contains all the required constraint wrenches.

5.3 Systematic Type Synthesis of V= Parallel Mechanisms

Based on the above conditions and the results of the preceding chapters, a general systematic procedure can now be proposed for the type synthesis of V= PMs. The proposed procedure can be divided into four main steps, namely:

Step 1. Decomposition of the wrench system of a V= PKC.
Step 2. Type synthesis of legs for PKCs.

Here, a leg for V= PKCs refers to a leg which has a specified leg-wrench system and which, together with a virtual chain, can form an $\mathcal{F}$-DOF single-loop KC. In order to achieve this, two sub-steps are proposed.

Step 2a. Type synthesis of $\mathcal{F}$-DOF single-loop KCs that involve a virtual chain and have a specified leg-wrench system.

Step 2b. Generation of types of legs for V= PKCs by removing the virtual chain from the $\mathcal{F}$-DOF single-loop KCs obtained in Step 2a.

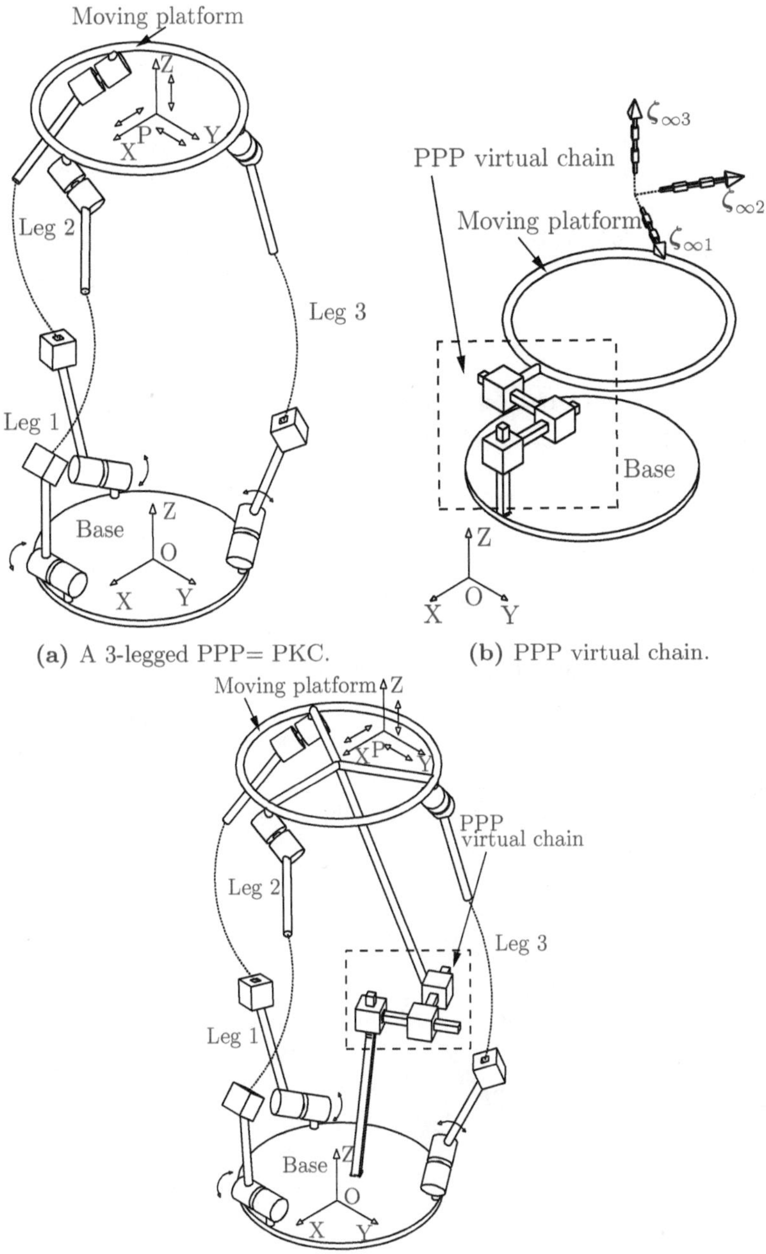

(a) A 3-legged PPP= PKC.

(b) PPP virtual chain.

(c) A 3-legged PPP= PKC with a PPP virtual chain added.

Fig. 5.1. PPP= PKC

Step 3. Assembly of legs for V= PKCs.
V= PKCs can be generated by assembling m legs for V= PKCs, obtained in Step 2, such that (1) each leg and a same virtual chain can form a $\mathcal{F}$-DOF single-loop kinematic chain, and (2) the linear combination of the leg-wrench systems is the same as that of the virtual chain (Condition (2) for V= PKCs). These conditions can be easily satisfied by inspection.

Step 4. Selection of actuated joints.
V= PMs can be generated by selecting the actuated joints for each V= PKC obtained in Step 3.

The above procedure will be elaborated upon in the following sections.

5.4 Step 1: Decomposition of the Wrench System of a Parallel Kinematic Chain

The decomposition of the wrench system of an m-legged V= PKC consists in finding all the leg-wrench systems and all the combinations of m leg-wrench systems for the V= PKC and a specified number of overconstraints Δ. Step 1 can be performed using (2.24) and (2.28).

5.4.1 Determination of the Leg-Wrench Systems

As the wrench system of a PKC is the linear combination of those of all its legs in a general configuration, any leg-wrench system in a V= PKC is a sub-system of the V= PKC's wrench system. Following the procedure proposed in Sect. 2.1.3, all the leg-wrench systems for a V= PKC can be identified without difficulty.

It is noted that in the type synthesis of PMs, one is only interested in the leg-wrench systems in which there is a set of basis wrenches with 0- and/or ∞-pitches.[1]

Example 5.1. Determine the leg-wrench systems for a PPP= PKC.

In any general configuration, the wrench system of a PPP= PKC is the same as that of its PPP virtual chain, i.e., the 3-ζ_∞-system, which is composed of all the ζ_∞.

It then follows that any leg-wrench system with order $c^i > 0$ of a PPP= PKC is either the 3-ζ_∞-system, a 2-ζ_∞-system, or a 1-ζ_∞-system in a general configuration (Fig. 5.2).

Example 5.2. Determine the leg-wrench systems for a PPR= PKC.

In any general configuration, the wrench system of a PPR= PKC is the same as that of its PPR virtual chain [Fig. 4.1d], i.e., a 2-ζ_∞-1-ζ_0-system, which is

[1] There exist leg-wrench systems with a non-zero finite pitch for PMs, such as the PPR= PMs (see Chap. 8), which have wrench systems composed of wrenches of different pitches. However, such leg-wrench systems are not presented in this book for clarity. The associated types of legs can be obtained following the procedure proposed in this book. Each of these legs contains one or more inactive joints than some types of legs listed in this book.

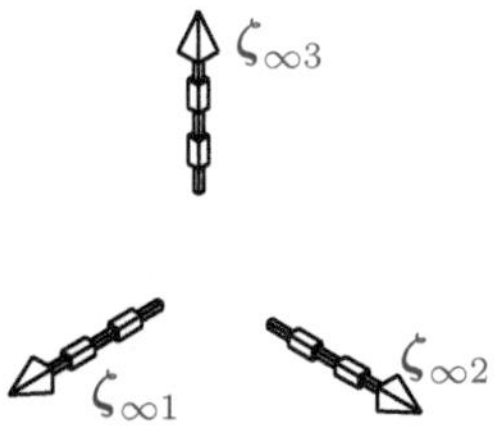

(a) 3-ζ_∞-system.

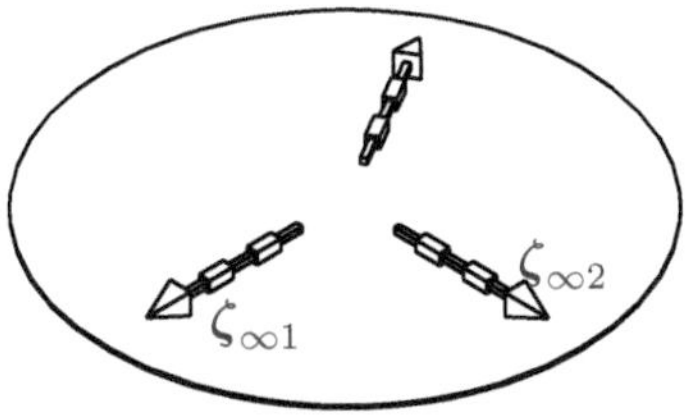

(b) 2-ζ_∞-system.

(c) 1-ζ_∞-system.

Fig. 5.2. Leg-wrench systems of PPP= PKCs

composed of (a) all the $\boldsymbol{\zeta}_\infty$ whose axes are perpendicular to the axis of the R joint and (b) all the $\boldsymbol{\zeta}_0$ whose axes are perpendicular to the directions of the P joints and coplanar with the axis of the R joint, and (c) other $\boldsymbol{\zeta}$ which are linear combinations of the above $\boldsymbol{\zeta}_\infty$ and $\boldsymbol{\zeta}_0$. One set of basis wrenches of the 2-$\boldsymbol{\zeta}_\infty$-1-$\boldsymbol{\zeta}_0$-system is $\boldsymbol{\zeta}_{01}$, $\boldsymbol{\zeta}_{\infty 2}$ and $\boldsymbol{\zeta}_{\infty 3}$.

It then follows that any leg-wrench system with order $c^i > 0$ of a PPR= PKC is either a 2-$\boldsymbol{\zeta}_\infty$-1-$\boldsymbol{\zeta}_0$-system, a 1-$\boldsymbol{\zeta}_\infty$-1-$\boldsymbol{\zeta}_0$-system, a 2-$\boldsymbol{\zeta}_\infty$-system, a 1-$\boldsymbol{\zeta}_0$-system or a 1-$\boldsymbol{\zeta}_\infty$-system in a general configuration (Fig. 5.3).

5.4.2 Determination of the Combinations of Leg-Wrench Systems

Wrench System Composed of Wrenches of the Same Pitch

For most of the commonly used motion patterns considered, all the wrenches in the wrench systems are of the same pitch. The leg-wrench systems are $c^i(0 \leq c^i \leq c)$-systems of the same pitch. The combination of leg-wrench systems can be simply represented by the combination of the orders, c^i, of leg-wrench systems.

The combinations of the orders, c^i, of leg-wrench systems can be determined by solving (2.28), which is recalled here for convenience

$$\sum_{i=1}^{m} c^i = 6 - \mathcal{F} + \Delta.$$

It should be noted that $0 \leq c^i \leq c$. Tables 5.1–5.5 show the set of c^i for $m(2 \leq m \leq (\mathcal{F}+1))$-legged V= PKCs. In Tables 5.1– 5.5, the sets of c^i corresponding to all possible values of Δ have been listed for completeness. For an m-legged $\mathcal{F}$-DOF V= PKC, Δ varies from 0 to $(m-1)(6-\mathcal{F})$.

Wrench System Composed of Wrenches of Different Pitches

For the motion patterns for which not all the wrenches in the wrench systems are of the same pitch, the combination of leg-wrench systems cannot be simply

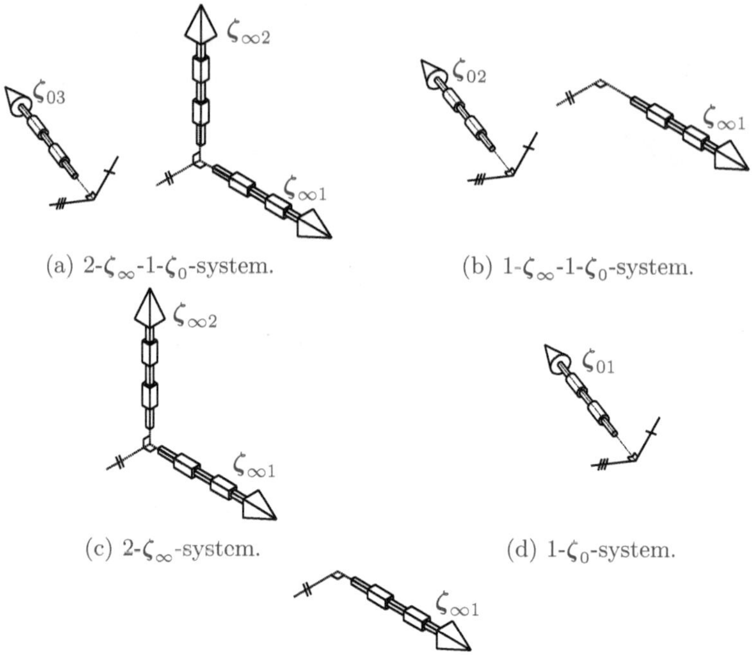

(a) 2-ζ_∞-1-ζ_0-system. (b) 1-ζ_∞-1-ζ_0-system.

(c) 2-ζ_∞-system. (d) 1-ζ_0-system.

(e) 1-ζ_∞-system.

Fig. 5.3. Leg-wrench systems of PPR= PKCs

represented by the combination of the orders, c^i, of leg-wrench systems. In the combination of leg-wrench systems, the types of leg-wrench systems should also be considered.

The decomposition of wrench systems composed of wrenches of different pitches can be performed in two steps, namely:

Step a. Determine the combinations of the orders, c^i, of leg-wrench systems by solving (2.28). This step is the same as in the case of wrench systems composed of wrenches of the same pitch.

Step b. For each combination of the orders, c^i, of leg-wrench systems, determine the combinations of the types of leg-wrench systems by solving (2.24).

Example 5.3. Determine the combinations of the leg-wrench systems of m-legged PPR= PKCs with $2 \leq m \leq 4$.

Step a. Determine the combinations of the orders, c^i, of leg-wrench systems by solving (2.28). The results are shown in Tables 5.1– 5.5.

Step b. For each combination of the orders, c^i, of leg-wrench systems, determine the combination of the types of leg-wrench systems by solving (2.24). The

Table 5.1. Combinations of c^i for m-legged 2-DOF PKCs (Case $2 \leq m \leq 3$)

m	c	Δ	c^1	c^2	c^3
2	4	4	4	4	
		3	4	3	
		2	4	2	
			3	3	
		1	4	1	
			3	2	
		0	4	0	
			3	1	
			2	2	
3	4	8	4	4	4
		7	4	4	3
		6	4	4	2
			4	3	3
		5	4	4	1
			4	3	2
			3	3	3
		4	4	4	0
			4	3	1
			4	2	2
			3	3	2
		3	4	3	0
			4	2	1
			3	3	1
			3	2	2
		2	4	2	0
			4	1	1
			3	3	0
			3	2	1
			2	2	2
		1	4	1	0
			3	2	0
			3	1	1
			2	2	1
		0	4	0	0
			3	1	0
			2	2	0
			2	1	1

Table 5.2. Combinations of c^i for m-legged 3-DOF PKCs (Case $2 \leq m \leq 4$)

m	c	Δ	c^1	c^2	c^3	c^4
2	3	3	3	3		
		2	3	2		
		1	3	1		
			2	2		
		0	3	0		
			2	1		
3	3	6	3	3	3	
		5	3	3	2	
		4	3	3	1	
			3	2	2	
		3	3	3	0	
			3	2	1	
			2	2	2	
		2	3	2	0	
			3	1	1	
			2	2	1	
		1	3	1	0	
			2	2	0	
			2	1	1	
		0	3	0	0	
			2	1	0	
			1	1	1	

m	c	Δ	c^1	c^2	c^3	c^4
4	3	9	3	3	3	3
		8	3	3	3	2
		7	3	3	3	1
			3	3	2	2
		6	3	3	3	0
			3	3	2	1
			3	2	2	2
		5	3	3	2	0
			3	3	1	1
			3	2	2	1
			2	2	2	2
		4	3	3	1	0
			3	2	2	0
			3	2	1	1
			2	2	2	1
		3	3	3	0	0
			3	2	1	0
			3	1	1	1
			2	2	2	0
			2	2	1	1
		2	3	2	0	0
			3	1	1	0
			2	2	1	0
			2	1	1	1
		1	3	1	0	0
			2	2	0	0
			2	1	1	0
			1	1	1	1
		0	3	0	0	0
			2	1	0	0
			1	1	1	0

combinations of leg-wrench systems which can form a 2-ζ_∞-1-ζ_0-system can be obtained using (2.24) and are listed in Tables 5.6–5.8.

5.5 Step 2: Type Synthesis of Legs

Once the different combinations of leg-wrench systems have been established, the kinematic chains that instantiate these systems can be determined. This operation is performed in two steps.

Table 5.3. Combinations of c^i for m-legged 4-DOF PKCs (Case $2 \le m \le 5$)

m	c	Δ	c^1	c^2	c^3	c^4	c^5
2	2	2	2	2			
		1	2	1			
		0	2	0			
3	2	4	2	2	2		
		3	2	2	1		
		2	2	2	0		
			2	1	1		
		1	2	1	0		
			1	1	1		
		0	2	0	0		
			1	1	0		
4	2	6	2	2	2	2	
		5	2	2	2	1	
		4	2	2	2	0	
			2	2	1	1	
		3	2	2	1	0	
			2	1	1	1	
		2	2	2	0	0	
			2	1	1	0	
			1	1	1	1	
		1	2	1	0	0	
			1	1	1	0	
		0	1	1	0	0	
5	2	8	2	2	2	2	2
		7	2	2	2	2	1
		6	2	2	2	2	0
			2	2	2	1	1
		5	2	2	2	1	0
			2	2	1	1	1
		4	2	2	2	0	0
			2	2	1	1	0
			2	1	1	1	1
		3	2	2	1	0	0
			2	1	1	1	0
			1	1	1	1	1
		2	2	2	0	0	0
			2	1	1	0	0
			1	1	1	1	0
		1	2	1	0	0	0
			1	1	1	0	0
		0	2	0	0	0	0
			1	1	0	0	0

Table 5.4. Combinations of c^i for m-legged 5-DOF PKCs (Case $2 \leq m \leq 6$)

m	c	Δ	c^1	c^2	c^3	c^4	c^5	c^6
2	1	1	1	1				
		0	1	0				
3	1	2	1	1	1			
		1	1	1	0			
		0	1	0	0			
4	1	3	1	1	1	1		
		2	1	1	1	0		
		1	1	1	0	0		
		0	1	0	0	0		
5	1	4	1	1	1	1	1	
		3	1	1	1	1	0	
		2	1	1	1	0	0	
		1	1	1	0	0	0	
		0	1	0	0	0	0	
6	1	5	1	1	1	1	1	1
		4	1	1	1	1	1	0
		3	1	1	1	1	0	0
		2	1	1	1	0	0	0
		1	1	1	0	0	0	0
		0	1	0	0	0	0	0

Table 5.5. Combinations of c^i for m-legged 6-DOF PKCs (Case $2 \leq m \leq 7$)

m	c	Δ	c^1	c^2	c^3	c^4	c^5	c^6	c^7
2	0	0	0	0					
3	0	0	0	0	0				
4	0	0	0	0	0	0			
5	0	0	0	0	0	0	0		
6	0	0	0	0	0	0	0	0	
7	0	0	0	0	0	0	0	0	0

5.5.1 Step 2a: Type Synthesis of $\mathcal{F}$-DOF Single-Loop Kinematic Chains That Involve a Virtual Chain and Have a Specified Leg-Wrench System

The type synthesis of $\mathcal{F}$-DOF single-loop KCs that involve a virtual chain and have a specified leg-wrench system can be performed as follows.

First, perform the type synthesis of single-loop KCs that involve a virtual chain and have a specified leg-wrench system (See Chap. 3). Secondly, discard

Table 5.6. Combinations of leg-wrench systems for 2-legged PPR= PKCs

Δ	2-ζ_∞-1-ζ_0	1-ζ_∞-1-ζ_0	2-ζ_∞	1-ζ_0	1-ζ_∞	0-system
3	2	0	0	0	0	0
2	1	1	0	0	0	0
	1	0	1	0	0	0
1	1	0	0	1	0	0
	1	0	0	0	1	0
	0	2	0	0	0	0
	0	1	1	0	0	0
	0	0	2	0	0	0
0	1	0	0	0	0	1
	0	1	0	1	0	0
	0	1	0	0	1	0
	0	0	1	1	0	0

those single-loop KCs in which the twists of all the joints but the virtual chain are linearly dependent. This operation should be performed for each of the leg-wrench systems obtained in Step 1 of the type synthesis procedure.

For illustrative purposes, let us consider the type synthesis of 3-DOF single-loop KCs that involve a PPP virtual chain and have a 2-ζ_∞-system.

First, we perform the type synthesis of 3-DOF single-loop KCs that involve a PPP virtual chain and have a 2-ζ_∞-system. According to the results of Chap. 3, there are seven joints in a 3-DOF single-loop KC that involves a PPP virtual chain and has a 2-ζ_∞-system. Such a 3-DOF single-loop KC is composed of one spatial Parallelaxis compositional unit (Fig. 5.4). It is pointed out that the P joints can be placed anywhere in the single-loop KC. For brevity, we list only the 3-DOF single-loop KCs from which all 3-DOF single-loop KCs can be obtained through a permutation of the joints. For example, an $(\text{RRPRV})_A$ single-loop KC can be obtained by changing the position of the P joint in the $(\text{PRRRV})_A$ single-loop KC shown in Fig. 5.4b. The 2-ζ_∞-system is composed of all the ζ_∞ whose directions are perpendicular to all the axes of the R joints.

Secondly, we discard those single-loop KCs in which the twists of all the joints but the virtual chain are linearly dependent. The $(\text{RRRRV})_A$ KC [Fig. 5.4a] should be discarded since the twists of all the four R-joints are linearly dependent. For each of the other KCs shown in Fig. 5.4, the KC composed of all the joints except the virtual chain must be a spatial Parallelaxis compositional unit.

Generally, there are many cases in which the twists of the joints within a same leg are linearly dependent. Fortunately, the conditions for the twists of the joints within a same leg to be linearly independent are not difficult to address in the type synthesis of PMs. This is because of the characteristics of the compositional units used to construct PMs and due to that we focus on the type synthesis of

Table 5.7. Combinations of leg-wrench systems for 3-legged PPR= PKCs

Δ	2-ζ_∞-1-ζ_0	1-ζ_∞-1-ζ_0	2-ζ_∞	1-ζ_0	1-ζ_∞	0-system
6	3	0	0	0	0	0
5	2	1	0	0	0	0
	2	0	1	0	0	0
4	2	0	0	1	0	0
	2	0	0	0	1	0
	1	2	0	0	0	0
	1	1	1	0	0	0
	1	0	2	0	0	0
3	2	0	0	0	0	1
	1	1	0	1	0	0
	1	1	0	0	1	0
	1	0	1	1	0	0
	1	0	1	0	1	0
	0	2	1	0	0	0
	0	1	2	0	0	0
2	1	1	0	0	0	1
	1	0	1	0	0	1
	1	0	0	2	0	0
	1	0	0	1	1	0
	1	0	0	0	2	0
	0	2	0	1	0	0
	0	2	0	0	1	0
	0	1	1	1	0	0
	0	1	1	0	1	0
	0	0	2	1	0	0
	0	0	2	0	1	0
1	1	0	0	1	0	1
	1	0	0	0	1	1
	0	2	0	0	0	1
	0	1	1	0	0	1
	0	0	2	0	0	1
	0	1	0	2	0	0
	0	1	0	1	1	0
	0	1	0	0	2	0
0	1	0	0	0	0	2
	0	1	0	1	0	1
	0	1	0	0	1	1
	0	0	1	1	0	1
	0	0	1	0	1	1
	0	0	0	1	2	0
	0	0	0	2	1	0

Table 5.8. Combinations of leg-wrench systems for 4-legged PPR= PKCs

Δ	2-ζ_∞-1-ζ_0	1-ζ_∞-1-ζ_0	2-ζ_∞	1-ζ_0	1-ζ_∞	0-system
9	4	0	0	0	0	0
8	3	1	0	0	0	0
	3	0	1	0	0	0
7	3	0	0	1	0	0
	3	0	0	0	1	0
6	3	0	0	0	0	1
	2	1	0	1	0	0
	2	1	0	0	1	0
	2	0	1	1	0	0
	2	0	1	0	1	0
	1	3	0	0	0	0
	1	2	1	0	0	0
	1	1	2	0	0	0
	1	0	3	0	0	0
5	2	1	0	0	0	1
	2	0	1	0	0	1
	2	0	0	2	0	0
	2	0	0	1	1	0
	2	0	0	0	2	0
	1	2	0	1	0	0
	1	2	0	0	1	0
	1	0	2	1	0	0
	1	0	2	0	1	0
4	2	0	0	1	0	1
	2	0	0	0	1	1
	1	2	0	0	0	1
	1	1	1	0	0	1
	1	0	2	0	0	1
	1	1	0	2	0	0
	1	1	0	0	2	0
	1	1	0	1	1	0
	1	0	1	2	0	0
	1	0	1	0	2	0
	1	0	1	1	1	0
3	2	0	0	0	0	2
	1	1	0	1	0	1
	1	1	0	0	1	1

continued on next page

CONTINUED FROM PREVIOUS PAGE

Δ	2-ζ_∞-1-ζ_0	1-ζ_∞-1-ζ_0	2-ζ_∞	1-ζ_0	1-ζ_∞	0-system
	1	0	1	1	0	1
	1	0	1	0	1	1
	0	2	0	2	0	0
	0	2	0	1	1	0
	0	2	0	0	2	0
	0	0	2	2	0	0
	0	0	2	1	1	0
	0	1	2	0	0	0
2	1	1	0	0	0	2
	1	0	1	0	0	2
	1	0	0	2	0	1
	1	0	0	1	1	1
	1	0	0	0	2	1
	0	2	0	1	0	1
	0	2	0	0	1	1
	0	1	1	1	0	1
	0	1	1	0	1	1
	0	0	2	1	0	1
1	1	0	0	1	0	2
	1	0	0	0	1	2
	0	2	0	0	0	2
	0	1	1	0	0	2
	0	1	0	2	0	1
	0	1	0	1	1	1
	0	1	0	0	2	1
	0	0	1	2	0	1
	0	0	1	1	1	1
	0	0	0	2	2	0
0	1	0	0	0	0	3
	0	1	0	1	0	2
	0	1	0	0	1	2
	0	0	1	1	0	2
	0	0	0	1	2	1
	0	0	0	2	1	1

kinematically non-redundant PMs. For brevity, such conditions are listed below without explanation:

(a) There are no coaxial R joints.
(b) There are no P joints along the same direction.
(c) The direction of at most one P joint is parallel to the axis of an R joints.
(d) At most three R joints have parallel axes.
(e) The axes of at most three R joints pass through the same point.
(f) The directions of at most two P joints are parallel to the same plane.
(g) The sum of the number of R joints with parallel axes and the number of P joints is not greater than four.
(h) If the directions of n_P P joints are perpendicular to the axes of n_R R joints with parallel axes, then $n_P + n_R \leq 3$.

5.5.2 Step 2b: Generation of Types of Legs

Once the types of $\mathcal{F}$-DOF single-loop kinematic chains that involve a virtual chain and that have a specified leg-wrench system have been obtained, the types of legs for V= PMs can be readily obtained by removing the virtual chains from the $\mathcal{F}$-DOF single-loop KCs involving a virtual chain.

In removing the virtual chain, the specific geometric conditions which guarantee the satisfaction of Condition (1) for V= PKCs — i.e., that each leg together

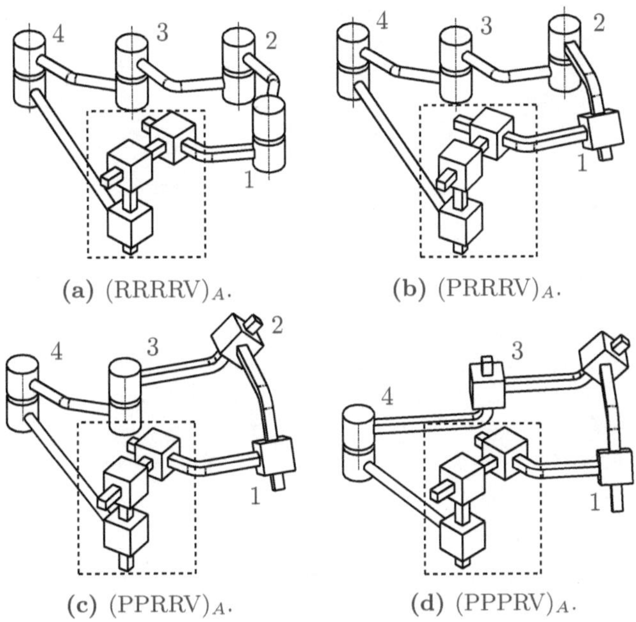

Fig. 5.4. 3-DOF single-loop KCs that involve a PPP virtual chain and have a 2-ζ_∞-system

with a same virtual chain should constitute an $\mathcal{F}$-DOF single-loop kinematic chain — should be clearly stated. These specific geometric conditions will be indicated by the notations that were introduced in Sect. 1.3 to represent the types of legs.

For instance, Fig. 5.5 shows a $(\text{PRRR})_A$ leg for PPP= PKC obtained from the 3-DOF $(\text{PRRRV})_A$ single-loop KC shown in Fig. 5.4. The wrench system of this leg is a 2-ζ_∞-system, which is composed of all the ζ_∞ whose directions are perpendicular to all the axes of the R joints.

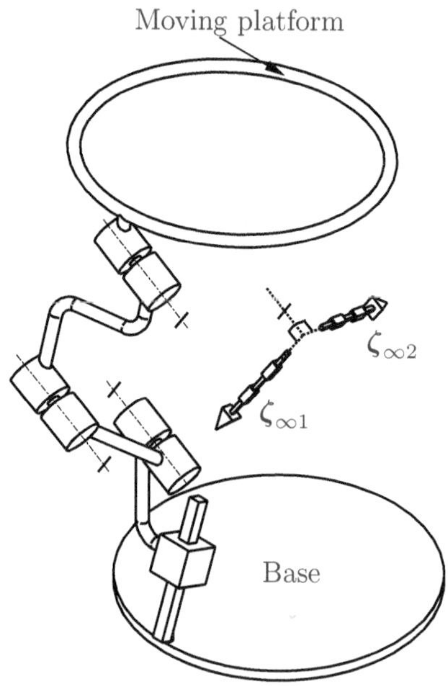

Fig. 5.5. $(\text{PRRR})_A$ leg for PPP= PKC

For some legs with $c^i > 0$ such as the $(\text{PRRR})_A$ leg shown in Fig. 5.5, the geometric conditions need to be satisfied only within a same leg and can therefore be derived directly from the associated single-loop kinematic chains. For other legs with $c^i > 0$, the geometric conditions among different legs should also be satisfied, in addition to geometric conditions within a same leg. This will become clearer in the next section and the following chapters where several families of PMs are synthesized.

5.6 Step 3: Assembly of Legs

The type synthesis of PKCs consists in obtaining the types of PKCs by assembling the legs obtained in Step 2 according to the combinations of m leg-wrench systems obtained in Step 1.

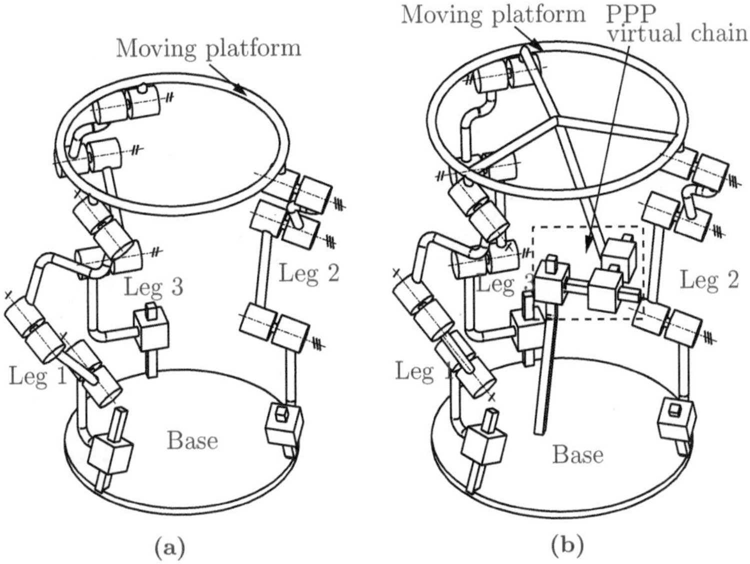

Fig. 5.6. (**a**) 3-$(PRRR)_A$ PPP= PKC and (**b**) 3-$(PRRR)_A$ PPP= PKC with a PPP virtual chain added

For the PKCs with invariant leg-wrench systems, i.e., the PKCs whose leg-wrench systems are all invariant with respect to the base and/or the moving platform, the condition which guarantees that the linear combination of all the leg-wrench systems constitutes the wrench system of the virtual chain (Conditions (2) for V= PKCs) will be revealed.

For those PKCs with varying leg-wrench systems, we make the assumption that the linear combination of the wrench systems of all the legs readily constitutes the desired wrench system.

For example, the 3-$(PRRR)_A$ PPP= PKC (Fig. 5.6) is a PKC with invariant leg-wrench systems. By imposing that not all the axes of the R joints on the moving platform are parallel, we guarantee that the linear combination of the three 2-ζ_∞-systems forms the 3-ζ_∞-system.

However, in many instances, the wrench system of a PKC varies with the configuration. The degeneracy of the wrench system of a PKC is referred to as a *constraint singularity* of a PM in [27, 133]. Since the constraint singularity analysis of a PM is input independent, it is more accurate to refer to it as the constraint singularity analysis of the PKC. The constraint singularity surfaces of a PKC comprise the open boundary of the workspace of the PMs corresponding to the PKC.

5.7 Step 4: Selection of the Actuated Joints

The selection of the actuated joints for V= PMs involves finding all the possible V= PMs for a given V= PKC and removing the cases for which the set of actuated joints is invalid. In the selection of the actuated joints, the following criteria are recommended:

(1) The actuated joints should be distributed among all the legs as evenly as possible.
(2) The actuated joints should preferably be on the base or close to the base.
(3) No unactuated active P joint should exist.

The first two criteria aim at optimizing the performance by minimizing the moving mass and inertia. The third criteria arises from practical considerations. Indeed, passive P joints tend to perform poorly, unless expensive guiding systems are used.

Since finding all the candidate V= PMs for a V= PKC is trivial, the validity detection of actuated joints of a candidate V= PM will be the focus of the selection of the actuated joints.

The selection of actuated joints for an $\mathcal{F}$-DOF PM should satisfy the validity condition for actuated joints of PMs proposed in Sect. 2.3. The validity detection of the actuated joints for PMs requires the computation of a 6×6 determinant. In fact, the validity detection of the actuated joints for an $\mathcal{F}$-DOF V= PM can be reduced to the calculation of an $\mathcal{F} \times \mathcal{F}$ determinant for most PMs discussed in this book, if an appropriate reference frame is selected.

The procedure for the validity detection of the actuated joints used in Chaps. 6–13 will now be outlined.

Following the validity condition of actuated joints for PMs (Sect. 2.3), we know that a set of $\mathcal{F}$ actuated joints for an $\mathcal{F}$-DOF PM is valid if and only if, in a general configuration, the actuation wrenches $\zeta^i_{\mathcal{D}j}$ $(i = 1, 2, \cdots, \mathcal{F})$, of the $\mathcal{F}$ actuated joints, together with a set of basis wrenches, ζ_k $(k = 1, 2, \cdots, c)$, of the wrench system, $\mathcal{W}$, of the PKC constitute a set of basis wrenches of the 6-system. We have

$$\left| \zeta^1_{\mathcal{D}j} \; \zeta^2_{\mathcal{D}j} \cdots \zeta^{\mathcal{F}}_{\mathcal{D}j} \; \zeta_1 \; \zeta_2 \cdots \zeta_c \right| \neq 0. \tag{5.1}$$

Since the linear dependency of screws is frame-independent, we can select a frame to represent a wrench system $\mathcal{W}$ in such a way that $(6-c)$ of the six scalar components in the same position are zero for each basis wrench of the wrench system $\mathcal{W}$. Let $\zeta^i_{t\mathcal{D}j}$ denote a vector consisting of the $(6-c)$ scalar components of the actuation wrench $\zeta^i_{\mathcal{D}j}$ in the same position as the $(6-c)$ vanishing scalar components of a basis wrench of the wrench system $\mathcal{W}$. $\zeta^i_{t\mathcal{D}j}$ is called the *t-component* of the actuation wrench $\zeta^i_{\mathcal{D}j}$, while the vector $\zeta^i_{w\mathcal{D}j}$ consisting of the other components of $\zeta^i_{\mathcal{D}j}$ is called the *w-component* of $\zeta^i_{\mathcal{D}j}$. For convenience, we make the assumption that the first $(6-c)$ of the six scalar components are all zero for each basis wrench of the wrench system $\mathcal{W}$. Equation (5.1) can then be expressed as

$$\begin{aligned} &\begin{vmatrix} \zeta^1_{t\mathcal{D}j} & \zeta^2_{t\mathcal{D}j} & \cdots & \zeta^{\mathcal{F}}_{t\mathcal{D}j} & \mathbf{0} & \mathbf{0} & \cdots & \mathbf{0} \\ \zeta^1_{w\mathcal{D}j} & \zeta^2_{w\mathcal{D}j} & \cdots & \zeta^{\mathcal{F}}_{w\mathcal{D}j} & \zeta_{w1} & \zeta_{w2} & \cdots & \zeta_{wc} \end{vmatrix} \\ &= \left| \zeta^1_{t\mathcal{D}j} \; \zeta^2_{t\mathcal{D}j} \cdots \zeta^{\mathcal{F}}_{t\mathcal{D}j} \right| \left| \zeta_{w1} \; \zeta_{w2} \cdots \zeta_{wc} \right| \neq 0. \end{aligned} \tag{5.2}$$

As $\left|\zeta_{w1}\ \zeta_{w2}\ \cdots\ \zeta_{wc}\right| \neq 0$, (5.2) can be reduced to

$$\left|\zeta^{1}_{t\not{D}j}\ \zeta^{2}_{t\not{D}j}\ \cdots\ \zeta^{\mathcal{F}}_{t\not{D}j}\right| \neq 0. \tag{5.3}$$

Then, the linear dependency of all the actuation wrenches and the wrench systems is equivalent to the linear dependency of the $\mathcal{F}(= 6 - c)$ t-components of the actuation wrenches. Thus, the validity detection of the actuated joints is reduced to the calculation of the determinant of an $\mathcal{F} \times \mathcal{F}$ matrix, each column of which is composed of the t-component of an actuation wrench.

The procedure for the validity detection of the actuated joints is proposed as follows:

Step 4a. If one or more of the actuated joints of a candidate PM are inactive, the set of actuated joints is invalid and the candidate PM should be discarded.

For the PKCs obtained using the virtual-chain approach proposed in this book, the inactive joints are revealed in the process of the type synthesis. For PKCs obtained using other approaches, the inactive joints may not be known. They can be revealed easily knowing that the actuation wrench of an inactive joint belongs to the wrench system of the PKC. This provides one way to detect inactive joints in a PKC.

As mentioned before, for a PKC with inactive joints and its kinematic equivalent PKC without inactive joints, the number of overconstraints as well as the reaction forces in the joints are different, although the inactive joints in a PKC make no contribution to the movement of the moving platform.

Step 4b. If the determinant of the $\mathcal{F} \times \mathcal{F}$ matrix composed of the t-components of all the actuation wrenches of actuated joints is always zero, the set of actuated joints for the PKC satisfying these conditions is invalid. In this case, the candidate PM should be discarded. Similarly, if the elements of the $\mathcal{F} \times \mathcal{F}$ matrix are constant and its determinant is only dependent on link parameters (Fig. 5.7), the conditions for its determinant to be zero should be revealed and PMs satisfying these conditions should also be discarded. The PMs for which the elements of the $\mathcal{F} \times \mathcal{F}$ matrix are constant and the determinant is not zero are proper and they have no constraint singularities. It is noted that in cases where there are no inactive joints among the set of actuated joints, we make the assumption that the set of actuated joints is valid if the elements of the $\mathcal{F} \times \mathcal{F}$ matrix are configuration dependent.

Example 5.4. Select actuated joints for the 3-$(\text{PRRR})_A$ PPP= PKC of Sect. 5.6.

Based on the recommended criteria for the selection of the actuated joints, we obtain one candidate 3-$(\underline{\text{P}}\text{RRR})_A$ PPP= PM [Fig. 5.7a] from the 3-$(\text{PRRR})_A$ PPP= PKC.

Now, let us perform the validity check for the actuated joints.

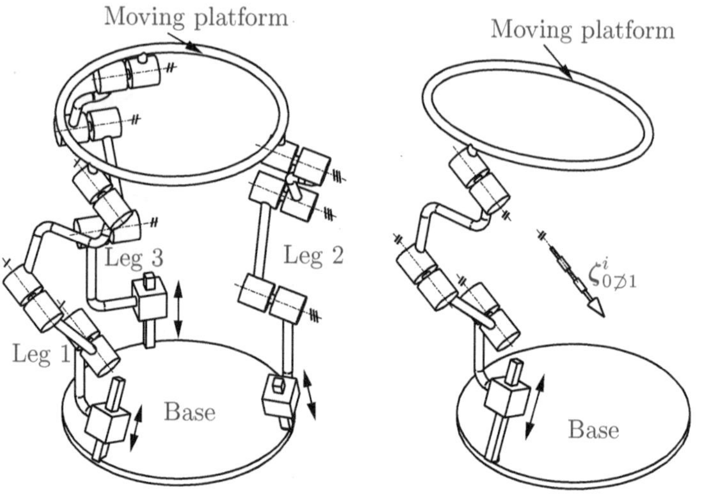

(a) 3-($\underline{\text{P}}$RRR)$_A$ PPP= PM.

(b) Actuation wrench of ($\underline{\text{P}}$RRR)$_A$ leg.

Fig. 5.7. Selection of actuated joints for 3-($\underline{\text{P}}$RRR)$_A$ PPP= PM

Step 4a. In the 3-(PRRR)$_A$ PPP= PKC, there are no inactive joints. Go to Step 4b.

Step 4b. In each ($\underline{\text{P}}$RRR)$_A$ leg [Fig. 5.7b] of the candidate 3-($\underline{\text{P}}$RRR)$_A$ PPP= PM [Fig. 5.7a], the first P joint is actuated. The actuation wrench is any ζ_0 whose axis is parallel to the axes of the three R joints within the same leg [Fig. 5.7b]. In any frame, $[\mathbf{0} \quad \mathbf{i}^T]^T$, $[\mathbf{0} \quad \mathbf{j}^T]^T$, $[\mathbf{0} \quad \mathbf{k}^T]^T$ can be used to denote a basis of $\mathcal{W}$. Here, **i**, **j** and **k** denote respectively the unit vectors along the X, Y and Z axes. Hence, the first three scalar components of all the basis wrenches of $\mathcal{W}$ are zero. Thus, the *t-component*, $\zeta^i_{t\not D j}$, of the $\zeta^i_{\not D j}$ is composed of the first three scalar components of the $\zeta^i_{\not D j}$, i.e.,

$$\zeta^i_{t\not D j} = \zeta^i_{F\not D j} \tag{5.4}$$

where $\zeta^i_{F\not D j}$ represents a unit vector which is parallel to the axes of the R joints within leg i.

The validity condition of actuated joints of a PPP= PM can be expressed as

$$\left| \zeta^1_{F\not D j} \; \zeta^2_{F\not D j} \; \zeta^3_{F\not D j} \right| \neq 0. \tag{5.5}$$

The elements of the 3×3 matrix for the 3-($\underline{\text{P}}$RRR)$_A$ PPP= PM are constant and its determinant is dependent only on link parameters. Hence, in order for the actuated joints to be valid, the PPP= PM should be such that all the axes of the R joints on the moving platform are not parallel to a plane.

5.8 Summary

Based on the results of Chaps. 2–4, this chapter proposed a systematic procedure for the type synthesis of PMs. Using the concept of virtual chain, two simple conditions were given to determine the validity of candidate PKCs. Then, a four-step procedure was presented for the type synthesis of PMs. The proposed approach is general and easy to use. It will be applied to the type synthesis of several families of PMs in the second part of this book.

Part II

Case Studies

学而时习之，不亦乐乎？
——孔子(公元前 551-479)

Isn't it a pleasure to study and practice what you have learned?
— Confucius (551–479 BC)

6. Three-DOF PPP= Parallel Mechanisms

As explained in Chap. 4, there exist many motion patterns for which PMs can be synthesized. One of the simplest — and yet one of the most commonly used — motion pattern is the one associated with translations. In this chapter, the type synthesis of 3-DOF PPP= PMs (also called translational PMs in the literature) is dealt with using the general approach proposed in Chap. 5. A PPP= PM is the parallel counterpart of the serial Cartesian robot and generates 3-DOF translational motion. It covers a wide range of applications. The four steps of the type synthesis of PPP= PMs are presented in detail.

6.1 Introduction

A PPP= PM (translational parallel mechanism) is a PM generating 3-DOF translational motion. PPP= PMs have a wide range of applications such as assembly and machining. Several types of PPP= PMs have been proposed [4, 13, 16, 28, 45, 47, 61, 70, 114, 115, 131]. A systematic approach is proposed in [45, 47] to generate PPP= PMs based on the displacement group theory. Systematic studies on the generation of 3-DOF PPP= PMs are performed using screw algebra and screw theory in [33] and [70] respectively.

In fact, most of the work reported to date on the systematic type synthesis of PPP= PMs [13, 33, 47, 70] deals mainly with the systematic type synthesis of translational parallel kinematic chains (PPP= PKCs). From a historical perspective, it is interesting to notice that the results on PPP= PKCs with 5-DOF legs reported in [13, 33, 47, 70] were in fact published in [58] in 1973. Hunt's work [58], which was performed in a different context (constant velocity couplings), is in fact not easy to read and has been largely ignored for many years. The selection of actuated joints for PPP= PMs was dealt with systematically in [70].

Using the general approach to the type synthesis of PMs proposed in Chap. 5, the type synthesis of PPP= PMs is dealt with in this chapter. The four steps for the type synthesis of PPP= PMs, i.e., (1) the decomposition of wrench systems of PPP= PKCs, (2) the type synthesis of legs for PPP= PKCs, (3) the assembly

X. Kong and C. Gosselin: Type Syn. of Parallel Mech., STAR 33, pp. 89–107, 2007.
springerlink.com

of legs to generate PPP= PKCs, and (4) the selection of actuated joints for PPP= PMs, are dealt with in Sects. 6.5, 6.6, 6.7, and 6.8, respectively.

6.2 Wrench System of a PPP= PKC

In any general configuration, the twist system of a PPP= PKC or its PPP virtual chain is the 3-$\boldsymbol{\xi}_\infty$-system. Since the virtual power developed by any $\boldsymbol{\zeta}_\infty$ along any twist within the 3-$\boldsymbol{\xi}_\infty$-system is 0, the wrench system of the PPP= PKC or its virtual chain is thus the 3-$\boldsymbol{\zeta}_\infty$-system (Fig. 6.1).

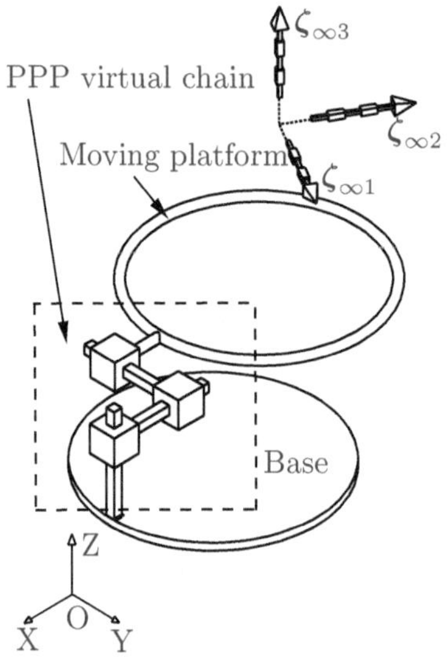

Fig. 6.1. Wrench system of a PPP= PKC

6.3 Conditions for a PKC to Be a PPP= PKC

When we connect the base and the moving platform of a PPP= PKC by a PPP virtual chain, the function of the PKC is not affected. Any of its legs and the PPP virtual chain will form a 3-DOF single loop kinematic chain. When the order of the leg-wrench system is greater than 0, the single-loop kinematic chain constructed must be an overconstrained kinematic chain.

Based on the concept of PPP virtual chain, it follows that a PKC is a PPP= PKC if it satisfies the two conditions below:

(1) Each leg of the PKC and a same PPP virtual chain constitute a 3-DOF single-loop kinematic chain.
(2) The wrench system of the PKC is the same as that of the PPP virtual chain, i.e., a 3-$\boldsymbol{\zeta}_\infty$-system, in any one general configuration.

6.4 Procedure for the Type Synthesis of PPP= PMs

A general procedure can thus be proposed for the type synthesis of PPP= PKCs as follows:

Step 1. Decomposition of the wrench system of 3-DOF PPP= PKCs.
Step 2. Type synthesis of legs for PPP= PKCs.
Here, a leg for PPP= PKCs refers to a leg satisfying Condition (1) for PPP= PKCs.
Step 3. Assembly of legs for PPP= PKCs.
PPP= PKCs can be generated by assembling two or more legs for PPP= PKCs, obtained in Step 1, such that Condition (2) for PPP= PKCs is satisfied.
Step 4. Selection of actuated joints.
PPP= PMs can be generated by selecting sets of actuated joints for each PPP= PKC obtained in Step 3.

The procedure proposed above will be elaborated upon in the following sections.

6.5 Step 1: Decomposition of the Wrench System of PPP= PKCs

The decomposition of the wrench system of an m-legged PPP= PKC consists in finding all the leg-wrench systems and all the combinations of m leg-wrench systems for the PPP= PKC and a specified number of overconstraints Δ. Step 1 can be performed using (2.24) and (2.28).

6.5.1 Determination of the Leg-Wrench Systems

As the wrench system of a PKC is the linear combination of those of all its legs in a general configuration, any leg-wrench system in a V= PKC is a sub-system of its wrench system. Following the procedure proposed in Sect. 2.1.3, all the leg-wrench systems for a V= PKC can be identified without difficulty.

In any general configuration, the wrench system of a PPP= PKC is the same as that of its PPP virtual chain, i.e., the 3-$\boldsymbol{\zeta}_\infty$-system, which is composed of all the $\boldsymbol{\zeta}_\infty$ (Fig. 6.1).

It then follows that any leg-wrench system with order $c^i > 0$ of a PPP= PKC is either the 3-$\boldsymbol{\zeta}_\infty$-system, a 2-$\boldsymbol{\zeta}_\infty$-system, or a 1-$\boldsymbol{\zeta}_\infty$-system in a general configuration (Fig. 6.2).

6.5.2 Determination of the Combinations of Leg-Wrench Systems

For the PPP motion pattern considered, all the wrenches in the wrench systems are of the same pitch. The leg-wrench systems are $c^i(0 \leq c^i \leq c)$-systems of ∞

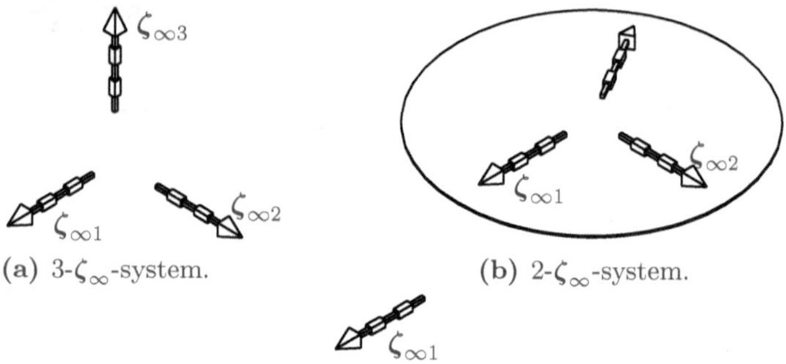

(a) 3-ζ_∞-system. (b) 2-ζ_∞-system.

(c) 1-ζ_∞-system.

Fig. 6.2. Leg-wrench systems of PPP= PKCs

pitch. The combination of leg-wrench systems can be simply represented by the combination of the orders, c^i, of leg-wrench systems.

The combinations of the orders, c^i, of leg-wrench systems have been determined by solving (2.28) and are shown in Tables 5.1–5.5.

6.6 Step 2: Type Synthesis of Legs

In this section, the type synthesis of legs for PPP= PKCs will be performed following the two substeps given in Sect. 5.5.

6.6.1 Step 2a: Type Synthesis of 3-DOF Single-Loop Kinematic Chains That Involve a Virtual Chain and Have a Specified Leg-Wrench System

The type synthesis of 3-DOF single-loop KCs that involve a PPP virtual chain and have a specified leg-wrench system can be performed as follows.

Firstly, we perform the type synthesis of single-loop KCs that involve a PPP virtual chain and have a specified leg-wrench system (see Chap. 3). Secondly, we discard those single-loop KCs in which the twists of all the joints but the PPP virtual chain are linearly dependent.

Cases with the 3-ζ_∞-System

The type synthesis of 3-DOF single-loop KCs that involve a PPP virtual chain and whose wrench system is the 3-ζ_∞-system is first performed. According to Chap. 3, there are six joints in a 3-DOF single-loop KC that involves a PPP virtual chain and has the 3-ζ_∞-system. Such a single-loop KC is composed of one spatial translational compositional unit (Fig. 6.3). We obtain one 3-DOF single-loop KC that involves a PPP virtual chain and has the 3-ζ_∞-system: PPPV. The 3-ζ_∞-system is composed of all the ζ_∞.

The single-loop KCs in which the twists of all the joints but the virtual chain are linearly dependent must then be discarded. For the PPPV KC shown in Fig. 6.3, the KC composed of all the joints except the virtual chain must be a spatial translational compositional unit. In other words, the directions of the three P joints are not parallel to a common plane. For brevity, we assume that the above geometric condition is satisfied.

Cases with a 2-ζ_∞-System

Similarly to the above case, the type synthesis of 3-DOF single-loop KCs that involve a PPP virtual chain and have a 2-ζ_∞-system is first performed. According to Chap. 3, there are seven joints in a 3-DOF single-loop KC that involves a PPP virtual chain and has a 2-ζ_∞-system. Such a single-loop KC is composed of one spatial parallelaxis compositional unit [Figs. 6.4a–6.4d]. The 2-ζ_∞-system is composed of all the ζ_∞ whose axes are perpendicular to all the axes of the R joints.

The single-loop KCs in which the twists of all the joints but the PPP virtual chain are linearly dependent must then be discarded. The R̀R̀R̀R̀V KC [Fig. 6.4a] should be discarded since the twists of all the four R joints are linearly dependent. For the KCs shown in Figs. 6.4b–6.4d, the KC composed of all the joints except the PPP virtual chain must be a spatial parallelaxis compositional unit. For brevity, we assume that the above geometric condition is satisfied.

Cases with a 1-ζ_∞-System

Finally, the cases with a 1-ζ_∞-system are considered. The type synthesis of 3-DOF single-loop KCs that involve a PPP virtual chain and have a 1-ζ_∞-system is first performed. According to Chap. 3, there are eight joints in a 3-DOF single-loop KC that involves a PPP virtual chain and has a 1-ζ_∞-system. Such a single-loop KC is composed of two spatial parallelaxis compositional units and/or planar compositional units or can be obtained by inserting one $(R)_L$ coaxial compositional unit in a 3-DOF single-loop KC that involves a PPP virtual chain and has a 2-ζ_∞-system. Some of the possible architectures are shown in Fig. 6.5. The 1-ζ_∞-system is the ζ_∞ whose direction is perpendicular to all the axes of the R joints.

The single-loop KCs in which the twists of all the joints but the virtual chain are linearly dependent must then be discarded. The R̀R̀R̀R̀ŔV KC [Fig. 6.5a] should be discarded since the twists of all the R joints except the Ŕ joint are linearly dependent. For the KCs shown in Figs. 6.5b–6.5h, we assume that the twists of all the joints but the virtual chain are linearly independent.

In the representation of the 3-DOF single-loop kinematic chains involving a PPP virtual chain and having a c^i-ζ_∞-system, the axes of the R joints denoted by the same symbols, R̀ or Ŕ, are parallel, while the axes of the R joints denoted by different symbols are not.

It is pointed out that the P and the only R joints within the $(R)_L$ compositional unit can be placed anywhere in the single-loop KC. For example, the

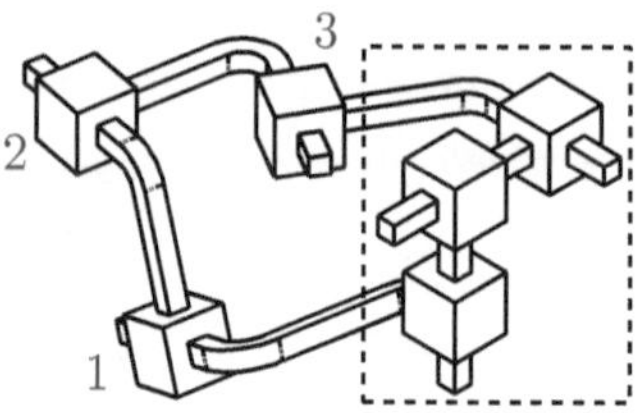

Fig. 6.3. Three-DOF single-loop KC involving a PPP virtual chain ($c^i = 3$): PPPV KC

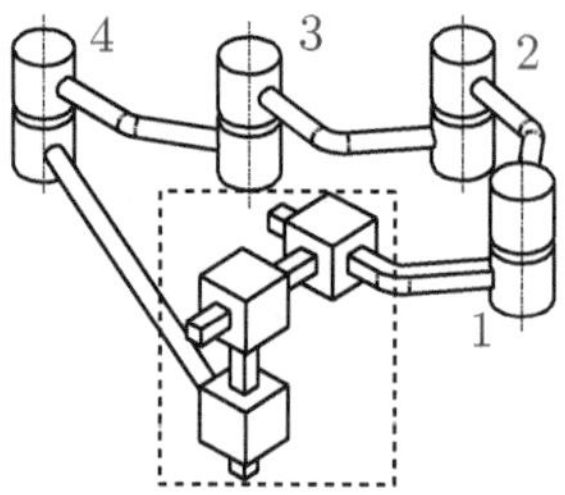

(a) R̀R̀R̀R̀V KC.

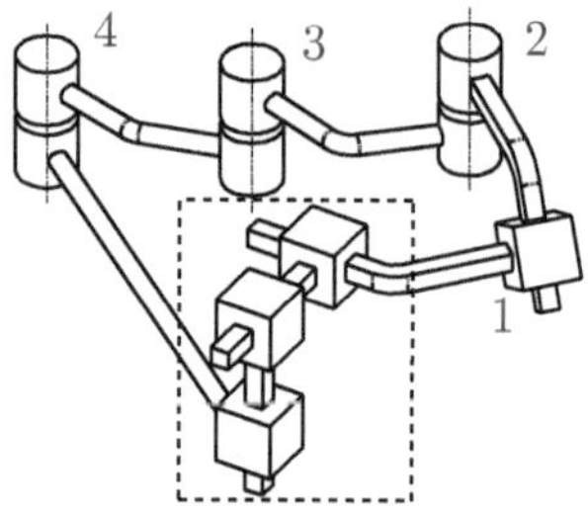

(b) PR̀R̀R̀V KC.

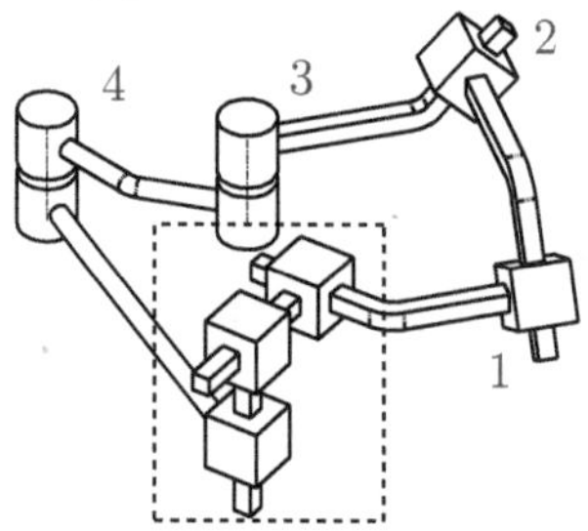

(c) PPR̀R̀V KC.

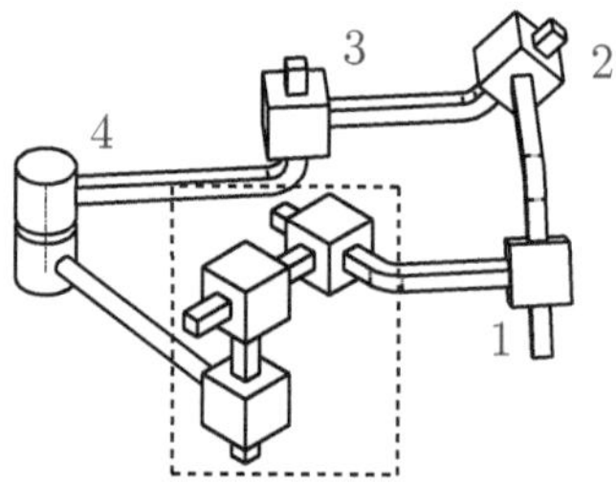

(d) PPPR̀V KC.

Fig. 6.4. Some 3-DOF single-loop KCs involving a PPP virtual chain ($c^i = 2$)

R̀R̀PR̀V single-loop KC can be obtained by changing the position of the P joint in the PR̀R̀R̀V single-loop KC [Fig. 6.4b]. For brevity, we list only the 3-DOF single-loop KCs from which all 3-DOF single-loop KCs can be obtained through the above operations.

6.6.2 Step 2b: Generation of Types of Legs

According to Condition (1) for the existence of PPP= PKCs (Sect. 6.3), the types of legs can be readily obtained from the 3-DOF single-loop kinematic chains obtained in Step 2a (Sect. 6.6.1) by removing the PPP virtual chain.

The specific geometric conditions which guarantee the satisfaction of Condition (1) for PPP= PKCs are clearly indicated by the notations that were

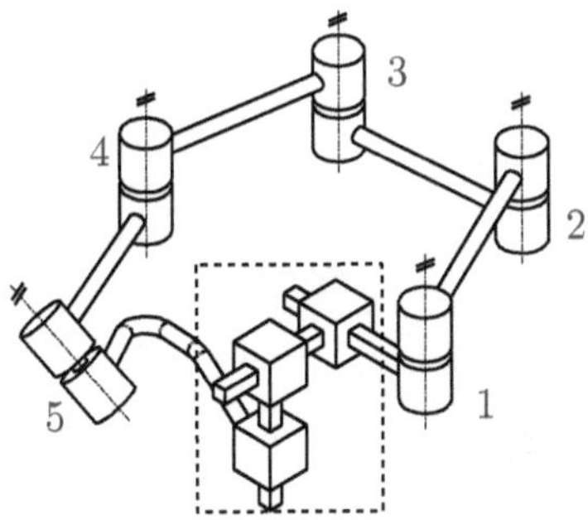

(a) R̀R̀R̀R̀ŔV KC.

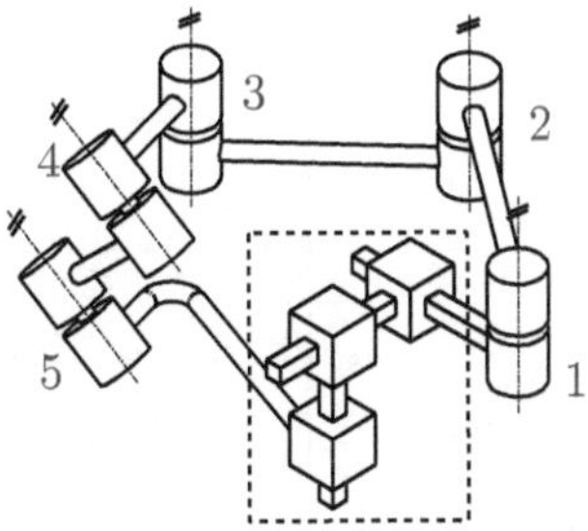

(b) R̀R̀R̀ŔŔV KC.

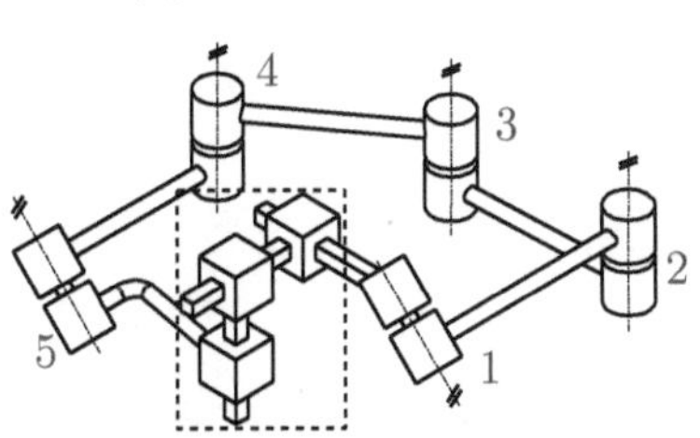

(c) R̀ŔŔŔR̀V KC.

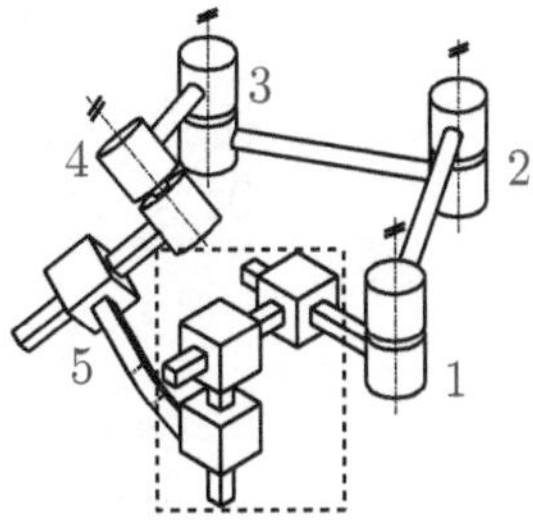

(d) R̀R̀R̀ŔPV KC.

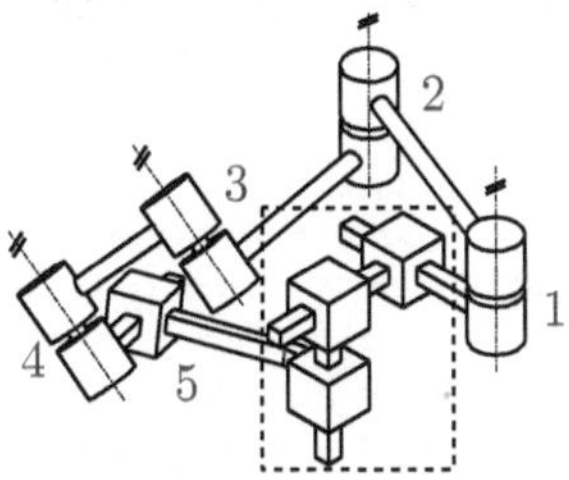

(e) R̀R̀ŔŔPV KC.

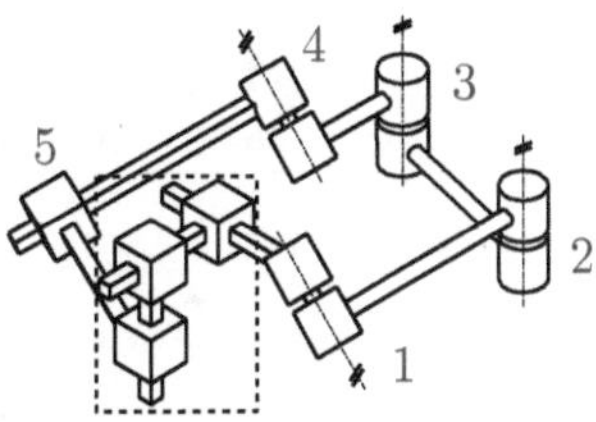

(f) R̀ŔŔR̀PV KC.

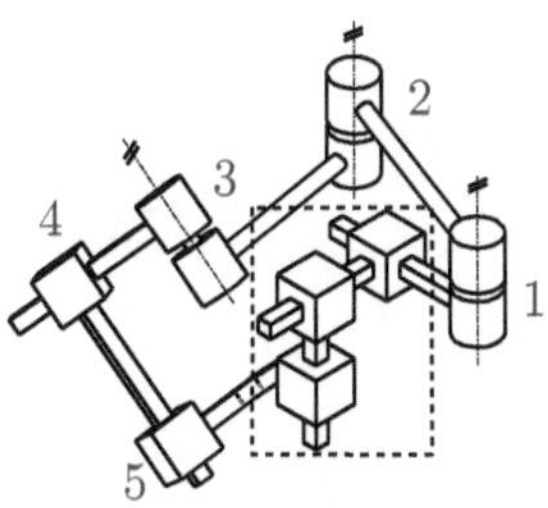

(g) R̀R̀ŔPPV KC.

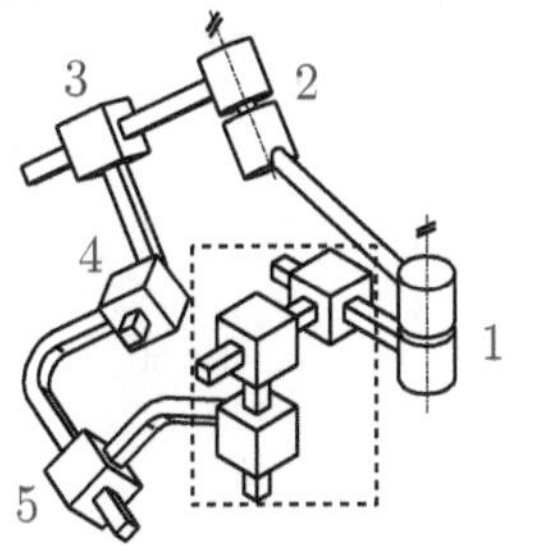

(h) R̀ŔPPPV KC.

Fig. 6.5. Some 3-DOF single-loop KCs involving a PPP virtual chain ($c^i = 1$)

introduced to represent the types of legs: the axes of the R joints denoted by the same symbols, R̀ or Ŕ, are parallel, while the axes of the R joints denoted by different symbols are not.

Table 6.1. Legs of PPP= PKCs

c^i	Class	No	Type	Description
3	3P	1	PPP	
2	3R-1P	2	R̀R̀R̀P	The axes of all the R joints are parallel
		3	R̀R̀PR̀	
		4	R̀PR̀R̀	
		5	PR̀R̀R̀	
	2R-2P	6	R̀R̀PP	
		7	R̀PR̀P	
		8	R̀PPR̀	
		9	PR̀R̀P	
		10	PR̀PR̀	
		11	PPR̀R̀	
	1R-3P	12	R̀PPP	The only R joint is inactive
		13	PR̀PP	
		14	PPR̀P	
		15	PPPR̀	
1	5R	16	ŔŔR̀R̀R̀	The axes of two or three successive R joints are parallel, while the axes of the other R joints are also parallel
		17	R̀R̀R̀ŔŔ	
		18	ŔR̀R̀R̀Ŕ	
		19	R̀ŔŔR̀R̀	
		20	R̀R̀ŔŔR̀	
	4R-1P	21	ŔŔR̀R̀P	The axes of two successive R joints or two R joints connected by the P joint are parallel to each other, the axes of the other two R joints are parallel to each other
		22	ŔŔR̀PR̀	
		23	ŔŔPR̀R̀	
		24	ŔPŔR̀R̀	
		25	PŔŔR̀R̀	
		26	ŔR̀R̀ŔP	
		27	ŔR̀R̀PŔ	
		28	ŔR̀PR̀Ŕ	
		29	ŔPR̀R̀Ŕ	
		30	PŔR̀R̀Ŕ	
		31-50	see Table 6.2	The axes of three R joints are parallel. The only R joint whose axis is not parallel to the axes of the other R joints is inactive
	3R-2P	51-80		The axes of two R joints are parallel to each other. The only R joint whose axis is not parallel to the axes of the other R joints is inactive
	2R-3P	81-90		Both R joints are inactive
0	Omitted	Omitted	Omitted	Omitted

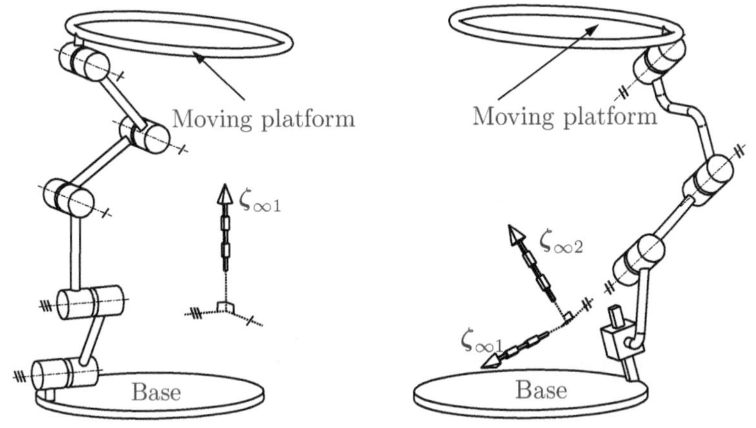

(a) ŔŔR̀R̀R̀ leg. (b) PR̀R̀R̀ leg.

Fig. 6.6. Some legs for PPP= PKCs

For example, by removing the virtual chain in a ŔŔR̀R̀R̀V kinematic chain [Fig. 6.5b], a ŔŔR̀R̀R̀ [Fig. 6.6a] leg can be obtained. In the ŔŔR̀R̀R̀ leg shown in Fig. 6.6a, the axes of the first two R joints are parallel to each other while the axes of the other three R joints are also parallel. This leg has a 1-$\boldsymbol{\zeta}_\infty$-system. The direction of the $\boldsymbol{\zeta}_\infty$ is perpendicular to the axes of all the R joints. By removing the virtual chain in the 3-DOF PŔŔŔV single-loop KC shown in Fig. 6.4b, the PŔŔŔ leg [Fig. 6.6b] is obtained. In this PŔŔŔ leg, the axes of all the R joints are parallel. This leg has a 2-$\boldsymbol{\zeta}_\infty$-system which is composed of all the $\boldsymbol{\zeta}_\infty$ whose axes are perpendicular to all the axes of the R joints.

All the types of legs for PPP= PKCs obtained are listed in Table 6.1. In order to provide a quick reference, a general description of the legs is given in the last column of Table 6.1. Meanwhile, the inactive joints, if any, are pointed out.

It is noted that the legs used to synthesize PPP= PKCs need to satisfy the geometric conditions only within a same leg. These conditions can be derived directly from the associated single-loop kinematic chains.

6.7 Step 3: Assembly of Legs

PPP= PKCs can be generated by assembling a set of legs for PPP= PKCs shown in Table 6.1 selected according to the combinations of the leg-wrench systems shown in Table 5.2. In assembling PPP= PKCs, the following condition should be met: the linear combination of the leg-wrench systems constitutes the 3-$\boldsymbol{\zeta}_\infty$-system (see Condition (2) for PPP= PKCs in Sect. 6.3).

For a PPP= PKC in which not all the leg-wrench systems are invariant with respect to the base or the moving platform, the linear combination of the leg-wrench systems usually constitutes the 3-$\boldsymbol{\zeta}_\infty$-system. For a PPP= PKC in which all the leg-wrench systems are invariant with respect to the base or the moving

Table 6.2. Legs of PPP= PKCs (No. 31–90)

No	Type
31-35	R̀ŔR̀R̀P R̀R̀ŔR̀P R̀ŔR̀PR̀ R̀R̀ŔPR̀ R̀R̀PŔR̀
36-40	R̀ŔPR̀R̀ R̀PŔR̀R̀ R̀PR̀ŔR̀ PR̀ŔR̀R̀ PR̀R̀ŔR̀
41-45	ŔPR̀R̀R̀ PŔR̀R̀R̀ PR̀R̀R̀Ŕ ŔR̀PR̀R̀ R̀PR̀R̀Ŕ
46-50	ŔR̀R̀PR̀ R̀R̀PR̀Ŕ ŔR̀R̀R̀P R̀R̀R̀ŔP R̀R̀R̀PŔ
51-55	ŔR̀R̀PP R̀R̀ŔPP R̀R̀PŔP R̀R̀PPŔ ŔR̀PR̀P
56-60	R̀PR̀ŔP R̀PR̀PŔ ŔR̀PPR̀ R̀PPR̀Ŕ ŔPR̀R̀P
61-65	PŔR̀R̀P PR̀R̀ŔP PR̀R̀PŔ ŔPR̀PR̀ PŔR̀PR̀
66-70	PR̀PR̀Ŕ ŔPPR̀R̀ PŔPR̀R̀ PPŔR̀R̀ PPR̀R̀Ŕ
71-75	PPR̀ŔR̀ PR̀ŔPR̀ PR̀PŔR̀ PR̀ŔR̀P R̀ŔPPR̀
76-80	R̀PŔPR̀ R̀PPŔR̀ R̀ŔPR̀P R̀PŔR̀P R̀ŔR̀PP
81-85	PR̀ŔPP PR̀PŔP PR̀PPŔ PPR̀ŔP PPR̀PŔ
86-90	PPPR̀Ŕ R̀ŔPPP R̀PŔPP R̀PPŔP R̀PPPŔ

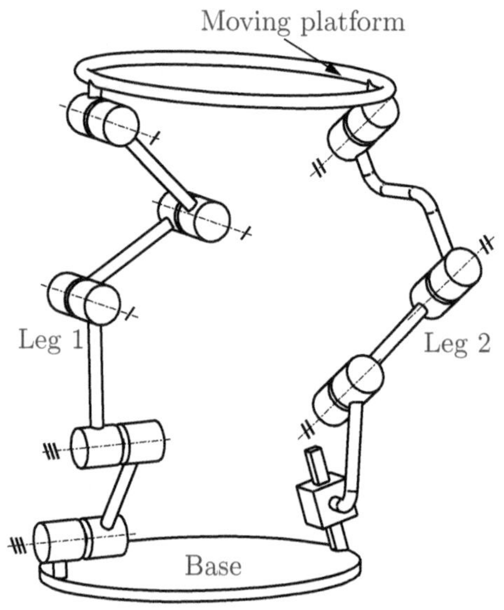

Fig. 6.7. ŔŔR̀R̀R̀-PR̀R̀R̀ PPP= PKC

platform, the geometry of the base or the moving platform should meet certain conditions to guarantee that the linear combination of the leg-wrench systems constitutes the 3-ζ_∞-system.

For example, by taking one ŔŔR̀R̀R̀ leg [Fig. 6.6a] and one PR̀R̀R̀ leg (Fig. 6.6b), a 2-legged ŔŔR̀R̀R̀-PR̀R̀R̀ PPP= PKC (Fig. 6.7) can be obtained. The two leg-wrench systems are both invariant with respect to the base or the moving platform,

Table 6.3. Three-legged PPP= PKCs with identical types of legs

Class	No	Type	Geometric condition
3P	1	3-PPP	
3R-1P	2-3	3-R̀R̀R̀P 3-R̀R̀PR̀	The axes of the R joints are not all parallel
	4-5	3-R̀PR̀R̀ 3-PR̀R̀R̀	
2R-2P	6-7	3-R̀R̀PP 3-R̀PR̀P	
	8-9	3-R̀PPR̀ 3-PR̀R̀P	
	10-11	3-PR̀PR̀ 3-PPR̀R̀	
1R-3P	12-13	3-R̀PPP 3-PR̀PP	
	14-15	3-PPR̀P 3-PPPR̀	
5R	16-17	3-ŔŔR̀R̀R̀ 3-R̀R̀R̀ŔŔ	Three lines each perpendicular to all the axes of the R joints within a leg are not parallel to a plane.
	18-19	3-ŔR̀R̀R̀Ŕ 3-R̀ŔŔR̀R̀	
	20	3-R̀R̀ŔŔR̀	
4R-1P	21-22	3-ŔŔR̀R̀P 3-ŔŔR̀PR̀	The same condition as types 16 and 17.
	23-24	3-ŔŔPR̀R̀ 3-ŔPŔR̀R̀	
	25-26	3-PŔŔR̀R̀ 3-ŔR̀R̀ŔP	
	27-28	3-ŔR̀R̀PŔ 3-ŔR̀PR̀Ŕ	
	29-30	3-ŔPR̀R̀Ŕ 3-PŔR̀R̀Ŕ	
	31-40	see Table 6.4	
	41-50		The same condition as types 16 and 17.
3R-2P	51-70		
	71-80		
2R-3P	81-90		The same condition as types 16 and 17.

hence the axes of all the R joints within the PPP= PKC should not be parallel to a plane in order to guarantee that the linear combination of their leg-wrench systems constitutes the 3-$\boldsymbol{\zeta}_\infty$-system.

Due to the large number of PPP= PKCs, only three-legged PPP= PKC with identical legs are listed in Table 6.3.

6.8 Step 4: Selection of Actuated Joints

6.8.1 t-Components of the Actuation Wrenches

In any frame, $[\mathbf{0}^T \quad \mathbf{i}^T]^T$, $[\mathbf{0}^T \quad \mathbf{j}^T]^T$, $[\mathbf{0}^T \quad \mathbf{k}^T]^T$ can be used to denote a basis of $\mathcal{W}$ of a PPP= PKC. Here, $\mathbf{i}$, $\mathbf{j}$ and $\mathbf{k}$ denote respectively the unit vectors along

Table 6.4. Types of PPP= PKCs (No. 31–90)

No	Type
31-35	3-R̀ŔR̀R̀P 3-R̀R̀ŔR̀P 3-R̀ŔR̀PR̀ 3-R̀R̀ŔPR̀ 3-R̀R̀PŔR̀
36-40	3-R̀ŔPR̀R̀ 3-R̀PŔR̀R̀ 3-R̀PR̀ŔR̀ 3-PR̀ŔR̀R̀ 3-PR̀R̀ŔR̀
41-45	3-ŔPR̀R̀R̀ 3-PŔR̀R̀R̀ 3-PR̀R̀R̀Ŕ 3-ŔR̀PR̀R̀ 3-R̀PR̀R̀Ŕ
46-50	3-ŔR̀R̀PR̀ 3-R̀R̀PR̀Ŕ 3-ŔR̀R̀R̀P 3-R̀R̀R̀ŔP 3-R̀R̀R̀PŔ
51-55	3-ŔR̀R̀PP 3-R̀R̀ŔPP 3-R̀R̀PŔP 3-R̀R̀PPŔ 3-ŔR̀PR̀P
56-60	3-R̀PR̀ŔP 3-R̀PR̀PŔ 3-ŔR̀PPR̀ 3-R̀PPR̀Ŕ 3-ŔPR̀R̀P
61-65	3-PŔR̀R̀P 3-PR̀R̀ŔP 3-PR̀R̀PŔ 3-ŔPR̀PR̀ 3-PŔR̀PR̀
66-70	3-PR̀PR̀Ŕ 3-ŔPPR̀R̀ 3-PŔPR̀R̀ 3-PPŔR̀R̀ 3-PPR̀R̀Ŕ
71-75	3-PPR̀ŔR̀ 3-PR̀ŔPR̀ 3-PR̀PŔR̀ 3-PR̀ŔR̀P 3-R̀ŔPPR̀
76-80	3-R̀PŔPR̀ 3-R̀PPŔR̀ 3-R̀ŔPR̀P 3-R̀PŔR̀P 3-R̀ŔR̀PP
81-85	3-PR̀ŔPP 3-PR̀PŔP 3-PR̀PPŔ 3-PPR̀ŔP 3-PPR̀PŔ
86-90	3-PPPR̀Ŕ 3-R̀ŔPPP 3-R̀PŔPP 3-R̀PPŔP 3-R̀PPPŔ

the X-, Y- and Z-axes, and $\mathbf{0}$ represents a 3×1 zero vector. Considering that the first three scalar components of all the basis wrenches of $\mathcal{W}$ are zero, the t-component, $\boldsymbol{\zeta}^i_{t\not{\mathcal{D}}j}$, of the $\boldsymbol{\zeta}^i_{\not{\mathcal{D}}j}$ is composed of the first three scalar components of the $\boldsymbol{\zeta}^i_{\not{\mathcal{D}}j}$, i.e.,

$$\boldsymbol{\zeta}^i_{t\not{\mathcal{D}}j} = \boldsymbol{\zeta}^i_{F\not{\mathcal{D}}j} \tag{6.1}$$

where $\boldsymbol{\zeta}^i_{F\not{\mathcal{D}}j}$ represents a unit vector which is parallel to the axes of the R joints within leg i.

Figure 6.8 shows the actuation wrenches of actuated joints in some legs for PPP= PMs. In the $\underline{\acute{R}}\acute{R}\grave{R}\grave{R}\grave{R}$ leg [Fig. 6.8a], the first R joint is actuated. The actuation wrench is any $\boldsymbol{\zeta}_0$ whose axis is parallel to the axes of the last three R joints and intersects the axis of the second R joint. In the $\acute{R}\underline{\acute{R}}\grave{R}\grave{R}\grave{R}$ leg [Fig. 6.8b], the second R joint is actuated. The actuation wrench is any $\boldsymbol{\zeta}_0$ whose axis is parallel to the axes of the last three R joints and intersects the axis of the first R joint. In the $\underline{P}\acute{R}\acute{R}\acute{R}$ leg [Fig. 6.8c], the first P joint is actuated. The actuation wrench is any $\boldsymbol{\zeta}_0$ whose axis is parallel to the axes of the three R joints. The t-component of each of the above actuation wrenches is the unit vector along its axis.

6.8.2 Procedure for the Validity Detection of Actuated Joints

The validity detection of actuated joints of PPP= PMs can be performed using the following steps.

Step 4a. If one or more of the actuated joints of a candidate PPP= PM are inactive, the set of actuated joints is invalid and the candidate PPP= PM should be discarded.

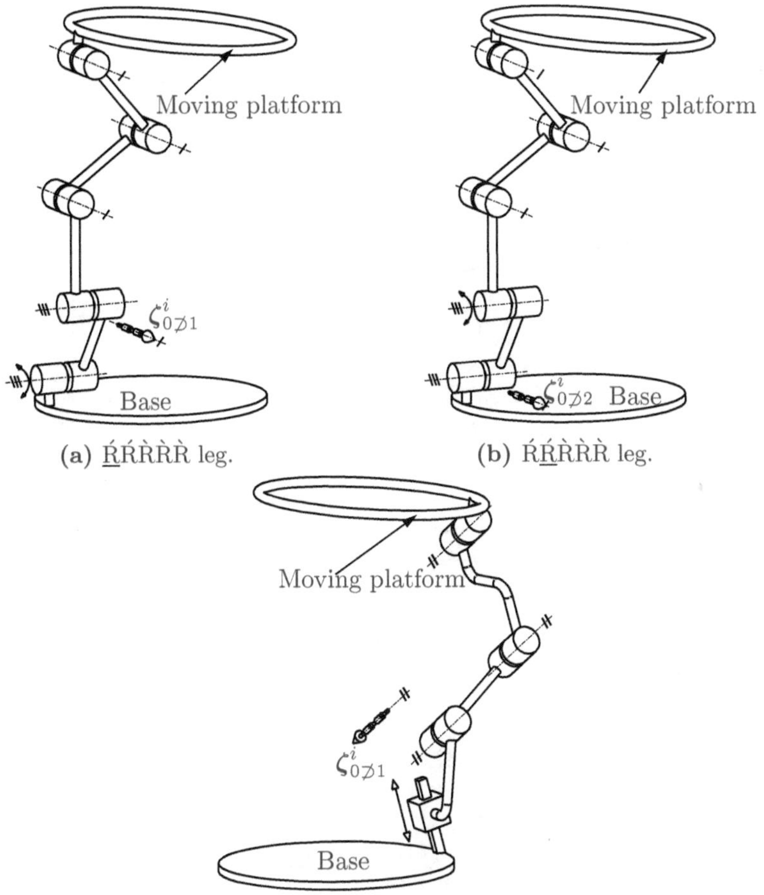

(a) $\underline{\acute{R}}\acute{R}\grave{R}\grave{R}\grave{R}$ leg. **(b)** $\acute{R}\underline{\acute{R}}\grave{R}\grave{R}\grave{R}$ leg.

(c) $\underline{P}\grave{R}\grave{R}\grave{R}$ leg.

Fig. 6.8. Actuation wrenches of some legs for PPP= PMs

Step 4b. If the t-components, i.e., the first vector components of all the actuation wrenches of actuated joints $\boldsymbol{\zeta}^i_{\not{D}j}$ are linearly dependent in a general configuration for a candidate PPP= PM, the set of actuated joints is invalid. In this case, the candidate PPP= PM should be discarded.

Following the criteria on the selection of actuated joints (Sect. 5.7) and the procedure for the detection of the validity of actuated joints, all the $m(m \geq 2)$-legged PPP= PMs corresponding to each PPP= PKC can be generated.

For example, the candidate PPP= PMs corresponding to the 2-legged PPP= PKC, $\acute{R}\acute{R}\grave{R}\grave{R}\grave{R}$-$P\grave{R}\grave{R}\grave{R}$, satisfying the above criteria are $\underline{\acute{R}\acute{R}}\grave{R}\grave{R}\grave{R}$-$\underline{P}\grave{R}\grave{R}\grave{R}$, $\underline{\acute{R}}\acute{R}\grave{R}\grave{R}\grave{R}$-$\underline{P\grave{R}}\grave{R}\grave{R}$ and $\acute{R}\underline{\acute{R}}\grave{R}\grave{R}\grave{R}$-$\underline{P\grave{R}}\grave{R}\grave{R}$ PPP= PMs (Fig. 6.9). Following the procedure for

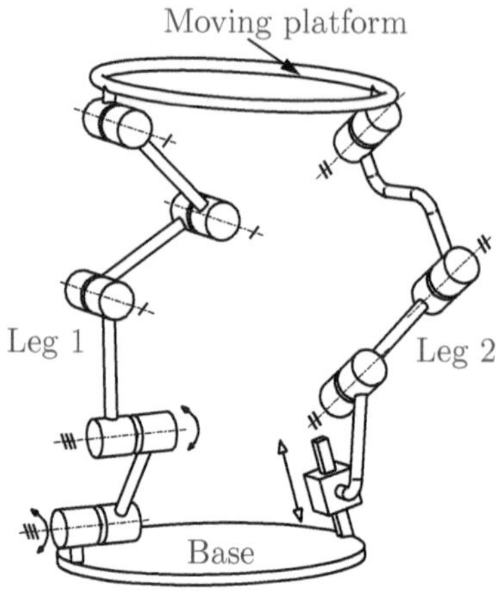

(a) $\underline{\acute{R}\acute{R}}\grave{R}\grave{R}\grave{R}\text{-}\underline{P}\grave{R}\grave{R}\grave{R}$ PPP= PM.

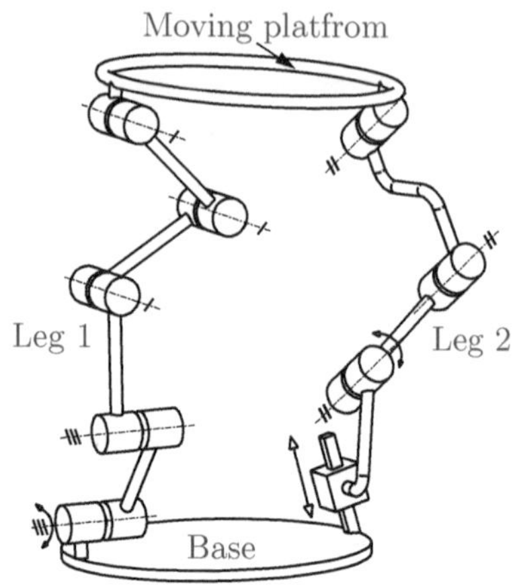

(b) $\underline{\acute{R}}\acute{R}\grave{R}\grave{R}\grave{R}\text{-}\underline{P\grave{R}}\grave{R}\grave{R}$ PPP= PM.

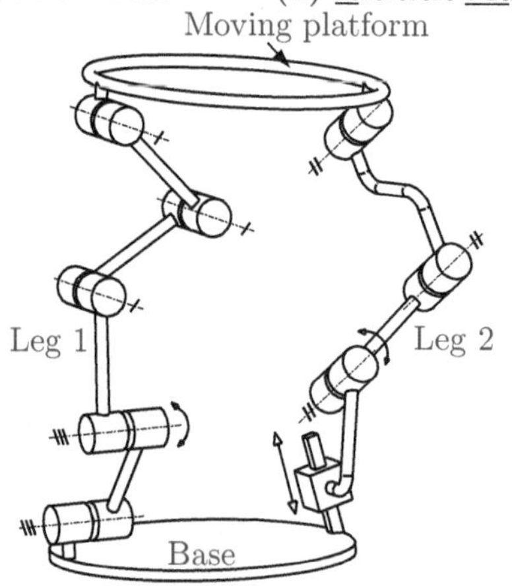

(c) $\acute{R}\underline{\acute{R}}\grave{R}\grave{R}\grave{R}\text{-}\underline{P\grave{R}}\grave{R}\grave{R}$ PPP= PM.

Fig. 6.9. Selection of actuated joints for the $P\grave{R}\grave{R}\grave{R}\text{-}\acute{R}\acute{R}\grave{R}\grave{R}\grave{R}$ PPP= PKC

the validity detection of actuated joints, the $\underline{\acute{R}\acute{R}}\grave{R}\grave{R}\grave{R}\text{-}\underline{P}\grave{R}\grave{R}\grave{R}$ PPP= PM should be discarded since the determinant of the matrix composed of the t-components of the three actuated joints is 0. Thus, the PPP= PMs corresponding to the 2-legged PPP= PKC, $\acute{R}\acute{R}\grave{R}\grave{R}\grave{R}\text{-}P\grave{R}\grave{R}\grave{R}$, are the $\underline{\acute{R}}\acute{R}\grave{R}\grave{R}\grave{R}\text{-}\underline{P\grave{R}}\grave{R}\grave{R}$ and $\acute{R}\underline{\acute{R}}\grave{R}\grave{R}\grave{R}\text{-}\underline{P\grave{R}}\grave{R}\grave{R}$ PPP= PMs.

Due to the large number of PPP= PKCs, a large number of PPP= PMs can be generated. Here, we only list 3-legged PPP= PMs with identical legs and whose actuated joints are all located on the base (Table 6.5). There are many new PPP= PMs among those listed in Table 6.5. Two of them, namely, the 3-$\underline{P}\grave{R}\grave{R}\grave{R}\acute{R}$ and 3-$\underline{P}\grave{R}\grave{R}\grave{R}$ PPP= PMs, are shown in Fig. 6.10.

Table 6.5. Three-legged PPP= PMs

Class	No	Type	Geometric condition
3P	1	3-P̲PP	Three lines each perpendicular to the axes of two unactuated P joints within a leg are not parallel to a plane.
3R-1P	2-3	3-R̲̀R̀R̀P 3-R̲̀R̀PR̀	
	4	3-R̲̀PR̀R̀	
	5	3-P̲R̀R̀R̀	All the axes of the R̀ joints are not parallel to a plane.
2R-2P	6	3-R̲̀R̀PP	The same condition as type 1.
	7-8	3-R̲̀PR̀P 3-R̲̀PPR̀	
	9-10	3-P̲R̀R̀P 3-P̲R̀PR̀	
	11	3-P̲PR̀R̀	
1R-3P	12-13	3-P̲PPR̀ 3-P̲PR̀P	The same condition as type 1.
	14-15	3-P̲R̀PP 3-R̀P̲PP	
5R	16	3-Ŕ̲ŔR̀R̀R̀	The same condition as type 5.
	17-18	3-R̲̀R̀R̀ŔŔ 3-Ŕ̲R̀R̀R̀Ŕ	
	19-20	3-R̲̀ŔŔR̀R̀ 3-R̲̀R̀ŔŔR̀	
4R-1P	21-22	3-Ŕ̲ŔR̀R̀P 3-Ŕ̲ŔR̀PR̀	
	23-24	3-Ŕ̲ŔPR̀R̀ 3-Ŕ̲PŔR̀R̀	
	25-26	3-P̲ŔŔR̀R̀ 3-Ŕ̲R̀R̀ŔP	
	27-28	3-Ŕ̲R̀R̀PŔ 3-Ŕ̲R̀PR̀Ŕ	
	29-30	3-Ŕ̲PR̀R̀Ŕ 3-P̲ŔR̀R̀Ŕ	
	31-38	see Table 6.6	
	39-43		The same condition as type 5.
	44-50		
3R-2P	51-54		The same condition as type 1.
	55-79		
	80		The same condition as type 1.
2R-3P	81-90		

It was revealed in [39, 72] that the 3-P̲R̀R̀R̀ and 3-P̲R̀R̀R̀Ŕ PPP= PMs have the following characteristics: (1) The forward displacement analysis can be performed by solving a set of linear equations. (2) The Jacobian matrix of the PPP= PM is constant. The inverse of the Jacobian matrix can be pre-calculated, and there is no need to calculate repeatedly the inverse of the Jacobian matrix in performing the forward displacement analysis and forward velocity analysis. (3) There is no

Table 6.6. Types of PPP= PMs (No. 31-90)

No	Type
31-35	3-R̲̀ŔR̀R̀P 3-R̲̀R̀ŔR̀P 3-R̲̀ŔR̀PR̀ 3-R̲̀R̀ŔPR̀ 3-R̲̀R̀PŔR̀
36-40	3-R̲̀ŔPR̀R̀ 3-R̲̀PŔR̀R̀ 3-R̲̀PR̀ŔR̀ 3-P̲R̀ŔR̀R̀ 3-P̲R̀R̀ŔR̀
41-45	3-ŔP̲R̀R̀R̀ 3-P̲ŔR̀R̀R̀ 3-P̲R̀R̀R̀Ŕ 3-ŔR̲̀PR̀R̀ 3-R̲̀PR̀R̀Ŕ
46-50	3-ŔR̲̀R̀PR̀ 3-R̲̀R̀PR̀Ŕ 3-ŔR̲̀R̀R̀P 3-R̲̀R̀R̀ŔP 3-R̲̀R̀R̀PŔ
51-55	3-ŔR̲̀R̀PP 3-R̲̀R̀ŔPP 3-R̲̀R̀PŔP 3-R̲̀R̀PPŔ 3-ŔR̲̀PR̀P
56-60	3-R̲̀PR̀ŔP 3-R̲̀PR̀PŔ 3-ŔR̲̀PPR̀ 3-R̲̀PPR̀Ŕ 3-ŔP̲R̀R̀P
61-65	3-P̲ŔR̀R̀P 3-P̲R̀R̀ŔP 3-P̲R̀R̀PŔ 3-ŔP̲R̀PR̀ 3-P̲ŔR̀PR̀
66-70	3-P̲R̀PR̀Ŕ 3-ŔP̲PR̀R̀ 3-P̲ŔPR̀R̀ 3-P̲PŔR̀R̀ 3-P̲PR̀R̀Ŕ
71-75	3-P̲PR̀ŔR̀ 3-P̲R̀ŔPR̀ 3-P̲R̀PŔR̀ 3-P̲R̀ŔR̀P 3-R̲̀ŔPPR̀
76-80	3-R̲̀PŔPR̀ 3-R̲̀PPŔR̀ 3-R̲̀ŔPR̀P 3-R̲̀PŔR̀P 3-R̲̀ŔR̀PP
81-85	3-P̲R̀ŔPP 3-P̲R̀PŔP 3-P̲R̀PPŔ 3-P̲PR̀ŔP 3-P̲PR̀PŔ
86-90	3-P̲PPR̀Ŕ 3-R̀ŔP̲PP 3-R̀P̲ŔPP 3-R̀P̲PŔP 3-R̀P̲PPŔ

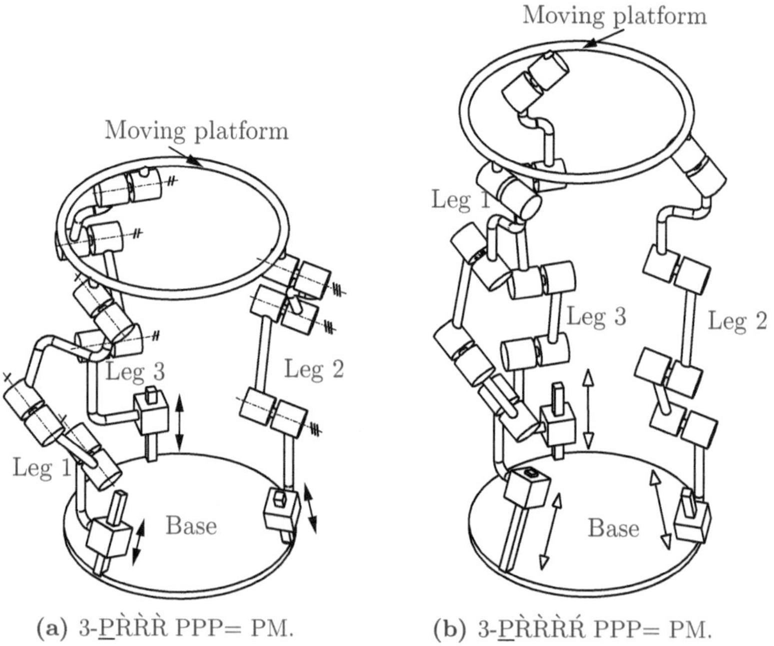

(a) 3-P̲R̀R̀R̀ PPP= PM. **(b)** 3-P̲R̀R̀R̀Ŕ PPP= PM.

Fig. 6.10. Some new PPP= PMs

constraint singularity. (4) There is no uncertainty singularity. Furthermore, these PPP= PMs can be rendered input-output decoupled (Fig. 6.11). The main characteristic of an input-output decoupled PPP= PM is that three translations of

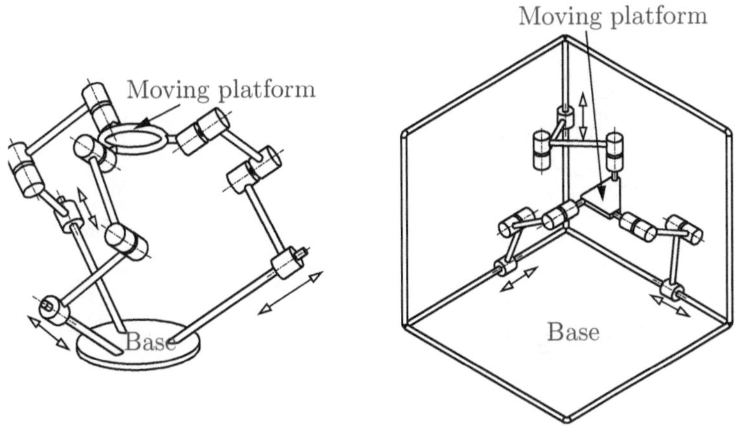

(a) 3-$\underline{P}\grave{R}\grave{R}\grave{R}$ (Case 1).

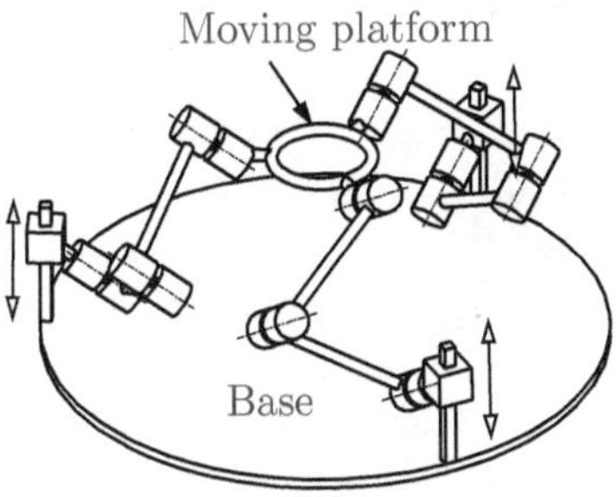

(b) 3-$\underline{P}\grave{R}\grave{R}\grave{R}$ (Case 2).

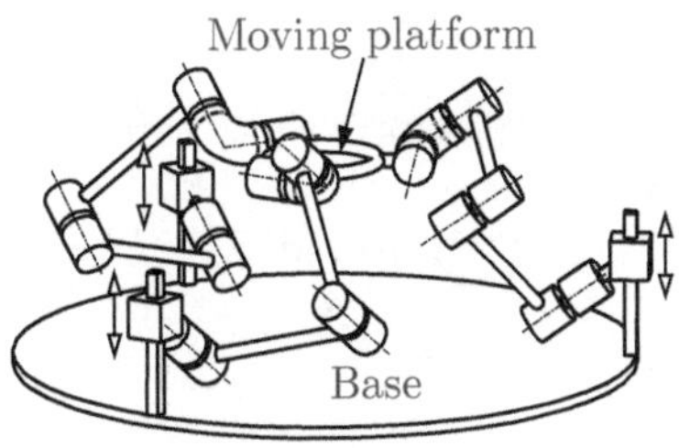

(c) 3-$\underline{P}\grave{R}\grave{R}\grave{R}\acute{R}$ (Case 1).

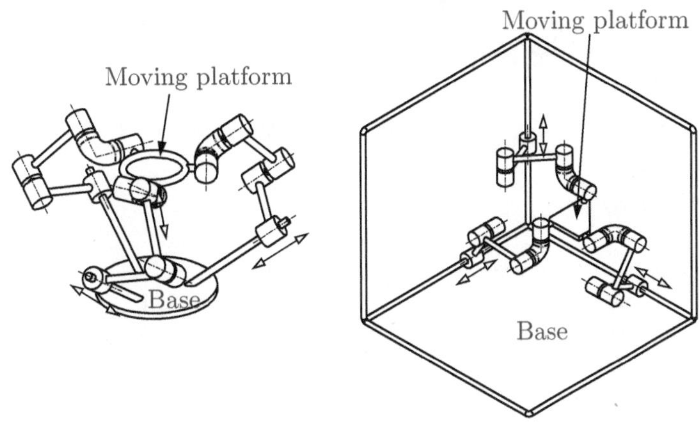

(d) 3-$\underline{P}\grave{R}\grave{R}\grave{R}\acute{R}$ (Case 2).

Fig. 6.11. Input-output decoupled PPP= PMs (with linear actuators)

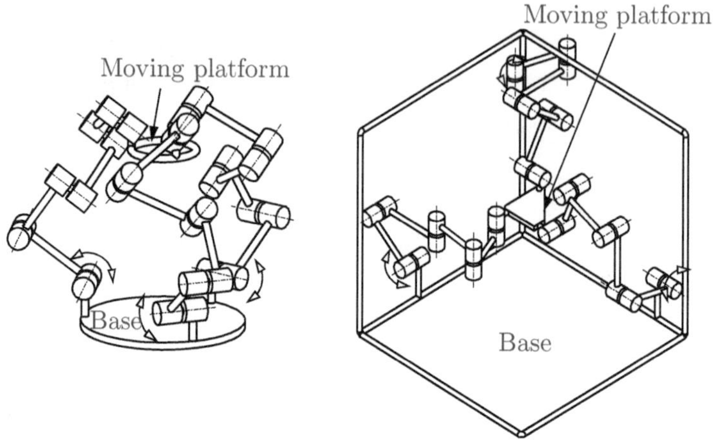

Fig. 6.12. Input-output decoupled 3-$\underline{\acute{R}}\acute{R}\grave{R}\grave{R}\grave{R}$ PPP= PMs (with rotary actuators)

the moving platform along three orthogonal directions can be controlled independently by three actuators. Several plastic models of mechanisms were built in the Robotics Laboratory at Laval University in order to illustrate the above results (See Fig. 6.13). One parallel manipulator based on the input-output decoupled 3-$\underline{P}\grave{R}\grave{R}\grave{R}$ PPP= PM, called the Tripteron [See Fig. 1.7a], was also built and tested at Laval University. The control of the Tripteron is very simple since its Jacobian matrix is the identity matrix. Using brushless DC motors, the prototype of the Tripteron performed trajectories involving accelerations of up to 8 g. For additional characteristics of the Tripteron, the reader is referred to [39, 40, 71, 72].

The type synthesis of PPP= PMs presented above allowed the identification of two promising classes of manipulators. The first is the class of input-output decoupled PPP= PMs (Figs. 6.11 and 6.12) [71], while the second is the class of PPP= PMs having a set of linear input-output equations (Fig. 6.10) [73, 79].

6.9 Summary

In this chapter, the type synthesis of PPP= PMs has been thoroughly solved using the virtual-chain approach proposed in Chap. 5. PPP= PKCs which were proposed in [47, 58] have been re-obtained. PPP= PKCs with inactive joints have also been obtained. Both overconstrained and non-overconstrained PPP= PKCs can be obtained. The validity detection of actuated joints of PPP= PMs has been reduced to the calculation of a 3×3 determinant. Some new PPP= PMs have also been revealed.

It should be pointed out that by replacing an R joint with a coaxial H (helical) joint in any PPP= PKC generated in this chapter, some PPP= PMs involving

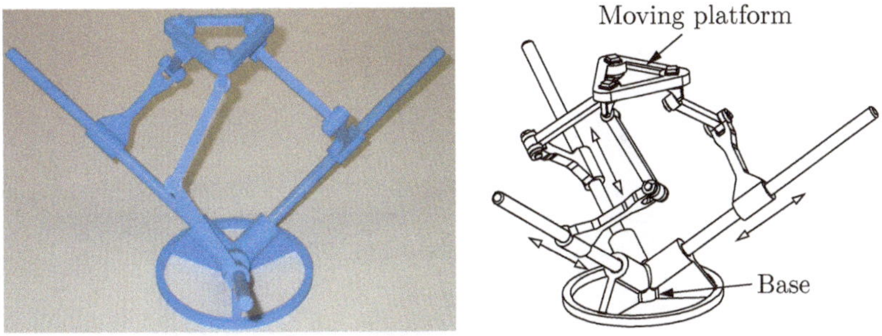

(a) 3-P̲R̀R̀R̀ (Case 1).

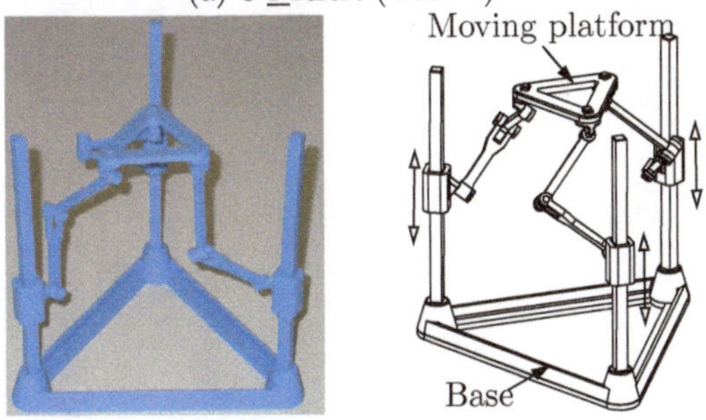

(b) 3-P̲R̀R̀R̀ (Case 2).

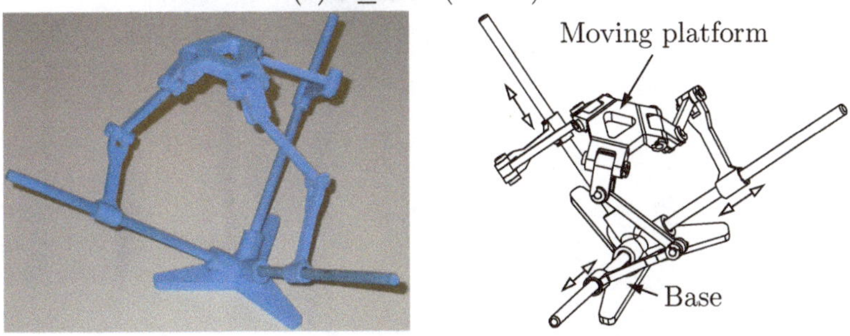

(c) 3-P̲R̀R̀R̀Ŕ (Case 2).

Fig. 6.13. Plastic models of input-output decoupled PPP= PMs (with linear actuators)

H joints can be obtained. Other PPP= PMs involving H joints can be obtained starting from 3-DOF single-loop KCs involving a PPP virtual chain and at least one H joint following the above procedure. Two cases can be considered: (1) 3-DOF single-loop KCs obtained from the R̀R̀R̀R̀V [Fig. 6.4a] KC by replacing one or more of the R̀ joints each with one coaxial H̀ joint and (2) 3-DOF single-loop KCs obtained from the R̀R̀R̀R̀ŔV [Fig. 6.5a], R̀R̀R̀ŔR̀V, R̀R̀ŔR̀R̀V, R̀ŔR̀R̀R̀V, ŔR̀R̀R̀R̀V KCs by replacing one or more of the R̀ (or Ŕ) joints each with one coaxial H̀ (or H́) joint. Finally, by replacing a P joint with a planar parallelogram [35, 47, 71, 94, 102] or a set of two consecutive P joints with a spatial parallelogram, additional variations of PPP= PMs can be obtained.

7. Three-DOF S= Parallel Mechanisms

Parallel mechanisms capable of performing arbitrary rotations about a fixed point are used in many applications including robotics, manufacturing and others. They represent an important family of devices and their type synthesis is a key issue. In this chapter, the type synthesis of 3-DOF S= PMs (also called spherical parallel mechanisms) is dealt with using the general approach proposed in Chap. 5. An S= PM refers to a 3-DOF PM generating 3-DOF spherical motion. The four steps of the type synthesis of S= PMs are presented in detail.

7.1 Introduction

An S= PM, also called spherical parallel mechanism, refers to a 3-DOF PM generating 3-DOF spherical motion. S= PMs have a wide range of applications such as orienting devices and wrists.

S= PMs proposed so far mainly include: (a) S= PMs in which each leg is composed of three R joints whose axes all pass through the centre of rotation [18, 19, 37], (b) S= PMs in which one of the legs is composed of only one unactuated S joint, (c) S= PMs in which each of the three legs is composed of two R joints whose axes pass through the centre of rotation of the moving platform and are connected by a kinematic chain equivalent to a planar joint [64], (d) S= PMs in which each of the three legs is composed of two R joints in series with parallel axes, and three R joints whose axes intersect at the centre of rotation of the moving platform [24], and (e) S= PMs in which each of the three legs is composed of an R, Π (planar parallelogram), and S joints [116].

An approach based on displacement group theory is proposed in [64] for the type synthesis of S= PMs. Some new types of S= PKCs and S= PMs have been obtained using this approach [25, 64]. Unfortunately, the number of overconstraints of S= PMs has not been revealed. The selection of the actuated joints has not been dealt with systematically either.

Using the general approach proposed in Chap. 5, the type synthesis of S= PMs is dealt with in this chapter. The decomposition of the wrench system of an S= PKC is dealt with in Sect. 7.5. The type synthesis of legs for S= PKCs is

X. Kong and C. Gosselin: Type Syn. of Parallel Mech., STAR 33, pp. 109–124, 2007.
springerlink.com

performed in Sect. 7.6. In Sect. 7.7, the assembly of legs to generate S= PKCs is dealt with while the selection of actuated joints for S= PMs is discussed in Sect. 7.8.

7.2 Wrench System of an S= PKC

In any general configuration, the twist system of an S= PKC or its S virtual chain is a 3-ξ_0-system whose centre is at the centre of rotation of the moving platform. It can be found without difficulty that the wrench system of an S= PKC is a 3-ζ_0-system whose centre is at the centre of rotation of the moving platform (Fig. 7.1).

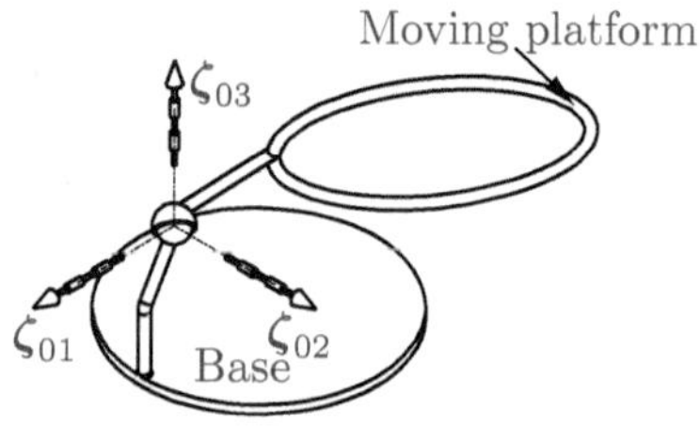

Fig. 7.1. Wrench system of an S= PKC

7.3 Conditions for a PKC to Be an S= PKC

When we connect the base and the moving platform of an S= PKC by an S virtual chain, the function of the PKC is not affected. Any of its legs and the S virtual chain will form a 3-DOF single loop kinematic chain. When the order of the leg-wrench system is greater than 0, the single-loop kinematic chain constructed must be an overconstrained kinematic chain.

Base on the concept of S virtual chain, it follows that a PKC is an S= PKC if it satisfies the two conditions below:

(1) Each leg of the PKC and a same S virtual chain constitute a 3-DOF single-loop kinematic chain.
(2) The wrench system of the PKC is the same as that of the S virtual chain, i.e., a 3-ζ_0-system, in any one general configuration.

7.4 Procedure for the Type Synthesis of S= PMs

A procedure can thus be proposed for the type synthesis of S= PKCs as follows:

Step 1. Decomposition of the wrench system of 3-DOF S= PKCs.
Step 2. Type synthesis of legs for S= PKCs.
Here, a leg for S= PKCs refers to a leg satisfying Condition (1) for S= PKCs.

Step 3. Assembly of legs for S= PKCs.
S= PKCs can be generated by taking two or more legs for S= PKCs, obtained in Step 1, such that Condition (2) for S= PKCs is satisfied.
Step 4. Selection of actuated joints.
S= PMs can be generated by selecting actuated joints in different ways for each S= PKC (Sect. 7.8), obtained in Step 3.

The procedure proposed above will be elaborated upon in the following sections.

7.5 Step 1: Decomposition of the Wrench System of S= PKCs

The decomposition of the wrench system of an m-legged S= PKC consists in finding all the leg-wrench systems and all the combinations of m leg-wrench systems for the S= PKC and a specified number of overconstraints Δ. Step 1 can be performed using (2.24) and (2.28).

7.5.1 Determination of the Leg-Wrench Systems

As the wrench system of a PKC is the linear combination of those of all its legs in a general configuration, any leg-wrench system in a S= PKC is a sub-system of its wrench system. Following the procedure proposed in Sect. 2.1.3, all the leg-wrench systems for an S= PKC can be identified without difficulty.

In any general configuration, the wrench system of an S= PKC is the same as that of its S virtual chain, i.e., a 3-$\boldsymbol{\zeta}_0$-system, which is composed of all the $\boldsymbol{\zeta}_0$ whose axes pass through the centre of the S virtual chain.

It then follows that any leg-wrench system with order $c^i > 0$ of an S= PKC is either a 3-$\boldsymbol{\zeta}_0$-system, a 2-$\boldsymbol{\zeta}_0$-system, or a 1-$\boldsymbol{\zeta}_0$-system in a general configuration (Fig. 7.2).

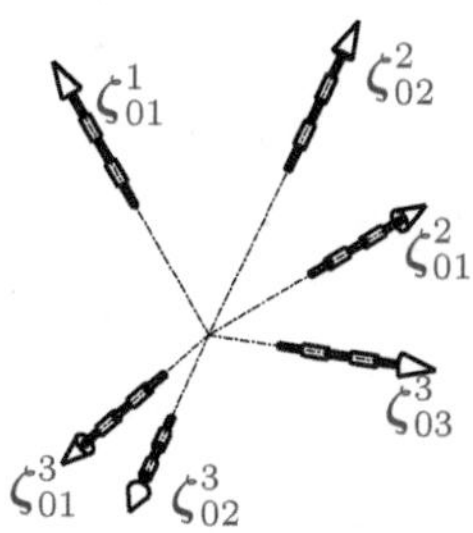

Fig. 7.2. Leg-wrench system of an S= PKC

7.5.2 Determination of the Combinations of Leg-Wrench Systems

For the S motion pattern considered, all the wrenches in the wrench system are of the same pitch. The leg-wrench systems are $c^i (0 \leq c^i \leq c)$-systems of 0 pitch. The combination of leg-wrench systems can be simply represented by the combination of the orders, c^i, of leg-wrench systems.

The combinations of the orders, c^i, of leg-wrench systems have been determined by solving (2.28) and are shown in Tables 5.1–5.5.

7.6 Step 2: Type Synthesis of Legs

In this section, the type synthesis of legs for S= PKCs will be performed using the two substeps described in Sect. 5.5.

7.6.1 Step 2a: Type Synthesis of 3-DOF Single-Loop Kinematic Chains That Involve an S Virtual Chain and Have a Specified Leg-Wrench System

The type synthesis of 3-DOF single-loop KCs that involve an S virtual chain and have a specified leg-wrench system can be performed as follows.

Firstly, perform the type synthesis of single-loop KCs that involve an S virtual chain and have a specified leg-wrench system (See Chap. 3). Secondly, discard those single-loop KCs in which the twists of all the joints but the virtual chain are linearly dependent.

Cases with a 3-ζ_0-System

Firstly, perform the type synthesis of 3-DOF single-loop KCs that involve an S virtual chain and have a 3-ζ_0-system. According to Chap. 3, there are six joints in a 3-DOF single-loop KC that involves an S virtual chain and has a 3-ζ_0-system. Such a single-loop KC is composed of one spherical compositional unit (Fig. 7.3). We obtain one 3-DOF single-loop KC that involves an S virtual chain and has a 3-ζ_0-system: $\hat{R}\hat{R}\hat{R}$S. The 3-ζ_0-system is composed of all the ζ_0 whose axes pass through the centre of the S virtual chain.

Secondly, discard those single-loop KCs in which the twists of all the joints but the virtual chain are linearly dependent. For brevity, we assume that the above condition is automatically satisfied.

Cases with a 2-ζ_0-System

Firstly, perform the type synthesis of 3-DOF single-loop KCs that involve an S virtual chain and have a 2-ζ_0-system. According to Chap. 3, there are seven joints in a 3-DOF single-loop KC that involves an S virtual chain and has a 2-ζ_0-system. Such a single-loop KC is formed by inserting one coaxial or codirectional compositional unit into one spherical compositional unit (Fig. 7.4).

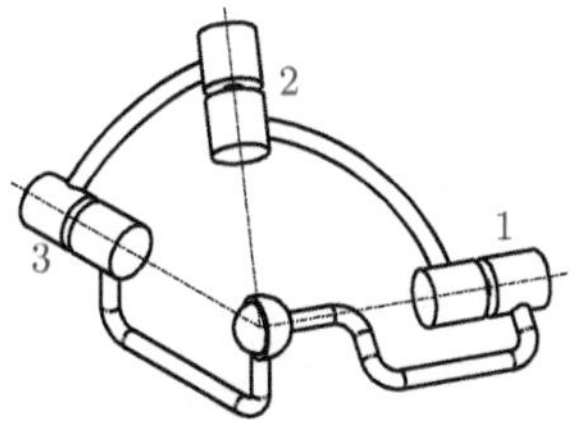

Fig. 7.3. Three-DOF single-loop KC involving a virtual chain and having a 3-$\boldsymbol{\zeta}_0$-system: $\hat{\text{R}}\hat{\text{R}}\hat{\text{R}}$S KC

The 2-$\boldsymbol{\zeta}_0$-system is composed of all the $\boldsymbol{\zeta}_0$ whose axes pass through the centre of the S virtual chain and intersect the axes of the R joints within the coaxial compositional unit or perpendicular to the directions of the P joints within the codirectional compositional unit.

Secondly, discard those single-loop KCs in which the twists of all the joints but the virtual chain are linearly dependent. This requires that there be only one R joint within the coaxial compositional unit or one P joint within the codirectional compositional unit.

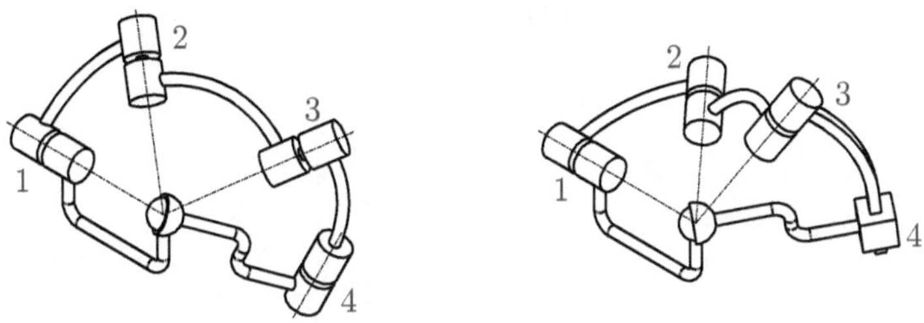

(a) $\hat{\text{R}}\hat{\text{R}}\hat{\text{R}}(\text{R})_L$S. **(b)** $\hat{\text{R}}\hat{\text{R}}\hat{\text{R}}(\text{P})_L$S.

Fig. 7.4. 3-DOF single-loop KCs involving an S virtual chain and having a 2-$\boldsymbol{\zeta}_0$-system

Cases with a 1-$\boldsymbol{\zeta}_0$-System

Firstly, perform the type synthesis of 3-DOF single-loop KCs that involve an S virtual chain and have a 1-$\boldsymbol{\zeta}_0$-system. According to Chap. 3, there are eight joints in a 3-DOF single-loop KC that involves an S virtual chain and has a 1-$\boldsymbol{\zeta}_0$-system. Such a single-loop KC is composed of (a) two spherical compositional units or (b) one spherical compositional unit and one planar compositional unit [Figs. 7.5a–7.5d] or can be obtained by inserting one coaxial or codirectional compositional unit into a 3-DOF single-loop KC that involves an S virtual chain and has a 2-$\boldsymbol{\zeta}_0$-system [Figs. 7.5e–7.5f]. The 1-$\boldsymbol{\zeta}_0$-system is the $\boldsymbol{\zeta}_0$ whose direction is perpendicular to all the axes of the R joints.

Table 7.1. Legs for S= PKCs

c^i	Class	No.	Type
3	3R	1	$\hat{R}\hat{R}\hat{R}$
2	4R	2	$\hat{R}\hat{R}\hat{R}R$
		3	$\hat{R}\hat{R}R\hat{R}$
		4	$\hat{R}R\hat{R}\hat{R}$
		5	$R\hat{R}\hat{R}\hat{R}$
	3R1P	6	$\hat{R}\hat{R}\hat{R}P$
		7	$\hat{R}\hat{R}P\hat{R}$
		8	$\hat{R}P\hat{R}\hat{R}$
		9	$P\hat{R}\hat{R}\hat{R}$
1	5R	10-19	Permutation of $\hat{R}\hat{R}\hat{R}RR$
		20	$\hat{R}\hat{R}(RRR)_E$
		21	$\hat{R}(RRR)_E\hat{R}$
		22	$(RRR)_E\hat{R}\hat{R}$
		23	$\hat{R}\hat{R}(RRR)_S$
		24	$\hat{R}(RRR)_S\hat{R}$
		25	$(RRR)_S\hat{R}\hat{R}$
	4R1P	26-45	Permutation of $\hat{R}\hat{R}\hat{R}RP$
		46	$\hat{R}\hat{R}(RRP)_E$
		47	$\hat{R}\hat{R}(RPR)_E$
		48	$\hat{R}\hat{R}(PRR)_E$
		49	$\hat{R}(RRP)_E\hat{R}$
		50	$\hat{R}(RPR)_E\hat{R}$
		51	$\hat{R}(PRR)_E\hat{R}$
		52	$(RRP)_E\hat{R}\hat{R}$
		53	$(RPR)_E\hat{R}\hat{R}$
		54	$(PRR)_E\hat{R}\hat{R}$
	3R2P	55-64	Permutation of $\hat{R}\hat{R}\hat{R}PP$
		65	$\hat{R}\hat{R}(RPP)_E$
		66	$\hat{R}\hat{R}(PRP)_E$
		67	$\hat{R}\hat{R}(PPR)_E$
		68	$\hat{R}(RPP)_E\hat{R}$
		69	$\hat{R}(PRP)_E\hat{R}$
		70	$\hat{R}(PPR)_E\hat{R}$
		71	$(RPP)_E\hat{R}\hat{R}$
		72	$(PRP)_E\hat{R}\hat{R}$
		73	$(PPR)_E\hat{R}\hat{R}$
0	omitted	omitted	omitted

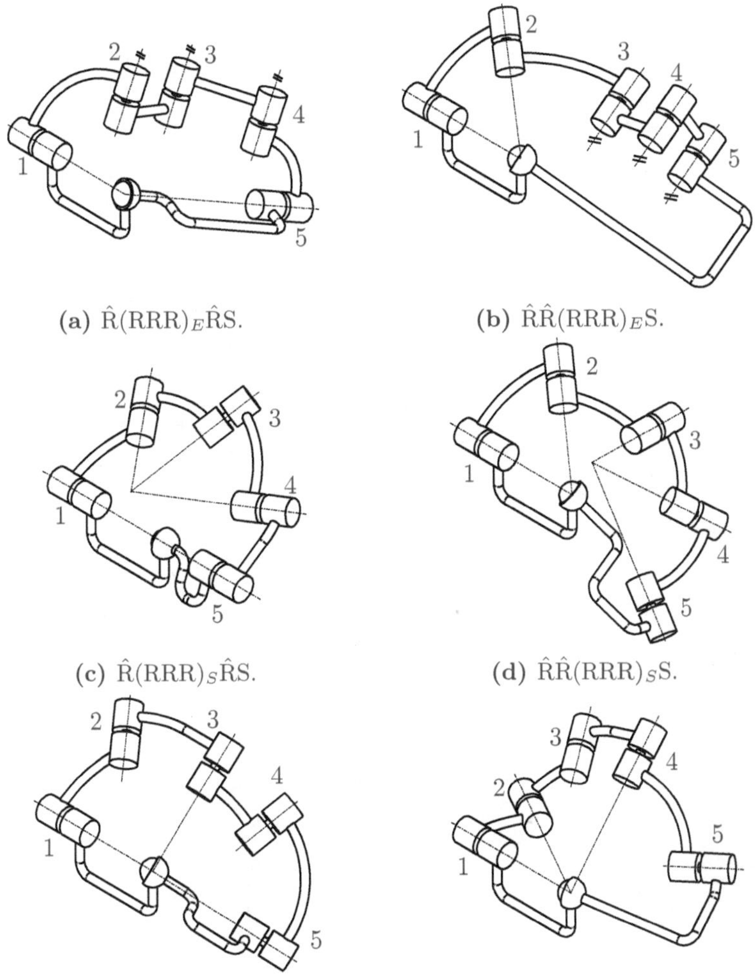

(a) $\hat{\mathrm{R}}(\mathrm{RRR})_E\hat{\mathrm{R}}\mathrm{S}$.

(b) $\hat{\mathrm{R}}\hat{\mathrm{R}}(\mathrm{RRR})_E\mathrm{S}$.

(c) $\hat{\mathrm{R}}(\mathrm{RRR})_S\hat{\mathrm{R}}\mathrm{S}$.

(d) $\hat{\mathrm{R}}\hat{\mathrm{R}}(\mathrm{RRR})_S\mathrm{S}$.

(e) $\hat{\mathrm{R}}(\mathrm{RRR})_B\hat{\mathrm{R}}\mathrm{S}$.

(f) $\hat{\mathrm{R}}\hat{\mathrm{R}}(\mathrm{RRR})_B\mathrm{S}$.

Fig. 7.5. 3-DOF single-loop KCs involving an S virtual chain and having a 1-$\boldsymbol{\zeta}_1$-system

Secondly, discard those single-loop KCs in which the twists of all the joints but the virtual chain are linearly dependent. This requires that there be only one R joint within the coaxial compositional unit or one P joint within the codirectional compositional unit.

In the representation of the 3-DOF single-loop kinematic chains involving an S virtual chain and having a c^i-$\boldsymbol{\zeta}_0$-system, $\hat{\mathrm{R}}$'s denote R joints whose axes intersect at the centre of the S virtual chain, $(\mathrm{XXX})_E$ (X represents an R or a P joint) denotes an equivalent planar joint formed by three successive joints, and $(\mathrm{RRR})_S$ denotes an equivalent spherical joint formed by three successive R joints, all the axes of the R joints within the $(\mathrm{XXX})_E$ or $(\mathrm{RRR})_S$ do not pass

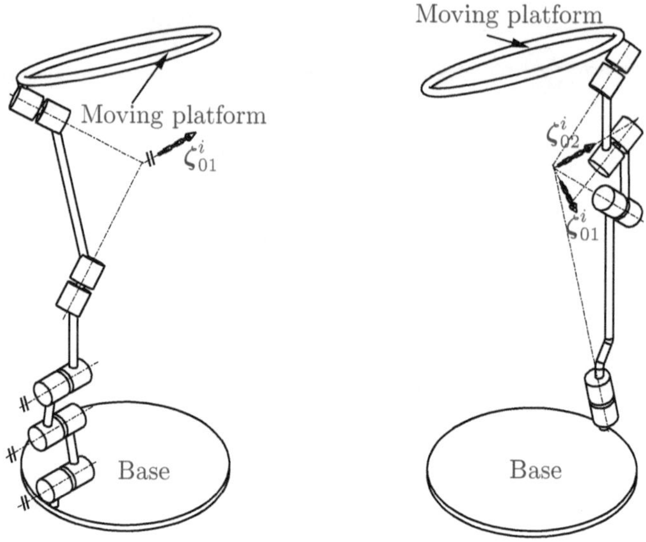

(a) $(RRR)_E\hat{R}\hat{R}$ leg. **(b)** $\hat{R}\hat{R}R\hat{R}$ leg.

Fig. 7.6. Some legs for S= PKCs

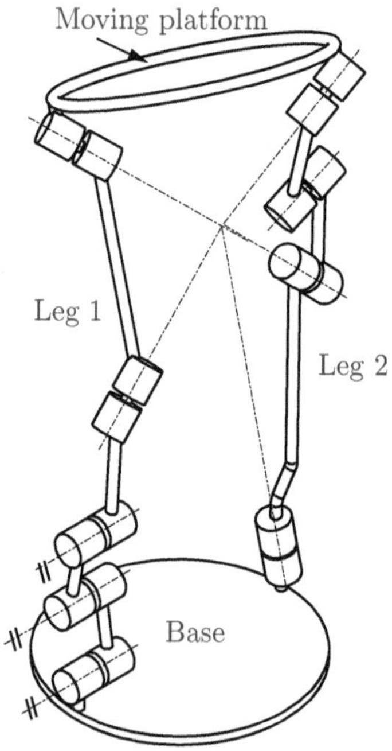

Fig. 7.7. $(RRR)_E\hat{R}\hat{R}$-$\hat{R}\hat{R}R\hat{R}$ S= PKC

through the intersection of the axes of the $\hat{R}$ joints, and $()_L$ denotes a coaxial or codirectional compositional unit.

It is pointed out that the $(P)_L$ and $(R)_L$ joints can be placed anywhere in the single-loop KC. For brevity, we list only the 3-DOF single-loop KCs from which all 3-DOF single-loop KCs can be obtained through the above operations. For example, the $\hat{R}\hat{R}P\hat{R}S$ single-loop KC can be obtained by changing the position of the P joint in $\hat{R}\hat{R}\hat{R}PS$ single-loop KC [Fig. 7.4].

7.6.2 Step 2b: Generation of Types of Legs

According to Condition (1) for the existence of S= PKCs (Sect. 7.3), the types of legs can be readily obtained from the 3-DOF single-loop kinematic chains obtained in Step 2a, Sect. 7.6.1, by removing the S virtual chain.

The specific geometric conditions which guarantee the satisfaction of Condition (1) for S= PKCs are clearly indicated by the notations that were introduced to represent the types of legs.

For example, by removing the S virtual chain in an $(RRR)_E\hat{R}\hat{R}S$ kinematic chain [Fig. 7.5b], an $(RRR)_E\hat{R}\hat{R}$ leg [Fig. 7.6a] can be obtained. In this leg, the axes of the first three R joints are parallel while the axes of the last two R joints intersect each other. This leg has a 1-$\boldsymbol{\zeta}_0$-system. The $\boldsymbol{\zeta}_0$ passes through the common point of the axes of two $\hat{R}$ joint and is parallel to the axes of the first three R joints.

All the types of legs for S= PKCs obtained are listed in Table 7.1. The legs used to synthesize S= PKCs need to satisfy not only the geometric conditions within a same leg, but also the geometric conditions among different legs.

7.7 Step 3: Assembly of Legs

S= PKCs can be generated by assembling a set of legs for S= PKCs shown in Table 7.1 selected according to the combinations of the leg-wrench systems shown in Tables 5.1–5.5. In assembling an S= PKC, the following condition should be satisfied: the linear combination of the leg-wrench systems constitutes a 3-$\boldsymbol{\zeta}_0$-system (see Condition (2) for S= PKCs in Sect. 7.3).

For an S= PKC in which not all the leg-wrench systems are invariant with respect to the base or the moving platform, the linear combination of the leg-wrench systems usually constitutes a 3-$\boldsymbol{\zeta}_0$-system. For an S= PKC in which all the leg-wrench systems are invariant with respect to the base or the moving platform, the geometry of the base or the moving platform should meet certain conditions to guarantee that the linear combination of the leg-wrench systems constitutes a 3-$\boldsymbol{\zeta}_0$-system.

Figure 7.6 shows two legs for S= PKCs. In the $(RRR)_E\hat{R}\hat{R}$ leg shown in Fig. 7.6a, the axes of the first three R joints are parallel while the axes of the last two R joints intersect each other. This leg has a 1-$\boldsymbol{\zeta}_0$-system. The $\boldsymbol{\zeta}_0$ passes through the common point of the axes of two $\hat{R}$ joint and is parallel to the axes of the first three R joints. In the $\hat{R}\hat{R}\hat{R}\hat{R}$ leg shown in Fig. 7.6b, all the axes of

Table 7.2. Three-legged S= PKCs with identical legs

c^i	Class	No.	Type	Number of overconstraints
3	3R	1	3-$\hat{R}\hat{R}\hat{R}$	6
2	4R	2	3-$\hat{R}\hat{R}\hat{R}R$	3
		3	3-$\hat{R}\hat{R}R\hat{R}$	
		4	3-$\hat{R}R\hat{R}\hat{R}$	
		5	3-$R\hat{R}\hat{R}\hat{R}$	
	3R1P	6	3-$\hat{R}\hat{R}\hat{R}P$	
		7	3-$\hat{R}\hat{R}P\hat{R}$	
		8	3-$\hat{R}P\hat{R}\hat{R}$	
		9	3-$P\hat{R}\hat{R}\hat{R}$	
1	5R	10-19	3-Permutation of $\hat{R}\hat{R}\hat{R}RR$	0
		20	3-$\hat{R}\hat{R}(RRR)_E$	
		21	3-$\hat{R}(RRR)_E\hat{R}$	
		22	3-$(RRR)_E\hat{R}\hat{R}$	
		23	3-$\hat{R}\hat{R}(RRR)_S$	
		24	3-$\hat{R}(RRR)_S\hat{R}$	
		25	3-$(RRR)_S\hat{R}\hat{R}$	
	4R1P	26-45	3-Permutation of $\hat{R}\hat{R}\hat{R}RP$	
		46	3-$\hat{R}\hat{R}(RRP)_E$	
		47	3-$\hat{R}\hat{R}(RPR)_E$	
		48	3-$\hat{R}\hat{R}(PRR)_E$	
		49	3-$\hat{R}(RRP)_E\hat{R}$	
		50	3-$\hat{R}(RPR)_E\hat{R}$	
		51	3-$\hat{R}(PRR)_E\hat{R}$	
		52	3-$(RRP)_E\hat{R}\hat{R}$	
		53	3-$(RPR)_E\hat{R}\hat{R}$	
		54	3-$(PRR)_E\hat{R}\hat{R}$	
	3R2P	55-64	3-Permutation of $\hat{R}\hat{R}\hat{R}PP$	
		65	3-$\hat{R}\hat{R}(RPP)_E$	
		66	3-$\hat{R}\hat{R}(PRP)_E$	
		67	3-$\hat{R}\hat{R}(PPR)_E$	
		68	3-$\hat{R}(RPP)_E\hat{R}$	
		69	3-$\hat{R}(PRP)_E\hat{R}$	
		70	3-$\hat{R}(PPR)_E\hat{R}$	
		71	3-$(RPP)_E\hat{R}\hat{R}$	
		72	3-$(PRP)_E\hat{R}\hat{R}$	
		73	3-$(PPR)_E\hat{R}\hat{R}$	

the R̂ joints intersect at a point. This leg has a 2-ζ_0-system which comprises all ζ_0 whose axes pass through the common point of all the axes of the R̂ joints and intersect the axis of the R joint. By taking one (RRR)$_E$R̂R̂ leg and one R̂R̂RR̂ leg, a 2-legged (RRR)$_E$R̂R̂-R̂R̂RR̂ S= PKC (Fig. 7.7) can be obtained. In the S= PKC, all the R̂ joints in the S= PKCs intersect at one point.

Due to the large number of S= PKCs, only the three-legged S= PKCs with legs of the same type are listed in Table 7.2.

The wrench system of each leg of the 3-(RRR)$_E$R̂R̂, 3-(RRR)$_S$R̂R̂, 3-(RRP)$_E$R̂R̂, 3-(RPR)$_E$R̂R̂, 3-(PRR)$_E$R̂R̂, 3-(RPP)$_E$R̂R̂, 3-(PRP)$_E$R̂R̂, 3-(PPR)$_E$R̂R̂ are invariant with respect to the base. To guarantee that the linear combination of the wrench systems of three legs within an S= PKC constitutes a 3-ζ_0-system, the base should be designed in such a way that the normals to the equivalent planar joints of all legs are not parallel to a plane or that the centre of rotation of the moving platform and the centres of the three equivalent spherical joints are not located on a same plane. The wrench system of each leg of the 3-R̂R̂(RRR)$_E$, 3-R̂R̂(RRR)$_S$, 3-R̂R̂(RRP)$_E$, 3-R̂R̂(RPR)$_E$, 3-R̂R̂(PRR)$_E$, 3-R̂R̂(RPP)$_E$, 3-R̂R̂(PRP)$_E$, 3-R̂R̂(PPR)$_E$ are invariant with respect to the moving platform. To guarantee that the linear combination of the wrench systems of three legs within an S= PKC constitutes a 3-ζ_0-system, the moving platform should be designed in such a way that the normals to the equivalent planar joints of all legs are not parallel to a plane or that the centre of rotation of the moving platform and the centres of the three equivalent spherical joints are not located on a same plane.

7.8 Step 4: Selection of Actuated Joints

7.8.1 t-Component of the Actuation Wrenches

Considering that the order of a screw system is coordinate free and for simplicity reasons, all the actuation wrenches and wrenches are expressed in a coordinate system with its origin at the centre of the wrench system. Thus, $[\mathbf{i}^T \quad \mathbf{0}^T]^T$, $[\mathbf{j}^T \quad \mathbf{0}^T]^T$, $[\mathbf{k}^T \quad \mathbf{0}^T]^T$ denote a basis of $\mathcal{W}$. Here, $\mathbf{i}$, $\mathbf{j}$ and $\mathbf{k}$ denote respectively the unit vectors along the X-, Y- and Z-axes, and $\mathbf{0}$ represents a 3×1 zero vector. The last three scalar components of all the basis wrenches of $\mathcal{W}$ are zero. Thus, the t-component, $\boldsymbol{\zeta}^i_{t\not{\mathcal{D}}j}$, of the $\boldsymbol{\zeta}^i_{\not{\mathcal{D}}j}$ is composed of the last three scalar components of the $\boldsymbol{\zeta}^i_{\not{\mathcal{D}}j}$, i.e.,

$$\boldsymbol{\zeta}^i_{t\not{\mathcal{D}}j} = \boldsymbol{\zeta}^i_{S\not{\mathcal{D}}j} \tag{7.1}$$

where $\boldsymbol{\zeta}^i_{S\not{\mathcal{D}}j}$ represents the second vector component of $\boldsymbol{\zeta}^i_{\not{\mathcal{D}}j}$.

Figure 7.8 shows the actuation wrenches of actuated joints in some legs for S= PMs. These two legs correspond to a same leg for S= PKCs. The wrench systems of these legs are both a 1-ζ_0-system. In the (<u>R</u>RR)$_E$R̂R̂ leg [Fig. 7.8a], the first R joint is actuated. $\boldsymbol{\zeta}^i_{0\not{\mathcal{D}}1}$ can be selected as the ζ_0 along the intersection of the plane defined by the axes of the second and third joints with the plane defined by the axes of the last two joints. In the (R<u>R</u>R)$_E$R̂R̂ leg [Fig. 7.8b], the

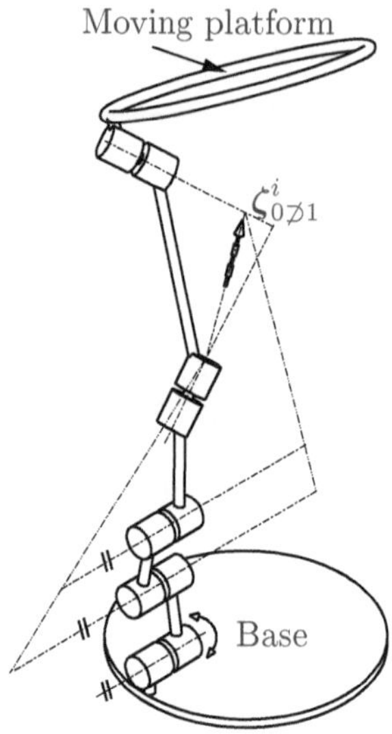

(a) $(\underline{\mathrm{R}}\mathrm{RR})_E\hat{\mathrm{R}}\hat{\mathrm{R}}$ leg.

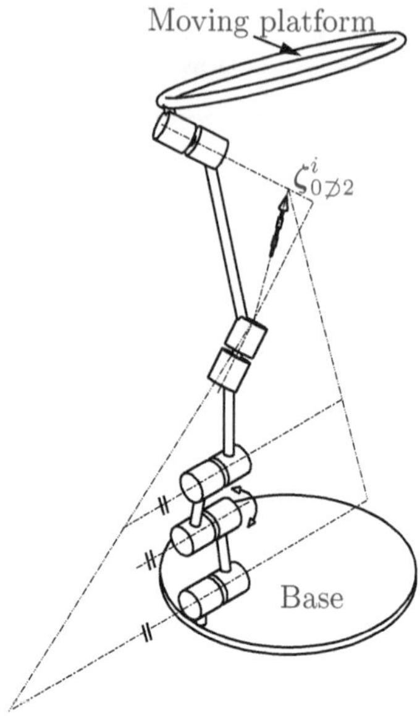

(b) $(\mathrm{R}\underline{\mathrm{R}}\mathrm{R})_E\hat{\mathrm{R}}\hat{\mathrm{R}}$ leg.

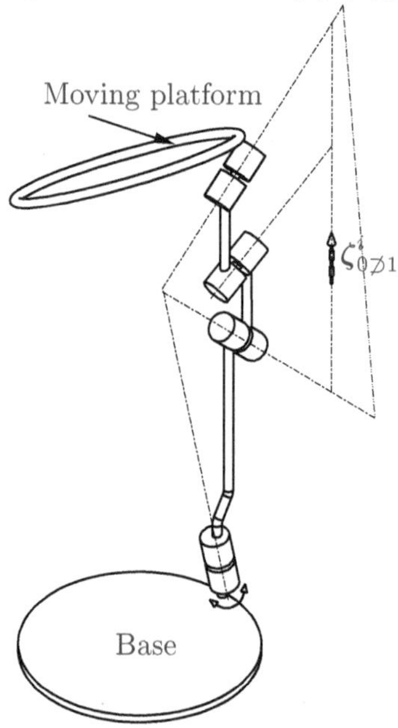

(c) $\underline{\hat{\mathrm{R}}}\hat{\mathrm{R}}\mathrm{R}\hat{\mathrm{R}}$ leg.

Fig. 7.8. Actuation wrenches of some legs for S= PKCs

second R joint is actuated. $\boldsymbol{\zeta}^{i}_{0\not{D}2}$ can be selected as the $\boldsymbol{\zeta}_0$ along the intersection of the plane defined by the axes of the first and third joints with the plane defined by the axes of the last two joints. In the $\underline{\hat{\text{R}}}\hat{\text{R}}\text{R}\hat{\text{R}}$leg [Fig. 7.8c], the first R joint is actuated. $\boldsymbol{\zeta}^{i}_{0\not{D}1}$ can be selected as the $\boldsymbol{\zeta}_0$ whose axis passes through the intersection of the axis of the R joint with the plane defined by the axes of the two unactuated $\hat{\text{R}}$ joints and does not pass through the intersection of the axes of all the $\hat{\text{R}}$ joints.

7.8.2 Procedure for the Validity Detection of Actuated Joints

The validity detection of actuated joints of S= PMs can thus be performed using the following steps.

Step 4a. If one or more of the actuated joints of a candidate S= PM are inactive, the set of actuated joints is invalid and the candidate S= PM should be discarded.

Although different approaches can be used to detect inactive joints, an alternative approach is proposed below.

A joint in an S= PKC is inactive if the second vector component of its actuation wrenches, $\boldsymbol{\zeta}^{i}_{S\not{D}j}$, is $\mathbf{0}$. In other words, a joint in an S= PKC is inactive if its actuation wrenches belong to the 3-$\boldsymbol{\zeta}_0$-system of the S= PKC. Physically speaking, the actuation wrenches of the inactive joint will not restrict the motion of the moving platform within its twist system.

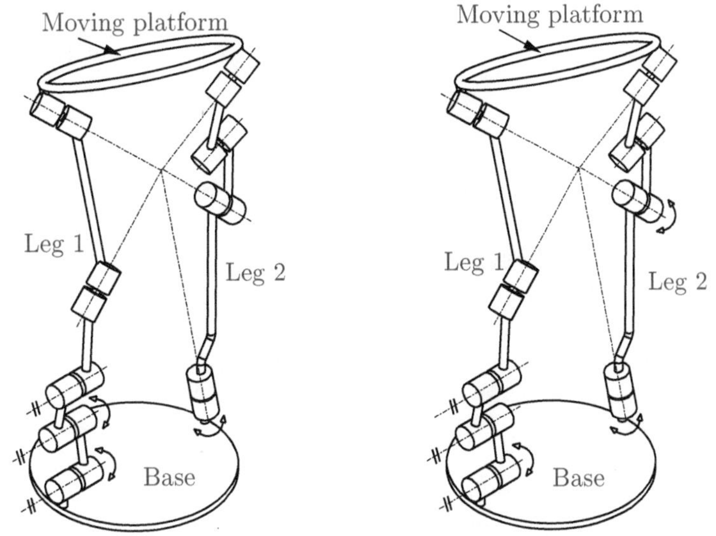

(a) $(\underline{\text{RR}}\text{R})_E\hat{\text{R}}\hat{\text{R}}$-$\underline{\hat{\text{R}}}\hat{\text{R}}\text{R}\hat{\text{R}}$ S= PM. **(b)** $(\underline{\text{R}}\text{RR})_E\hat{\text{R}}\hat{\text{R}}$-$\underline{\hat{\text{R}}\hat{\text{R}}}\text{R}\hat{\text{R}}$ S= PM.

Fig. 7.9. Selection of actuated joints for the $\text{RRR}\hat{\text{R}}\hat{\text{R}}$-$\hat{\text{R}}\hat{\text{R}}\text{R}\hat{\text{R}}$ S= PKC

Table 7.3. Three-legged S= PMs

c^i	Class	No.	Type
3	3R	1	3-$\underline{\hat{\mathrm{R}}}\hat{\mathrm{R}}\hat{\mathrm{R}}$
2	4R	2	3-$\underline{\hat{\mathrm{R}}}\hat{\mathrm{R}}\hat{\mathrm{R}}\mathrm{R}$
		3	3-$\underline{\hat{\mathrm{R}}}\hat{\mathrm{R}}\mathrm{R}\hat{\mathrm{R}}$
		4	3-$\underline{\hat{\mathrm{R}}}\mathrm{R}\hat{\mathrm{R}}\hat{\mathrm{R}}$
		5	3-$\mathrm{R}\underline{\hat{\mathrm{R}}}\hat{\mathrm{R}}\hat{\mathrm{R}}$
1	5R	6-15	3-Permutation of $\underline{\hat{\mathrm{R}}}\hat{\mathrm{R}}\hat{\mathrm{R}}\mathrm{RR}$
		16	3-$\underline{\hat{\mathrm{R}}}\hat{\mathrm{R}}(\mathrm{RRR})_E$
		17	3-$\underline{\hat{\mathrm{R}}}(\mathrm{RRR})_E\hat{\mathrm{R}}$
		18	3-$(\underline{\mathrm{R}}\mathrm{RR})_E\hat{\mathrm{R}}\hat{\mathrm{R}}$
		19	3-$\underline{\hat{\mathrm{R}}}\hat{\mathrm{R}}(\mathrm{RRR})_S$
		20	3-$\underline{\hat{\mathrm{R}}}(\mathrm{RRR})_S\hat{\mathrm{R}}$
		21	3-$(\underline{\mathrm{R}}\mathrm{RR})_S\hat{\mathrm{R}}\hat{\mathrm{R}}$
	4R-1P	22	3-$(\mathrm{RR}\underline{\mathrm{P}})_E\hat{\mathrm{R}}\hat{\mathrm{R}}$
		23	3-$(\mathrm{R}\underline{\mathrm{P}}\mathrm{R})_E\hat{\mathrm{R}}\hat{\mathrm{R}}$
		24	3-$(\underline{\mathrm{P}}\mathrm{RR})_E\hat{\mathrm{R}}\hat{\mathrm{R}}$

For example, the actuation wrenches of the $\hat{\mathrm{R}}\hat{\mathrm{R}}\underline{\mathrm{R}}\hat{\mathrm{R}}$ leg for S= PMs when the R joint is selected as actuated joint are the $\boldsymbol{\zeta}_0$ whose axes pass through the common points of the axes of the three $\hat{\mathrm{R}}$ joints and do not pass through the axis of the R joint. The second vector component of all its actuation wrenches is $\mathbf{0}$, thus the R joint is inactive and cannot be selected as actuated joint. In fact, for a leg involving three $\hat{\mathrm{R}}$ joints, all the joints in the leg except the $\hat{\mathrm{R}}$ joints are inactive.

Step 4b. If the t-components, i.e., the second vector components of the actuation wrenches of actuated joints $\boldsymbol{\zeta}^i_{\mathcal{D}j}$ are linearly dependent in a general configuration for a candidate S= PM, the set of actuated joints is invalid. In this case, the candidate S= PM should be discarded.

Following the criteria on the selection of actuated joints (Sect. 5.7) and the procedure for the detection of the validity of actuated joints, all the $m(m \geq 2)$-legged S= PMs corresponding to each S= PKC can be generated.

For example, the candidate S= PMs corresponding to the 2-legged S= PKC, $(\mathrm{RRR})_E\hat{\mathrm{R}}\hat{\mathrm{R}}$-$\hat{\mathrm{R}}\hat{\mathrm{R}}\mathrm{R}\hat{\mathrm{R}}$ satisfying the above criteria are $(\underline{\mathrm{R}}\mathrm{RR})_E\hat{\mathrm{R}}\hat{\mathrm{R}}$-$\underline{\hat{\mathrm{R}}}\hat{\mathrm{R}}\mathrm{R}\hat{\mathrm{R}}$ and $(\underline{\mathrm{R}}\mathrm{RR})_E\hat{\mathrm{R}}\hat{\mathrm{R}}$-$\underline{\hat{\mathrm{R}}}\underline{\hat{\mathrm{R}}}\mathrm{R}\hat{\mathrm{R}}$ (Fig. 7.9). Following the procedure for the detection of the validity of actuated joints, the $(\underline{\mathrm{R}}\mathrm{RR})_E\hat{\mathrm{R}}\hat{\mathrm{R}}$-$\underline{\hat{\mathrm{R}}}\hat{\mathrm{R}}\mathrm{R}\hat{\mathrm{R}}$ S= PM should be discarded since the determinant of the matrix composed of the t-components of the three actuated joints is 0. Thus, there is only one S= PM corresponding to the 2-legged S= PKC, i.e., the $(\underline{\mathrm{R}}\mathrm{RR})_E\hat{\mathrm{R}}\hat{\mathrm{R}}$-$\underline{\hat{\mathrm{R}}}\underline{\hat{\mathrm{R}}}\mathrm{R}\hat{\mathrm{R}}$ S= PM.

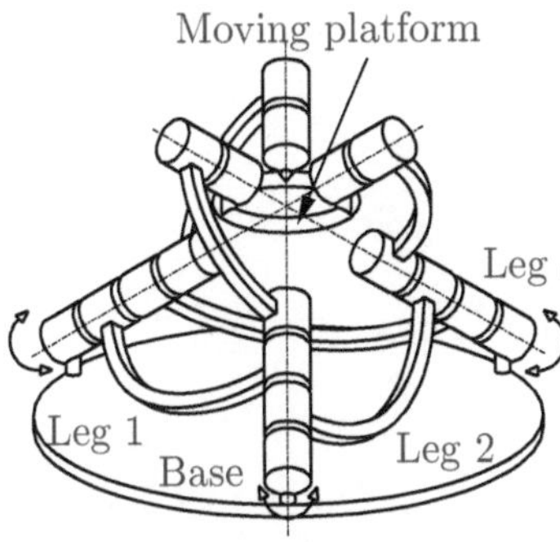

(a) 3-$\underline{\hat{R}}\hat{R}\hat{R}$ S= PM.

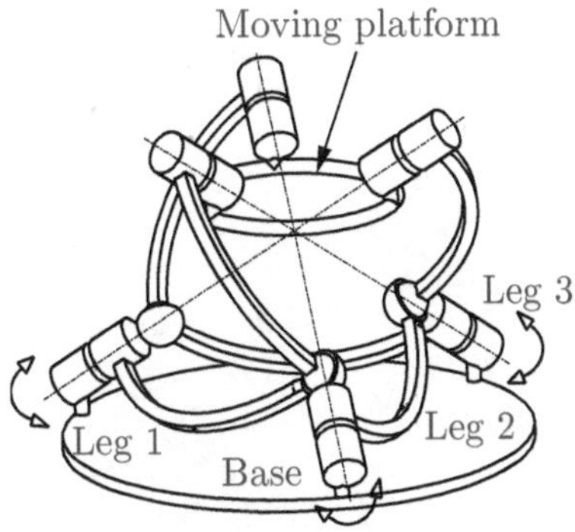

(b) 3-$\underline{\hat{R}}S\hat{R}$ S= PM.

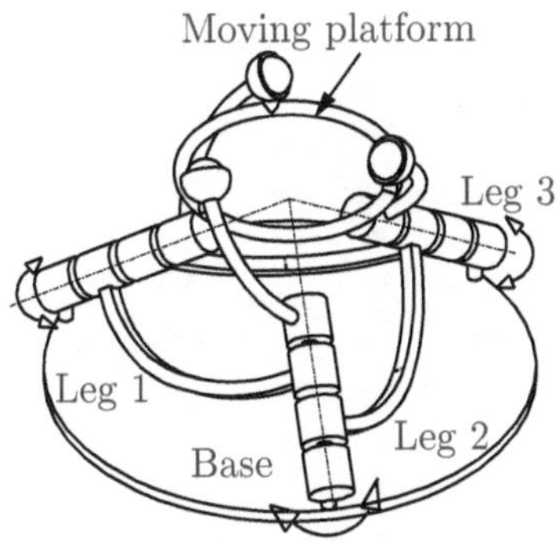

(c) 3-$\underline{\hat{R}}\hat{R}S$ S= PM.

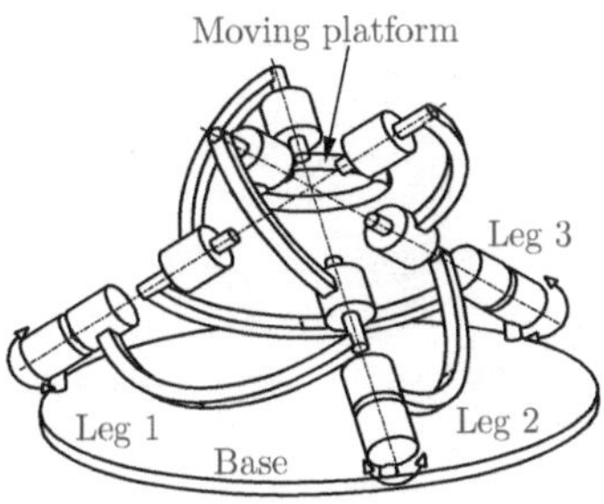

(d) 3-$\underline{\hat{R}}\hat{C}\hat{C}$ S= PM.

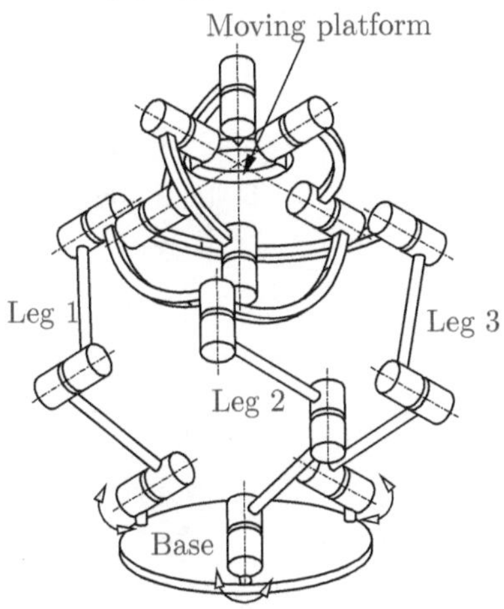

(e) 3-$(\underline{R}RR)_E\hat{R}\hat{R}$ S= PM.

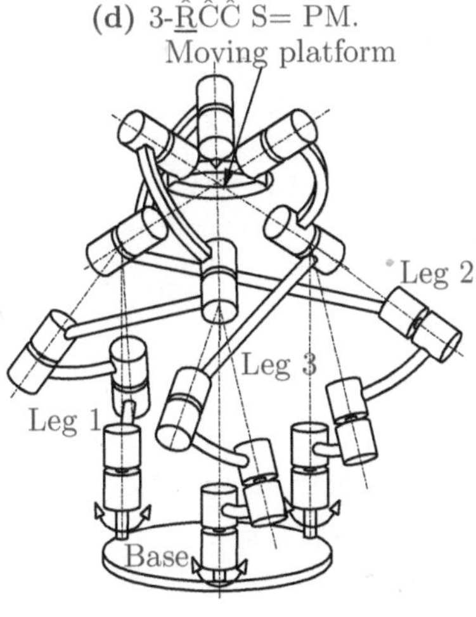

(f) 3-$(\underline{R}RR)_S\hat{R}\hat{R}$ S= PM.

Fig. 7.10. Six S= PMs shown in an isotropic configuration

Due to the large number of S= PMs, only 3-legged S= PMs with all legs of the same type satisfying the above criteria are listed in Table 7.3. There are 11 new S= PMs in Table 7.3 (see No. 16, 18–27).

7.9 Summary

In this chapter, the type synthesis of S= PMs has been thoroughly solved using the general type synthesis approach proposed in Chap. 5. S= PKCs which were proposed in [47, 58] have been re-obtained. S= PKCs with inactive joints have also been obtained. Both overconstrained and non-overconstrained S= PKCs can be obtained and some new S= PMs have also been revealed.

By substituting a combination of an R joint and a P joint with parallel axes, a combination of two R joints whose axes are not parallel and a combination of three R joints whose axes are not parallel with a C, U, and S joint respectively, all the special cases of S= PMs can be obtained. For example, the RUU S= PM [24] is a special case of the 3-$\hat{\underline{R}}\hat{R}RR\hat{R}$ S= PM. To make the conditions for S= PMs clear in their representation and for simplicity reasons, these special cases are not listed in Table 7.3.

Four of the new S= PMs, namely the 3-$(\underline{R}RR)_E\hat{R}\hat{R}$, 3-$(\underline{R}RR)_S\hat{R}\hat{R}$, 3-$\hat{\underline{R}}S\hat{R}$ and 3-$\hat{\underline{R}}\hat{R}S$ S= PMs, are shown in Fig. 7.10. Based on the above four new S= PMs, some variations of the Agile eye [38] can be proposed (Fig. 7.10). Compared with the agile eye, the arrangement of the location of the actuated joints on the base is more flexible for the 3-$(\underline{R}RR)_E\hat{R}\hat{R}$ [Fig. 7.10e] and 3-$(\underline{R}RR)_S\hat{R}\hat{R}$ [Fig. 7.10f] S= PMs. In the case of the 3-$(\underline{R}RR)_S\hat{R}\hat{R}$ S= PM, the axes of three actuated joints can be parallel [Fig 7.10f]. However, these two new S= PMs are more complex in structure than the Agile eye. Compared with the Agile eye, which is overconstrained, the S= PMs shown in Figs. 7.10b and 7.10c are not overconstrained.

The 3-$\hat{\underline{R}}S\hat{R}$ S= PM can also be regarded as a special case of the 3-$\hat{\underline{R}}\hat{R}RR\hat{R}$, 3-$\hat{\underline{R}}R\hat{R}R\hat{R}$ or 3-$\hat{\underline{R}}RR\hat{R}\hat{R}$ S= PMs, while the 3-$\hat{\underline{R}}\hat{R}S$ S= PM can also be regarded as a special case of the 3-$\hat{\underline{R}}\hat{R}\hat{R}RR$, 3-$\hat{\underline{R}}\hat{R}R\hat{R}R$ or 3-$\hat{\underline{R}}\hat{R}RR\hat{R}$ S= PMs.

8. Three-DOF PPR= Parallel Mechanisms

The previous chapters addressed the type synthesis of PMs that can perform either pure translations or pure translations. Another type of 3-DOF motion pattern that can be used in several applications is the PPR motion pattern. This motion pattern involves two translations and one rotation. In this chapter, the type synthesis of 3-DOF PPR= PMs (also called cylindrical parallel mechanisms) is dealt with using the general approach proposed in Chap. 5. A PPR= PM refers to a 3-DOF PM generating 3-DOF PPR motion. The four steps of the type synthesis of PPR= PMs are presented in detail.

8.1 Introduction

PPR= PMs are the parallel counterparts of the 3-DOF PPR serial robots, which are composed of two P joints and one R joint. The moving platform of a PPR= PM can rotate arbitrarily about an axis undergoing a planar translation. PMs in this class have application potential mainly in assembly and machining. Compared with the type synthesis of planar PMs [59], spherical PMs[46, 59, 75] and translational PMs [13, 31, 33, 47, 58, 65, 70, 79, 121, 123], few researchers studied the type synthesis of PPR= PMs. Up to now, only a few PPR= PMs [95, 123] have been proposed[1]. This chapter aims at performing a systematic study on the type synthesis of PPR= PMs.

Using the general approach proposed in Chap. 5, the type synthesis of PPR= PMs is dealt with in this chapter. The decomposition of the wrench system of a PPR= PKC is dealt with in Sect. 8.5. The type synthesis of legs for PPR= PKCs is performed in Sect. 8.6. In Sect. 8.7, the assembly of legs to generate PPR= PKCs is presented while the selection of actuated joints for PPR= PMs is discussed in Sect. 8.8.

[1] In [95, 123], different names were used for the PPR= PMs.

X. Kong and C. Gosselin: Type Syn. of Parallel Mech., STAR 33, pp. 125–140, 2007.
springerlink.com

8.2 Wrench System of a PPR= PKC

In any general configuration, the twist system of a PPR= PKC is a 2-$\boldsymbol{\xi}_\infty$-1-$\boldsymbol{\xi}_0$-system. Therefore, the wrench system of a PPR= PKC or its PPR virtual chain (Fig. 8.1) is a 2-$\boldsymbol{\zeta}_\infty$-1-$\boldsymbol{\zeta}_0$-system, which is composed of (a) all the $\boldsymbol{\zeta}_\infty$ whose axes are perpendicular to the axis of the R joint and (b) all the $\boldsymbol{\zeta}_0$ whose axes are perpendicular to the directions of the P joints and coplanar with the axis of the R joint, and (c) other $\boldsymbol{\zeta}$ which are linear combinations of the above $\boldsymbol{\zeta}_\infty$ and $\boldsymbol{\zeta}_0$. One set of basis wrenches of the 2-$\boldsymbol{\zeta}_\infty$-1-$\boldsymbol{\zeta}_0$-system is $\boldsymbol{\zeta}_{\infty 1}$, $\boldsymbol{\zeta}_{\infty 2}$ and $\boldsymbol{\zeta}_{03}$ (Fig. 8.1).

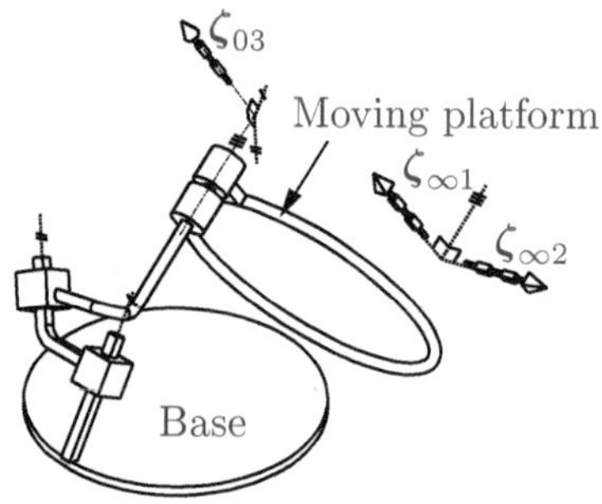

Fig. 8.1. Wrench system of a PPR= PKC

8.3 Conditions for a PKC to Be a PPR= PKC

When we connect the base and the moving platform of a PPR= PKC by a PPR virtual chain, the function of the PKC is not affected. Any of its legs and the PPR virtual chain will form a 3-DOF single loop kinematic chain. When the order of the leg-wrench system is greater than 0, the single-loop kinematic chain constructed must be an overconstrained kinematic chain.

Based on the concept of PPR virtual chain, it follows that a PKC is a PPR= PKC if it satisfies the two conditions below:

(1) Each leg of the PKC and a same PPR virtual chain constitute a 3-DOF single-loop kinematic chain.
(2) The wrench system of the PKC is the same as that of the PPR virtual chain, i.e., a 2-$\boldsymbol{\zeta}_\infty$-1-$\boldsymbol{\zeta}_0$-system, in any one general configuration.

8.4 Procedure for the Type Synthesis of PPR= PMs

A procedure can thus be proposed for the type synthesis of PPR= PKCs as follows:

Step 1. Decomposition of the wrench system of 3-DOF PPR= PKCs.
Step 2. Type synthesis of legs for PPR= PKCs.
Here, a leg for PPR= PKCs refers to a leg satisfying Condition (1) for PPR= PKCs.

Step 3. Assembly of legs for PPR= PKCs.
PPR= PKCs can be generated by taking two or more legs for PPR= PKCs, obtained in Step 1, such that Condition (2) for PPR= PKCs is satisfied.
Step 4. Selection of actuated joints.
PPR= PMs can be generated by selecting actuated joints in different ways for each PPR= PKC (Sect. 8.8), obtained in Step 3.

The procedure proposed above will be elaborated upon in the following sections.

8.5 Step 1: Decomposition of the Wrench System of PPR= PKCs

The decomposition of the wrench system of an m-legged PPR= PKC consists in finding all the leg-wrench systems and all the combinations of m leg-wrench systems for the PPR= PKC and a specified number of overconstraints Δ. Step 1 can be performed using (2.24) and (2.28).

8.5.1 Determination of the Leg-Wrench Systems

As the wrench system of a PKC is the linear combination of the leg-wrench systems in a general configuration, any leg-wrench system in a PPR= PKC is a sub-system of its wrench system. Following the procedure proposed in Sect. 2.1.3, all the leg-wrench systems for a PPR= PKC can be identified without difficulty.

In any general configuration, the wrench system of a PPR= PKC is the same as that of its PPR virtual chain, i.e., a 2-ζ_∞-1-ζ_0-system. It then follows that any leg-wrench system with order $c^i > 0$ of a PPR= PKC is either a 2-ζ_∞-1-ζ_0-system, a 1-ζ_∞-1-ζ_0-system, a 2-ζ_∞-system, a 1-ζ_0-system or a 1-ζ_∞-system in a general configuration (Fig. 8.2). Here, c^i denotes the order of leg-wrench system of leg i.

8.5.2 Determination of the Combinations of Leg-Wrench Systems

For the PPR motion pattern considered, not all the wrenches in the wrench system are of the same pitch. The combinations of leg-wrench systems have been determined by solving (2.28) and are shown in Table 5.7.

8.6 Step 2: Type Synthesis of Legs

In this section, the type synthesis of legs for PPR= PKCs will be performed using the two substeps described in Sect. 5.5.

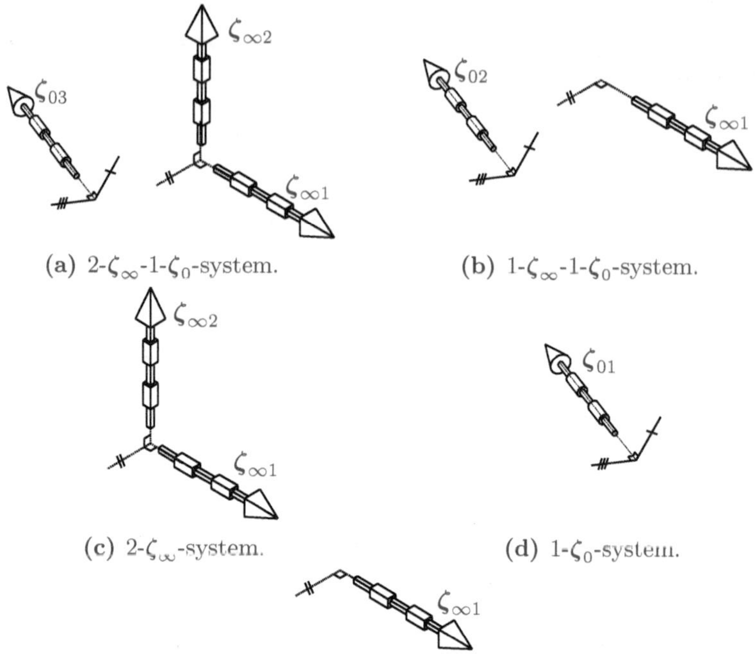

(a) 2-ζ_∞-1-ζ_0-system. (b) 1-ζ_∞-1-ζ_0-system.

(c) 2-ζ_∞-system. (d) 1-ζ_0-system.

(e) 1-ζ_∞-system.

Fig. 8.2. Leg-wrench systems of PPR= PKCs

8.6.1 Step 2a: Type Synthesis of 3-DOF Single-Loop Kinematic Chains That Involve a PPR Virtual Chain and Have a Specified Leg-Wrench System

The type synthesis of 3-DOF single-loop KCs that involve a PPR virtual chain and have a specified leg-wrench system can be performed as follows.

Firstly, perform the type synthesis of single-loop KCs that involve a PPR virtual chain and have a specified leg-wrench system (See Chap. 3). Secondly, discard those single-loop KCs in which the twists of all the joints but the virtual chain are linearly dependent.

Cases with a 2-ζ_∞-1-ζ_0-System

Firstly, perform the type synthesis of 3-DOF single-loop KCs that involve a PPR virtual chain and have a 2-ζ_∞-1-ζ_0-system. According to Chap. 3, there are six joints in a 3-DOF single-loop KC that involves a PPR virtual chain and has a 2-ζ_∞-1-ζ_0-system. Such a single-loop KC is formed by inserting one coaxial compositional unit into one planar translational compositional unit (Fig. 8.3). One set of basis wrenches of the 2-ζ_∞-1-ζ_0-system is composed of $\zeta_{\infty 1}$, $\zeta_{\infty 2}$ and ζ_{03}. The directions of the $\zeta_{\infty 1}$, $\zeta_{\infty 2}$ are perpendicular to the

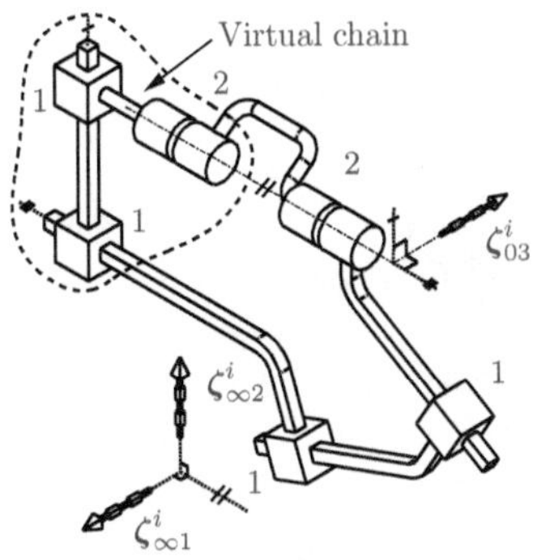

Fig. 8.3. Three-DOF single-loop KC involving a virtual chain and having a 2-ζ_∞-1-ζ_0-system: $(\mathrm{PP})^{\|}_E\ddot{\mathrm{R}}\mathrm{V}$ KC

axes of the R joints, and the axis of the ζ_{03} is perpendicular to the directions of all the P joints and intersects the axes of the R joints.

Secondly, discard those single-loop KCs in which the twists of all the joints but the virtual chain are linearly dependent. This requires that there be only two R joints within the coaxial compositional unit.

We obtain one 3-DOF single-loop KC that involves a PPR virtual chain and has a 2-ζ_∞-1-ζ_0-system: $(\mathrm{PP})^{\|}_E\ddot{\mathrm{R}}\mathrm{V}$.

Cases with a 1-ζ_∞-1-ζ_0-System

Firstly, perform the type synthesis of 3-DOF single-loop KCs that involve a PPR virtual chain and have a 1-ζ_∞-1-ζ_0-system. According to Chap. 3, there are seven joints in a 3-DOF single-loop KC that involves a PPR virtual chain and has a 1-ζ_∞-1-ζ_0-system. It is formed either by inserting one coaxial compositional unit into one planar compositional unit (Fig. 8.4) or by inserting one coaxial compositional unit or codirectional compositional unit into a 3-DOF single-loop KC that involves a PPR virtual chain and has a 1-ζ_∞-1-ζ_0-system. One set of basis wrenches of the 1-ζ_∞-1-ζ_0-system is composed of $\zeta_{\infty 1}$ and ζ_{02}. The direction of the $\zeta_{\infty 1}$ is perpendicular to the axes of all the R joints while the axis of the ζ_{02} is perpendicular to the directions of all the P joints and intersects the axes of the R joints.

Secondly, discard those single-loop KCs in which the twists of all the joints but the virtual chain are linearly dependent. This requires that there be only two R joints within the coaxial compositional unit in the first case or that there be only one joint within the coaxial or codirectional compositional unit in the second case.

We obtain several 3-DOF single-loop KCs that involve a PPR virtual chain and have a 1-ζ_∞-1-ζ_0-system: $(\mathrm{XXX})^{\|}_E\ddot{\mathrm{R}}\mathrm{V}$.

Cases with a 2-ζ_∞-System

Firstly, perform the type synthesis of 3-DOF single-loop KCs that involve a PPR virtual chain and have a 2-ζ_∞-system. According to Chap. 3, there are seven joints in a 3-DOF single-loop KC that involves a PPR virtual chain and has a 2-ζ_∞-system. Such a single-loop KC is formed by one parallelaxis compositional unit (Fig. 8.4). The 2-ζ_∞-system is composed of all the ζ_∞ whose directions are perpendicular to the axes of the R joints within the parallelaxis compositional unit.

Secondly, discard those single-loop KCs in which the twists of all the joints but the virtual chain are linearly dependent.

We obtain several 3-DOF single-loop KCs that involve a PPR virtual chain and have a 1-ζ_∞-1-ζ_0-system: $(\mathrm{XXX})_E^{\parallel}\ddot{\mathrm{R}}\mathrm{V}$.

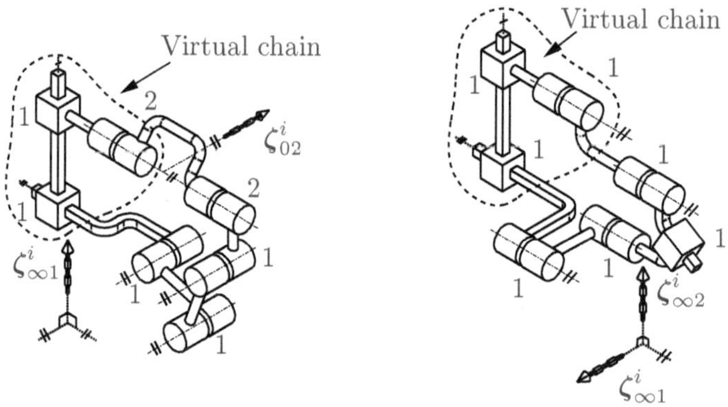

(a) $(\mathrm{RRR})_E^{\parallel}\ddot{\mathrm{R}}\mathrm{V}$. **(b)** R̋R̋PR̋V.

Fig. 8.4. 3-DOF single-loop kinematic chains involving a PPR virtual chain and having a 2-ζ-system

Cases with a 1-ζ_0-System

Firstly, perform the type synthesis of 3-DOF single-loop KCs that involve a PPR virtual chain and have a 1-ζ_0-system. According to Chap. 3, there are eight joints in a 3-DOF single-loop KC that involves a PPR virtual chain and has a 1-ζ_0-system. Such a single-loop KC is composed of (a) one planar translational compositional unit and one spherical compositional unit or (b) one planar compositional unit and one spherical compositional unit [Fig. 8.5a] or can be obtained by inserting one coaxial compositional unit into a 3-DOF single-loop KC that involves a PPR virtual chain and has a 1-ζ_∞-1-ζ_0-system. The 1-ζ_0-system is the ζ_0 whose axis passes through the intersection of the axes of the R joints within the spherical compositional unit and perpendicular to the plane associate with the planar compositional unit.

Secondly, discard those single-loop KCs in which the twists of all the joints but the virtual chain are linearly dependent. This requires that there be only two R joints within the coaxial compositional unit.

Cases with a 1-$\boldsymbol{\zeta}_\infty$-System

Firstly, perform the type synthesis of 3-DOF single-loop KCs that involve a PPR virtual chain and have a 1-$\boldsymbol{\zeta}_\infty$-system. According to Chap. 3, there are eight joints in a 3-DOF single-loop KC that involves a PPR virtual chain and has a 1-$\boldsymbol{\zeta}_\infty$-system. Such a KC is composed of (a) two spatial parallelaxis compositional units or (b) one spatial compositional unit and one planar compositional unit [Figs. 8.5b–8.5c]. It can also be obtained by inserting one coaxial compositional unit into a 3-DOF single-loop KC that involves a PPR virtual chain and has a 2-$\boldsymbol{\zeta}_\infty$-system [Fig. 8.4a] or by inserting one codirectional compositional unit into a 3-DOF single-loop KC that involves a PPR virtual chain and has a 1-$\boldsymbol{\zeta}_\infty$-1-$\boldsymbol{\zeta}_0$-system [Fig. 8.4b]. The 1-$\boldsymbol{\zeta}_\infty$-system is the $\boldsymbol{\zeta}_\infty$ whose direction is perpendicular to all the axes of the R joints.

Secondly, discard those single-loop KCs in which the twists of all the joints but the virtual chain are linearly dependent. For brevity, we make the assumption that this condition is always satisfied.

In the representation of the 3-DOF single-loop kinematic chains involving a PPR virtual chain and having a c^i-$\boldsymbol{\zeta}_0$-system, $()^{\|}_{E}$ denotes a kinematic chain in which the joints are arranged in such a way that all the links move along parallel planes that are parallel to a same plane, Ṙ denote R joints with intersecting axes within a same leg, Ŕ denote R joints with parallel axes within a same leg, R̈ denote R joints with coaxial axes, passing through the intersections of the axes of the Ṙ joints if any, within a PM, and R̋ denote R joints with parallel axes, being parallel to the axes of the R̈ joints or a line passing through at least two intersections of the axes of the Ṙ joints if any, within a PM.

8.6.2 Step 2b: Generation of Types of Legs

According to Condition (1) for the existence of PPR= PKCs (Sect. 8.3), the types of legs can be readily obtained from the 3-DOF single-loop kinematic chains obtained in Step 2a, Sect. 8.6.1, by removing the PPR virtual chain.

The specific geometric conditions which guarantee the satisfaction of Condition (1) for PPR= PKCs are clearly indicated by the notations that were introduced to represent the types of legs.

For example, by removing the virtual chain in an (RRR)$^{\|}_{E}$ṘṘV kinematic chain [Fig. 8.5c], an (RRR)$^{\|}_{E}$ṘṘ leg [Fig. 8.6d] can be obtained. Figure 8.6 shows some legs for PPR= PKCs and their leg-wrench systems. The leg-wrench system of the (RRR)$^{\|}_{E}$R̈ leg [Fig. 8.6b] is a 1-$\boldsymbol{\zeta}_\infty$-1-$\boldsymbol{\zeta}_0$-system. Its basis can be represented by a $\boldsymbol{\zeta}_\infty$ whose axis is perpendicular to the axes of all the R joints within a same leg and a $\boldsymbol{\zeta}_0$ whose axis intersects the axes of the R̈ joint and is parallel to the axes of the R joints within (RRR)$^{\|}_{E}$. The leg-wrench systems of the R̋ŔŔŔR̋ leg [Fig. 8.6e] and the ŔŔR̋R̋R̋ leg [Fig. 8.6f] are both a 1-$\boldsymbol{\zeta}_\infty$-system. The axis of the basis wrench $\boldsymbol{\zeta}_\infty$ is perpendicular to the axes of all the R joints within a same leg.

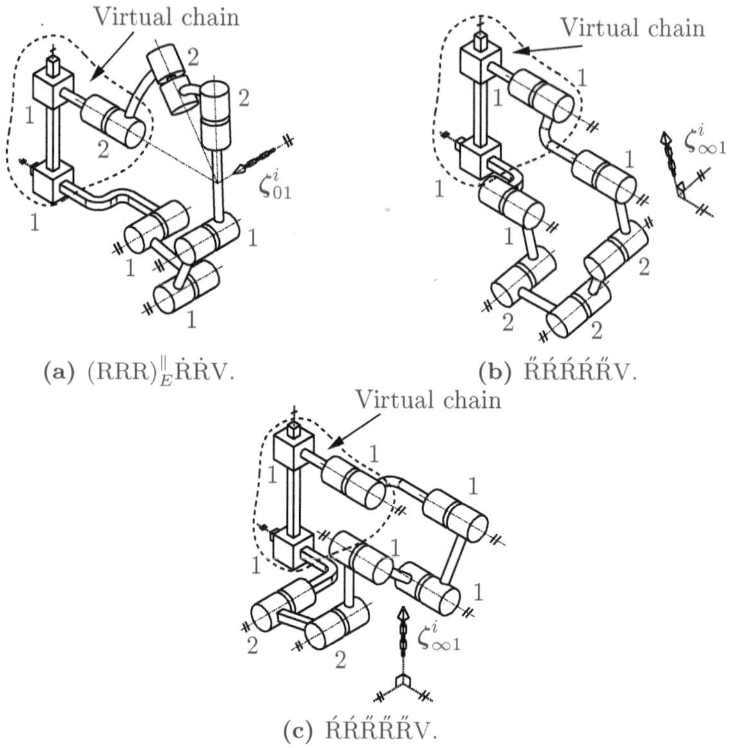

(a) $(\mathrm{RRR})_E^{\|}\dot{\mathrm{R}}\dot{\mathrm{R}}\mathrm{V}$. (b) $\H{\mathrm{R}}\acute{\mathrm{R}}\acute{\mathrm{R}}\acute{\mathrm{R}}\H{\mathrm{R}}\mathrm{V}$. (c) $\acute{\mathrm{R}}\acute{\mathrm{R}}\H{\mathrm{R}}\H{\mathrm{R}}\H{\mathrm{R}}\mathrm{V}$.

Fig. 8.5. 3-DOF single-loop KCs involving a PPR virtual chain and having a 1-ζ-system

All the types of legs for PPR= PMs obtained are listed in Table 8.1. For legs with $c^i = 0$, one is interested in legs with simple structures: RUS, PUS and UPS legs[59]. The legs used to synthesize PPR= PKCs need to satisfy not only the geometric conditions within a same leg, but also the geometric conditions among different legs.

8.7 Step 3: Assembly of Legs

PPR= PKCs can be generated by assembling a set of legs for PPR= PKCs shown in Table 8.1 selected according to the combinations of the leg-wrench systems shown in Table 5.7.In assembling PPR= PKCs, the following condition should be met: the linear combination of the leg-wrench systems constitutes a 2-ζ_∞-1-ζ_0-system (see Condition (2) for PPR= PKCs in Sect. 8.3).

Let us take PPR= PKCs of family 3 (Table 5.7) as an example. A PPR= PKC of family 3 has two legs with a 1-ζ_∞-1-ζ_0-system and one leg with a 1-ζ_∞-system. The required legs can be selected from Table 8.1. By assembling these

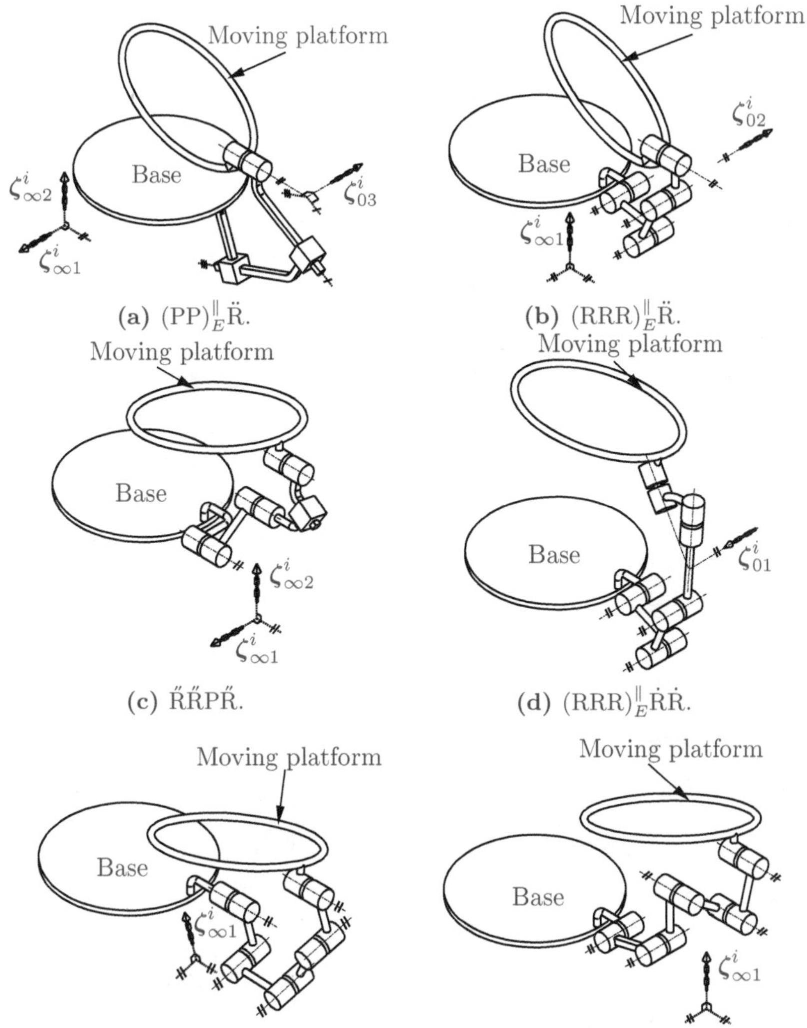

(a) $(\mathrm{PP})^{\parallel}_{E}\ddot{\mathrm{R}}$. **(b)** $(\mathrm{RRR})^{\parallel}_{E}\ddot{\mathrm{R}}$.

(c) R̋R̋PR̋. **(d)** $(\mathrm{RRR})^{\parallel}_{E}\dot{\mathrm{R}}\dot{\mathrm{R}}$.

(e) R̋ŔŔŔR̋. **(f)** ŔŔR̋R̋R̋.

Fig. 8.6. Some legs for PPR= PKCs

legs, we obtain PPR= PKCs of family 3. For example, a set of two $(\mathrm{RRR})^{\parallel}_{E}\ddot{\mathrm{R}}$ legs with a 1-ζ_{∞}-1-ζ_0-system and one R̋ŔŔŔR̋ leg with a 1-ζ_{∞}-system can be used to construct a 2-$(\mathrm{RRR})^{\parallel}_{E}\ddot{\mathrm{R}}$-R̋ŔŔŔR̋ PPR= PKC [Fig. 8.7(b)]. A set of two $(\mathrm{RRR})^{\parallel}_{E}\ddot{\mathrm{R}}$ legs with 1-ζ_{∞}-1-ζ_0-system and one ŔŔR̋R̋R̋ leg with a 1-ζ_{∞}-system can be used to construct a 2-$(\mathrm{RRR})^{\parallel}_{E}\ddot{\mathrm{R}}$-ŔŔR̋R̋R̋ PPR= PKC (Fig. 8.8).

It can be observed from Table 5.7 that among the combinations of sets of leg-wrench systems, there is only one combination, the combination of three

Table 8.1. Legs for PPR= PKCs

c^i	leg-wrench system	No.	Type
3	$2\text{-}\zeta_\infty\text{-}1\text{-}\zeta_0$	1	$(PP)_E^{\|}\ddot{R}$
2	$1\text{-}\zeta_\infty\text{-}1\text{-}\zeta_0$	2	$(XXX)_E^{\|}\ddot{R}$
	$2\text{-}\zeta_\infty$	3	$(XXXX)_A$
1	$1\text{-}\zeta_0$	4	$(XXX)_E^{\|}\dot{R}\dot{R}$
		5	$(XX)_E^{\|}\dot{R}\dot{R}\dot{R}$
	$1\text{-}\zeta_\infty$	6	$(XX)_A(XXX)_B$
		7	$\ddot{R}(XXX)_B\ddot{R}$
		8	$P(XXX)_B\ddot{R}$
		9	$\ddot{R}(XXX)_BP$
		10	$(XXX)_B(XX)_A$
		11	$(XXX)_A\acute{R}\acute{R}$
		12	$\acute{R}\acute{R}(XXX)_A$
0	omitted	omitted	omitted

$2\text{-}\zeta_\infty\text{-}1\text{-}\zeta_0$, in which all the leg-wrench system are of the same type. From Table 8.1, there is one type of leg, $(PP)_E^{\|}\ddot{R}$, with a $2\text{-}\zeta_\infty\text{-}1\text{-}\zeta_0$ wrench system. Thus there is only one 3-legged PPR= PKC, 3-$(PP)_E^{\|}\ddot{R}$ (Fig. 8.9), with identical type of legs.

It is noted that there exist no constraint singularities [133] for PPR= PKCs of families 1, 2, 4, 5, 6, 9, 10, 11 and 12. For a PPR= PM, constraint singularities occur at configurations in which the order of the wrench system of the PPR= PKC is less than 3.

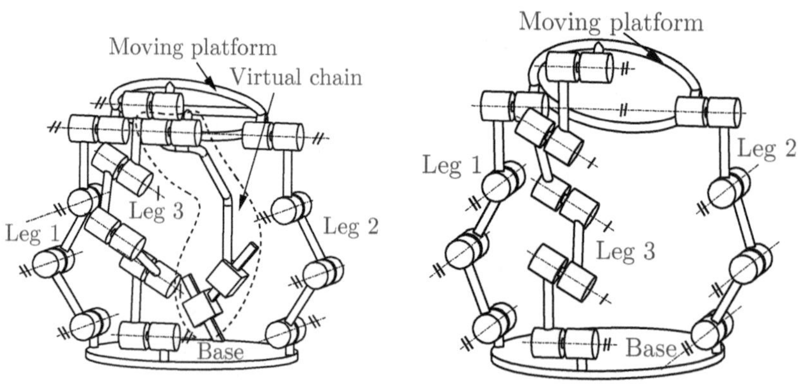

(a) (b)

Fig. 8.7. (**a**) 2-$(RRR)_E^{\|}\ddot{R}$-$\text{\H{R}}\acute{R}\acute{R}\acute{R}\text{\H{R}}$ PPR= PKC with a PPR virtual chain added and (**b**) 2-$(RRR)_E^{\|}\ddot{R}$-$\text{\H{R}}\acute{R}\acute{R}\acute{R}\text{\H{R}}$ PPR= PKC

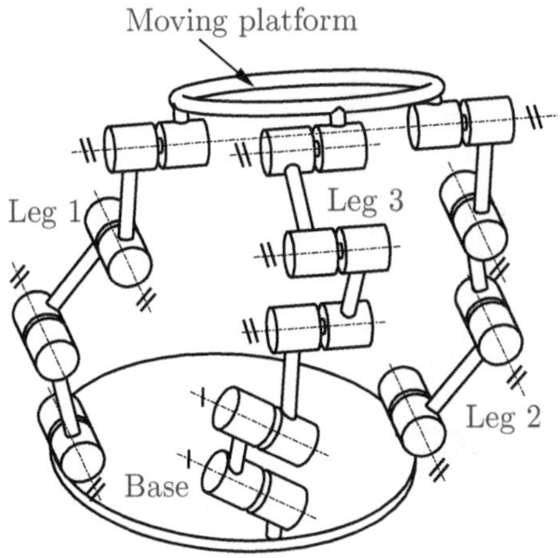

Fig. 8.8. 2-(RRR)$^{\|}_{E}$R̈-ŔŔR̋R̋R̋ PPR= PKCs

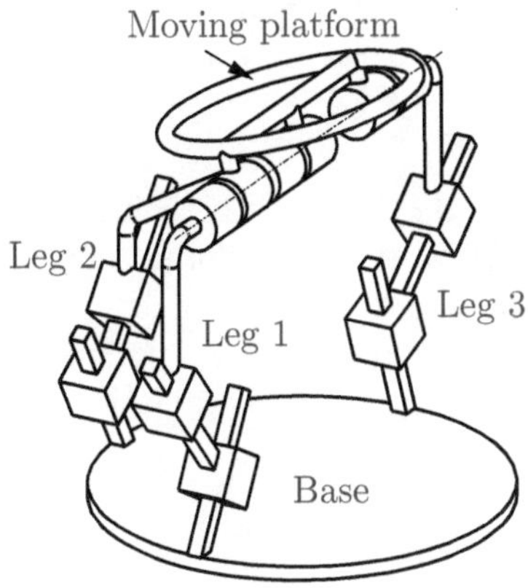

Fig. 8.9. 3-(PP)$^{\|}_{E}$R̈ PPR= PKCs

8.8 Step 4: Selection of Actuated Joints

In this section, we focus on the PPR= PMs associated with a PPR virtual chain in which the axis of the R joint and the directions of the two P joints are parallel to the same plane.

8.8.1 t-Component of the Actuation Wrenches

Considering that the order of a screw system is coordinate free and for simplicity reasons, all the actuation wrenches and wrenches are expressed in a coordinate system with its X-axis perpendicular to the directions of the two P joints and its Y-axis along the axis of the R joint within the PPR virtual chain.

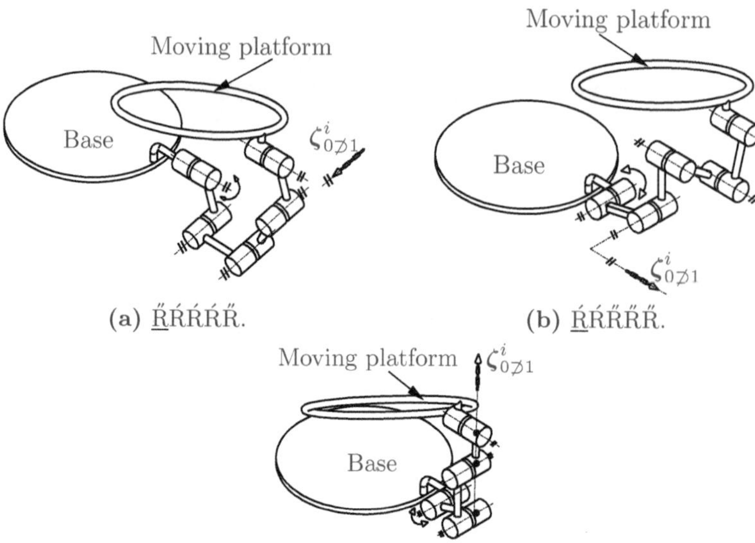

(a) R̲̋ŔŔŔR̋. (b) R̲ŔR̋R̋R̋. (c) $(\underline{\text{R}}\text{RR})^{\|}_{E}\ddot{\text{R}}$.

Fig. 8.10. Actuation wrenches of some legs for PPR= PKCs

In this way, the second, third and fifth scalar components of all the basis wrenches of $\mathcal{W}$ are zero. Thus, the t-component, $\zeta^i_{t\not{D}j}$, of the $\zeta^i_{\not{D}j}$ is composed of the second, third and fifth scalar components of the $\zeta^i_{\not{D}j}$.

Figure 8.10 shows the actuation wrenches of actuated joints in some legs for PPR= PMs. In the R̲̋ŔŔŔR̈ leg [Fig. 8.10(a)], the first R joint is actuated. $\zeta^i_{\not{D}1}$ can be chosen as any ζ_0 whose axis intersects the axis of the unactuated R̈ joint and is parallel to the axes of the R joints within ŔŔŔ. In the R̲ŔR̋R̋R̋ leg [Fig. 8.10(b)], the first R joint is actuated. $\zeta^i_{\not{D}1}$ can be chosen as any ζ_0 whose axis intersects the axis of the unactuated R joint within ŔŔ and is parallel to the axes of the R joints within R̋R̋R̋. In the $(\underline{\text{R}}\text{RR})^{\|}_{E}\ddot{\text{R}}$ leg [Fig. 8.10(c)], the first R joint is actuated. $\zeta^i_{\not{D}1}$ can be chosen as any ζ_0 whose axis intersects the axes of the three unactuated R joints and is not parallel to the axes of the R joints within $(\text{RRR})^{\|}_{E}$.

8.8.2 Procedure for the Validity Detection of Actuated Joints

The validity detection of actuated joints of PPR= PMs can thus be performed using the following steps.

Step 4a. If one or more of the actuated joints of a candidate PPR= PM are inactive, the set of actuated joints is invalid and the candidate PPR= PM should be discarded.

Step 4b. If the second vector components of the actuation wrenches of actuated joints $\zeta^{i}_{\not{D}j}$ are linearly dependent in a general configuration for a candidate PPR= PM, the set of actuated joints is invalid. In this case, the candidate PPR= PM should be discarded.

Following the criteria on the selection of actuated joints (Sect. 5.7) and the procedure for the detection of the validity of actuated joints, all the $m(m \geq 2)$-legged PPR= PMs corresponding to each PPR= PKC can be generated.

For example, the candidate 3-legged PPR= PM corresponding to the 2-(RRR)$^{\|}_{E}$R̈-ŔŔR̋R̋R̋ PPR= PKC (Fig. 8.8) is the 2-(R̲RR)$^{\|}_{E}$R̈-Ŕ̲ŔR̋R̋R̋ PPR= PM [Fig. 8.11(a)]. We have

$$\begin{vmatrix} \zeta^{1}_{t,\not{D}1} & \zeta^{2}_{t,\not{D}1} & \zeta^{3}_{t,\not{D}1} \end{vmatrix} = \begin{vmatrix} * & * & * \\ * & * & * \\ 0 & 0 & 0 \end{vmatrix} = 0.$$

where $*$ denotes an arbitrary number.

According to the validity condition of actuated joints, the set of actuated joints is invalid. The candidate 2-(R̲RR)$^{\|}_{E}$R̈-Ŕ̲ŔR̋R̋R̋ PPR= PM is thus discarded

The candidate 3-legged PPR= PM corresponding to the 2-(RRR)$^{\|}_{E}$R̈-R̋ŔŔŔR̋ PPR= PKC [Fig. 8.7(b)] is the 2-(R̲RR)$^{\|}_{E}$R̈-R̲̋ŔŔŔR̋ PPR= PM [Fig. 8.11(b)]. Following the procedure for the detection of the validity of actuated joints, it can be verified that the set of actuated joints is valid.

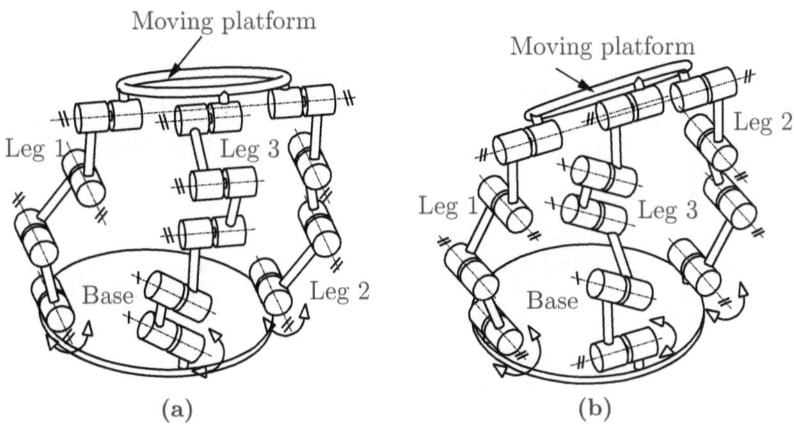

Fig. 8.11. Some candidate PPR= PMs: (**a**) 2-(R̲RR)$^{\|}_{E}$R̈-Ŕ̲ŔR̋R̋R̋ and (**b**) 2-(R̲RR)$^{\|}_{E}$R̈-R̲̋ŔŔŔR̋

Due to the large amount of PPR= PMs and space limitation, only several PPR= PMs are presented (Figs. 8.11(a), 8.12 8.13, and 8.14).

It is noted that for the PMs corresponding to the $3\text{-}(PP)^{\parallel}_{E}\ddot{R}$ PPR= PKC (Fig. 8.9) — the sole PPR= PKC with identical type of legs, the three P joints located on the base do not constitute a valid set of actuated joints. Thus, there are no 3-DOF PPR= PMs with identical types of legs and fixed actuators.

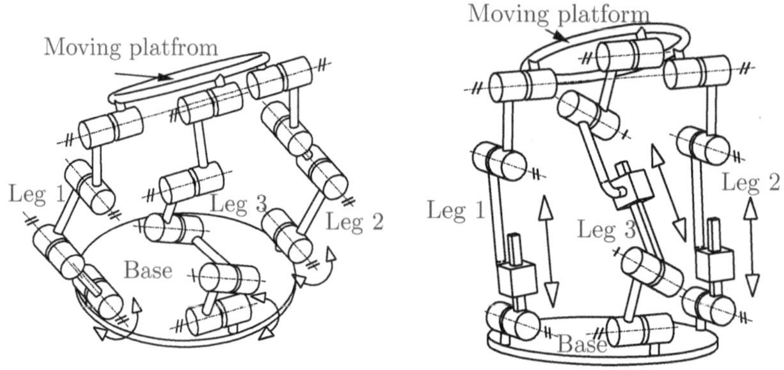

(a) (b)

Fig. 8.12. Some 3-DOF PPR= PMs of family 3: (**a**) $2\text{-}(\underline{R}RR)^{\parallel}_{E}\ddot{R}\text{-}\underline{\text{\H{R}}}\acute{R}\acute{R}\text{\H{R}}\text{\H{R}}$ and (**b**) $2\text{-}(R\underline{P}R)^{\parallel}_{E}\ddot{R}\text{-}\text{\H{R}}\acute{R}\underline{P}\acute{R}\text{\H{R}}$

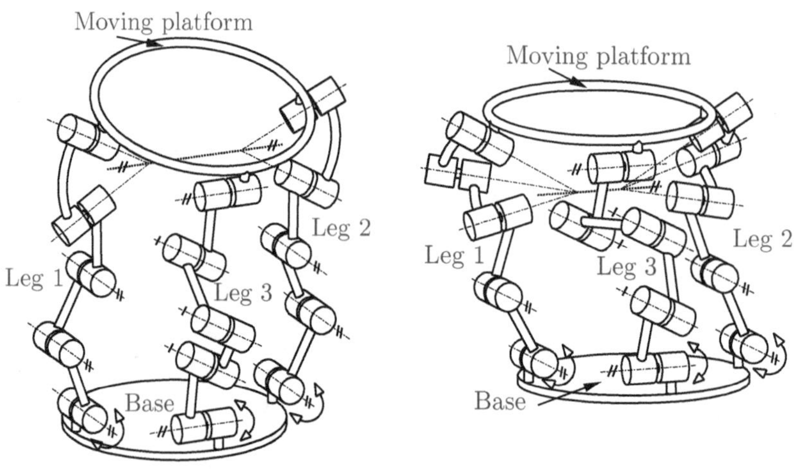

(a) (b)

Fig. 8.13. Some 3-DOF PPR= PMs of family 13: (**a**) $2\text{-}(\underline{R}RR)^{\parallel}_{E}\dot{R}\dot{R}\text{-}\underline{\text{\H{R}}}\acute{R}\acute{R}\acute{R}\text{\H{R}}$ and (**b**) $2\text{-}(\underline{R}R)^{\parallel}_{E}\dot{R}\dot{R}\dot{R}\text{-}\underline{\text{\H{R}}}\acute{R}\acute{R}\acute{R}\text{\H{R}}$

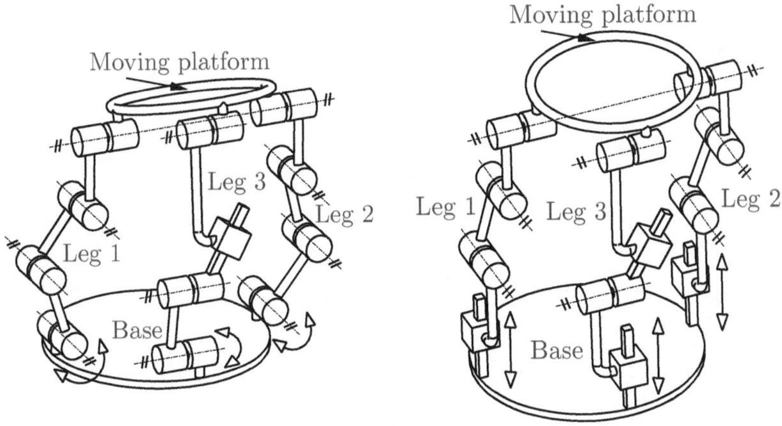

(a) (b)

Fig. 8.14. Some 3-DOF PPR= PMs of family 2: (**a**) $2\text{-}(\underline{\text{R}}\text{RR})^{\parallel}_{E}\ddot{\text{R}}\text{-}\underline{\text{\H{R}}}\text{\H{R}P\H{R}}$ and (**b**) $2\text{-}(\underline{\text{P}}\text{RR})^{\parallel}_{E}\ddot{\text{R}}\text{-}\underline{\text{P}}\text{\H{R}P\H{R}}$

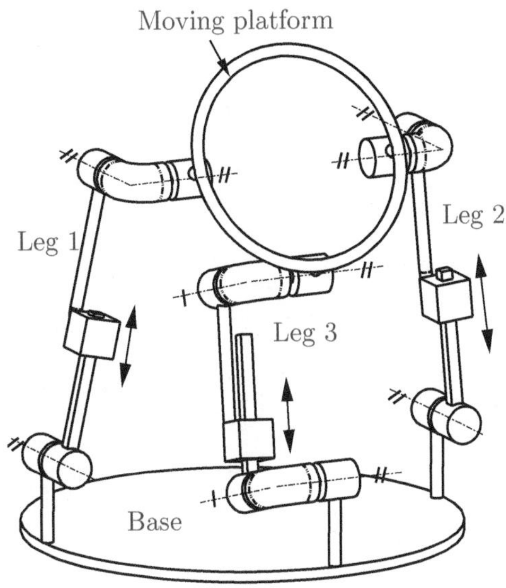

Fig. 8.15. $2\text{-R}\underline{\text{P}}\text{U-U}\underline{\text{P}}\text{U}$ PPR= PMs

8.9 Summary

In this chapter, the type synthesis of PPR= PMs has been thoroughly solved using the general type synthesis approach proposed in Chap. 5. PPR= PKCs which were proposed in [47, 58] have been re-obtained. PPR= PKCs with inactive joints have also been obtained. Either overconstrained or non-overconstrained PPR= PKCs can be obtained. Some new PPR= PMs have also been revealed.

From these PPR= PMs obtained, a number of variations can be obtained using the following techniques:

(a) Substitute a combination of one R and one P joints with parallel axes with a C joint or one of its equivalents [47];
(b) Substitute a combination of two successive R joints with non-parallel axes with a U joint;
(c) Substitute a combination of three successive R joints with concurrent axes with an S joint;
(d) Substitute one or more P joints each with a planar parallelogram [123?];
(e) Substitute two of three successive P joints with a spatial parallelogram [35];
(f) Substitute one or more R joints in the leg with a c^i-ζ_∞-system each with a coaxial H joint.

For instance, Fig. 8.15 shows a 2-R$\underline{\text{P}}$U-U$\underline{\text{P}}$U PPR= PM. This PM is obtained from the 2-$(\text{R}\underline{\text{P}}\text{R})^{\|}_{E}\ddot{\text{R}}$-$\H{\text{R}}\acute{\text{R}}\underline{\text{P}}\acute{\text{R}}\H{\text{R}}$ PPR= PM shown in Fig. 8.12(b) by substituting a combination of two successive R joints with non-parallel axes with a U joint. As compared with the original PM, the variation has fewer links.

It is also noted that PPR= PMs proposed in [95] are in fact variations of the PPR= PMs shown in Fig. 8.14, which are obtained by a) substituting a combination of two successive R joints with non-parallel axes with a U joint and b) replacing the unactuated P joints each with a planar parallelogram.

9. Four-DOF PPPR= Parallel Mechanisms

In many robotic applications, it is required to move objects in all Cartesian directions as well as orient these objects around an axis having a given direction. A typical example of such tasks is the assembly of computer circuit boards on which electronic components must be mounted. Rotations are required only around an axis orthogonal to the board while translations in all directions must be performed to pick up and place the components. Serial SCARA robots were developed specifically for this family of tasks in the 1970s. SCARA robots have become very popular and are widely spread in industry. The motion pattern of SCARA robots can be described using a PPPR virtual chain. In this chapter, the type synthesis of 4-DOF PPPR= PMs (also called 3T1R PMs in the literature) is dealt with using the general approach proposed in Chap. 5. A PPPR= PM is the parallel counterpart of the serial SCARA robot and generates 4-DOF PPPR motion (also called Schönflies motion). As mentioned above, PMs with a PPPR motion pattern cover a wide range of applications. The four steps of the type synthesis of PPPR= PMs are presented in detail.

9.1 Introduction

SCARA robots [97] are widely used 4-DOF serial robots. The end-effector of a SCARA robot generates PPPR motion (also called Schönflies motion) which refers to a rotation about any axis with a given direction in conjunction with 3-DOF translations. PPPR= PMs are the parallel counterparts of the SCARA robots. In a PPPR= PM, the moving platform generates PPPR motions which are controlled by four actuated joints distributed in different legs.

Several PPPR parallel kinematic chains (PPPR= PKCs) were proposed in [59]. However, no PPPR= PMs with all the actuated joints located on the base can be obtained from these PPPR= PKCs. PPPR= PMs proposed in [16, 17, 110] contain either four legs having different structures or S joints. A 3-UPU 4-DOF PPPR= PM was proposed in [54]. Recently, a systematic study on the type synthesis of PPPR= PMs was presented based on the units of single-opened-chain in [126]. Three types of 4-legged PPPR= PMs with legs of the same type which have

X. Kong and C. Gosselin: Type Syn. of Parallel Mech., STAR 33, pp. 141–157, 2007.
springerlink.com

no unactuated P joint were proposed. It was also claimed that all actuated joints could be located on the base. However, only two types of the parallel manipulators proposed in [126] are functional while the other one in which the four translational degrees of freedom of C joints are actuated is flawed. In the latter case, the four translational degrees of freedom of C joints cannot control the translational degree of freedom along the direction perpendicular to the axes of the four C joints. The reason of this misconception is that the selection of actuated joints has not been well solved. In addition, the number of over-constraints (also redundant constraints or passive constraints) of PPPR= PMs has not been revealed. Several PPPR= PMs were also proposed in [132].

In fact, most works on the systematic type synthesis of PPPR= PMs [126] deal mainly with the systematic type synthesis of PPPR= PKCs. Also, the results on PPPR= PKCs with 5-DOF legs published in [126] are incomplete. Recently, the type synthesis of PPPR= PMs, including the selection of actuated joints, was dealt with systematically in [78].

Using the general approach to the type synthesis of PMs proposed in Chap. 5, the type synthesis of PPPR= PMs is dealt with in this chapter. Four steps for the type synthesis of PPPR= PMs, i.e., (1) the decomposition of wrench systems of PPPR= PKCs, (2) the type synthesis of legs for PPPR= PKCs, (3) the assembly of legs to generate PPPR= PKCs, and (4) the selection of actuated joints for PPPR= PMs, are dealt with in Sects. 9.5, 9.6, 9.7, and 9.8, respectively.

9.2 Wrench System of a PPPR= PKC

In any general configuration, the twist system of a PPPR= PKC or its PPPR virtual chain is a 3-$\boldsymbol{\xi}_\infty$-1-$\boldsymbol{\xi}_0$-system. Since the virtual power developed by any $\boldsymbol{\zeta}_\infty$ whose direction is perpendicular to the axis of the R joint within the PPPR virtual chain along any twist within the 3-$\boldsymbol{\xi}_\infty$-1-$\boldsymbol{\xi}_0$-system is 0, the wrench system of the PPPR= PKC or its virtual chain is thus a 2-$\boldsymbol{\zeta}_\infty$-system (Fig. 9.1) in which the directions of all the $\boldsymbol{\xi}_\infty$ are perpendicular to the axis of the R joint within the PPPR virtual chain.

9.3 Conditions for a PKC to Be a PPPR= PKC

When we connect the base and the moving platform of a PPPR= PKC by an appropriate PPPR virtual chain, the function of the PKC is not affected. Any of its legs and the PPPR virtual chain will form a 4-DOF single loop KC. When the order of the leg-wrench system is greater than 0, the single-loop KC formed must be an overconstrained KC.

Based on the concept of PPPR virtual chain, it follows that a PKC is a PPPR= PKC if it satisfies the two conditions below:

(1) Each leg of the PKC and a same PPPR virtual chain form a 4-DOF single-loop KC.
(2) The wrench system of the PKC is the same as that of the PPPR virtual chain, i.e., a 2-$\boldsymbol{\zeta}_\infty$-system, in any one general configuration.

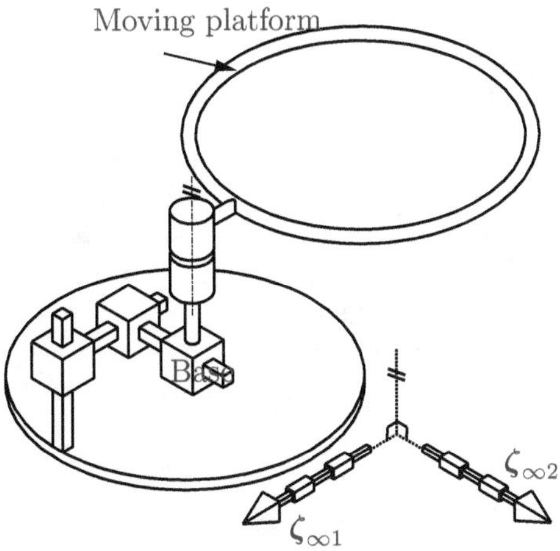

Fig. 9.1. Wrench system of a PPPR= PKC

9.4 Procedure for the Type Synthesis of PPPR= PMs

A general procedure can thus be proposed for the type synthesis of PPPR= PKCs as follows:

Step 1. Decomposition of the wrench system of 4-DOF PPPR= PKCs.
Step 2. Type synthesis of legs for PPPR= PKCs.
Here, a leg for PPPR= PKCs refers to a leg satisfying Condition (1) for PPPR= PKCs.
Step 3. Assembly of legs for PPPR= PKCs.
PPPR= PKCs can be generated by assembling two or more legs for PPPR= PKCs, obtained in Step 1, such that Condition (2) for PPPR= PKCs is satisfied.
Step 4. Selection of actuated joints.
PPPR= PMs can be generated by selecting actuated joints in different ways for each PPPR= PKC (Sect. 3.1), obtained in Step 3.

The procedure proposed above will be elaborated upon in the following sections.

9.5 Step 1: Decomposition of the Wrench System of PPPR= PKCs

The decomposition of the wrench system of an m-legged PPPR= PKC consists in finding all the leg-wrench systems and all the combinations of m leg-wrench systems for the PPPR= PKC and a specified number of overconstraints Δ. Step 1 can be performed using (2.24) and (2.28).

9.5.1 Determination of the Leg-Wrench Systems

As the wrench system of a PKC is the linear combination of the leg-wrench systems in a general configuration, any leg-wrench system in a V= PKC is a subsystem of its wrench system. Following the procedure proposed in Sect. 2.1.3, all the leg-wrench systems for a V= PKC can be identified without difficulty.

In any general configuration, the wrench system of a PPPR= PKC is the same as that of its PPPR virtual chain, i.e., a 2-ζ_∞-system, which is composed of all the ζ_∞ whose directions are perpendicular to the axis of the R joint within the PPPR virtual chain (Fig. 9.1). It then follows that any leg-wrench system with order $c^i > 0$ of a PPPR= PKC is either a 2-ζ_∞-system or a 1-ζ_∞-system in a general configuration (Fig. 9.2).

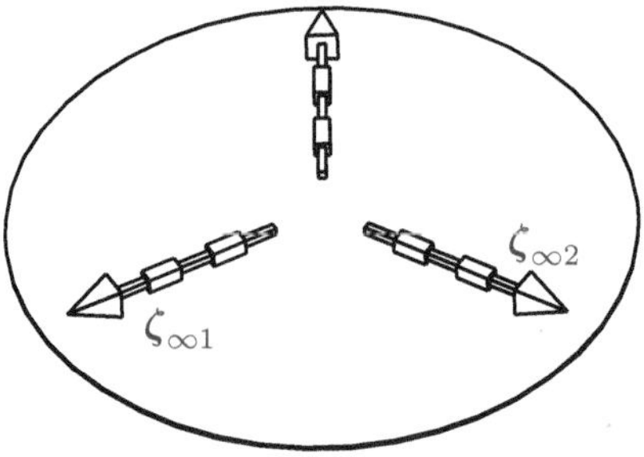

Fig. 9.2. Leg-wrench system of a PPPR= PKC

9.5.2 Determination of the Combinations of Leg-Wrench Systems

For the PPPR motion pattern considered, all the wrenches in the wrench systems are of the same pitch. The leg-wrench systems are $c^i (0 \leq c^i \leq 2)$-systems of ∞ pitch. The combination of leg-wrench systems can be simply represented by the combination of the orders, c^i, of leg-wrench systems.

The combinations of the orders, c^i, of leg-wrench systems have been determined by solving (2.28) and are shown in Tables 5.1–5.5.

9.6 Step 2: Type Synthesis of Legs

In this section, the type synthesis of legs for PPPR= PKCs will be performed following the two substeps given in Sect. 5.5.

9.6.1 Step 2a: Type Synthesis of 4-DOF Single-Loop KCs That Involve a PPPR Virtual Chain and Have a Specified Leg-Wrench System

The type synthesis of 4-DOF single-loop KCs that involve a PPPR virtual chain and have a specified leg-wrench system can be performed as follows.

Firstly, perform the type synthesis of single-loop KCs that involve a PPPR virtual chain and have a specified leg-wrench system (See Chap. 3). Secondly, discard those single-loop KCs in which the twists of all the joints but the PPPR virtual chain are linearly dependent.

Cases with a 2-ζ_∞-System

Firstly, perform the type synthesis of 4-DOF single-loop KCs that involve a PPPR virtual chain and have a 2-ζ_∞-system. According to Chap. 3, there are eight joints in a 4-DOF single-loop KC that involves a PPPR virtual chain and has a 2-ζ_∞-system. Such a single-loop KC is composed of one spatial parallelaxis compositional unit (Fig. 9.3). The 2-ζ_∞-system is composed of all the ζ_∞ whose axes are perpendicular to the axes of all the R joints.

Secondly, discard those single-loop KCs in which the twists of all the joints but the virtual chain are linearly dependent. The $\overset{''}{R}\overset{''}{R}\overset{''}{R}\overset{''}{R}\overset{''}{V}$ [Fig. 9.3a] 4-DOF single-loop KC should be discarded since the twists of all the four $\overset{''}{R}$ joints are linearly dependent. For the KCs shown in Figs. 9.3b–9.3d, the KC composed of all the joints except the PPPR virtual chain must be a spatial parallelaxis compositional unit. For brevity, we assume that the above condition is automatically satisfied.

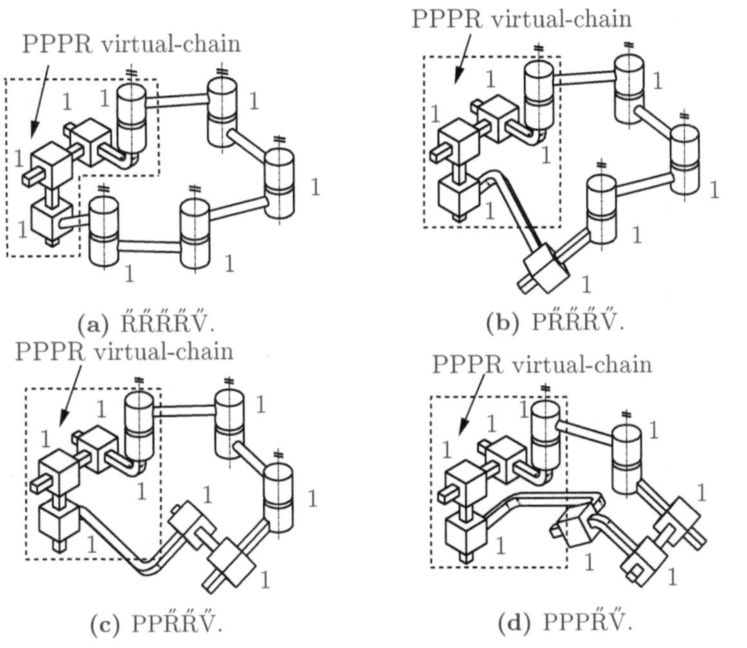

(a) $\overset{''}{R}\overset{''}{R}\overset{''}{R}\overset{''}{R}\overset{''}{V}$. (b) $P\overset{''}{R}\overset{''}{R}\overset{''}{R}\overset{''}{V}$. (c) $PP\overset{''}{R}\overset{''}{R}\overset{''}{V}$. (d) $PPP\overset{''}{R}\overset{''}{V}$.

Fig. 9.3. Some 4-DOF single-loop KCs involving a PPPR virtual chain ($c^i = 2$)

Cases with a 1-ζ_∞-System

Firstly, perform the type synthesis of 4-DOF single-loop KCs that involve a PPPR virtual chain and have a 1-ζ_∞-system. According to Chap. 3, there are eight joints in a 4-DOF single-loop KC that involves a PPPR virtual chain

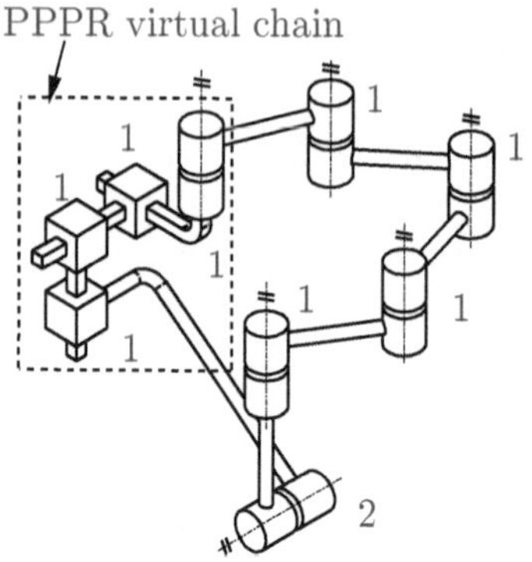

(a) ŔR̋R̋R̋R̋V̋.

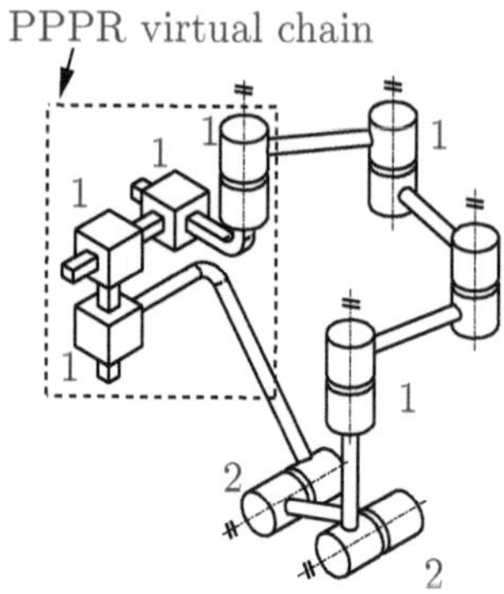

(b) ŔŔR̋R̋R̋V̋.

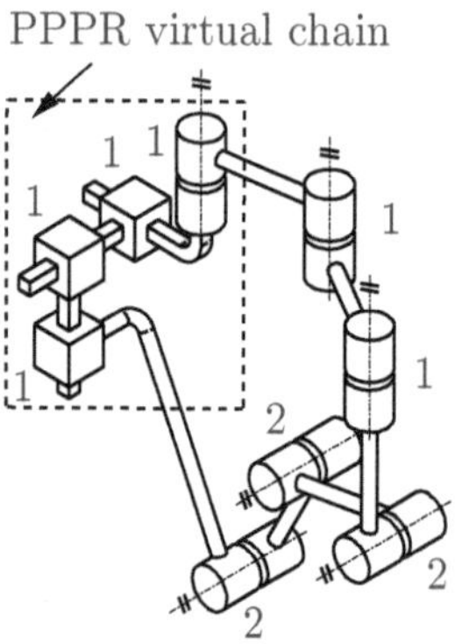

(c) ŔŔŔR̋R̋V̋.

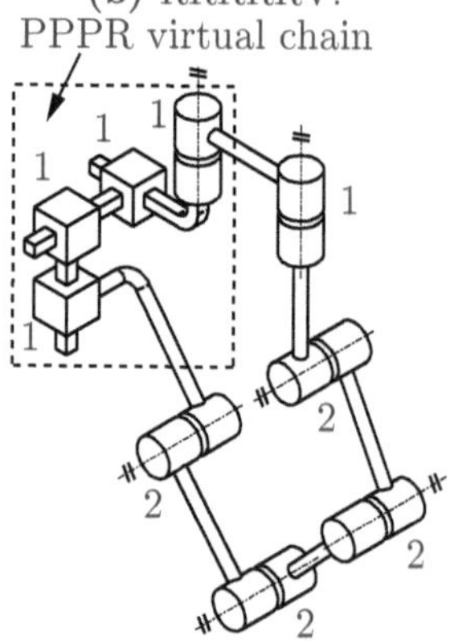

(d) ŔŔŔŔR̋V̋.

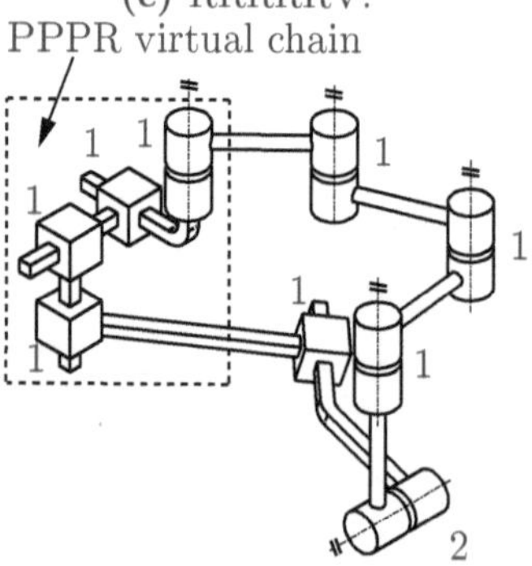

(e) PŔR̋R̋R̋V̋.

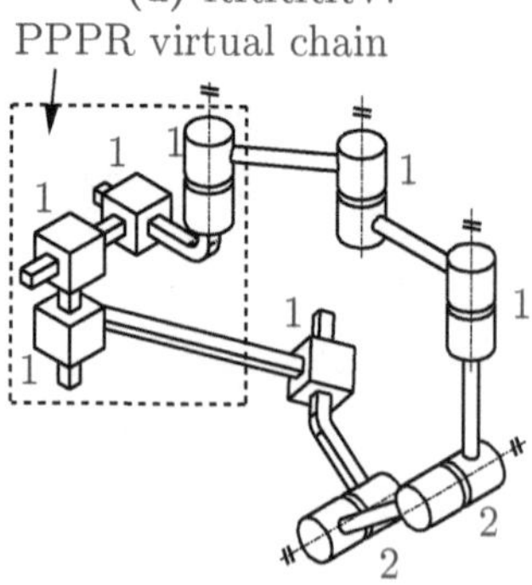

(f) PŔŔR̋R̋V̋.

Fig. 9.4. Some 4-DOF single-loop KCs involving a PPPR virtual chain ($c^i = 1$)

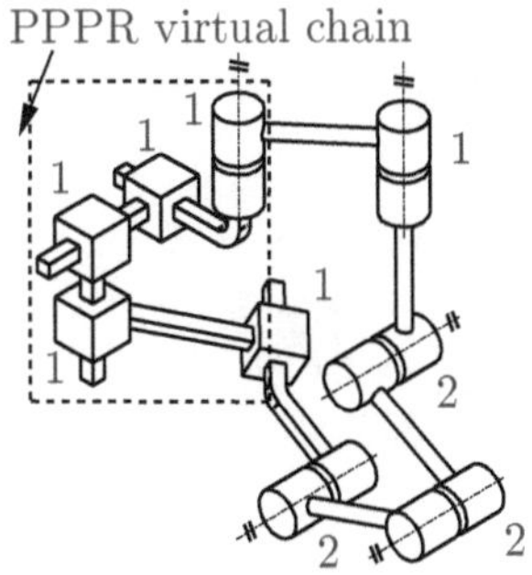

(g) PŔŔŔR̋V̋.

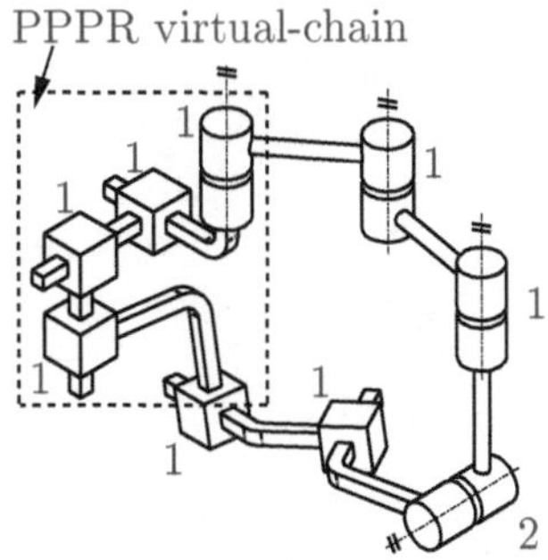

(h) PPŔR̋R̋V̋.

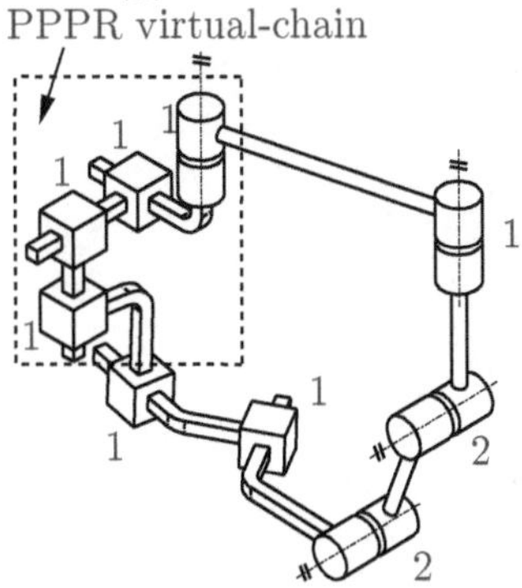

(i) PPŔŔR̋V̋.

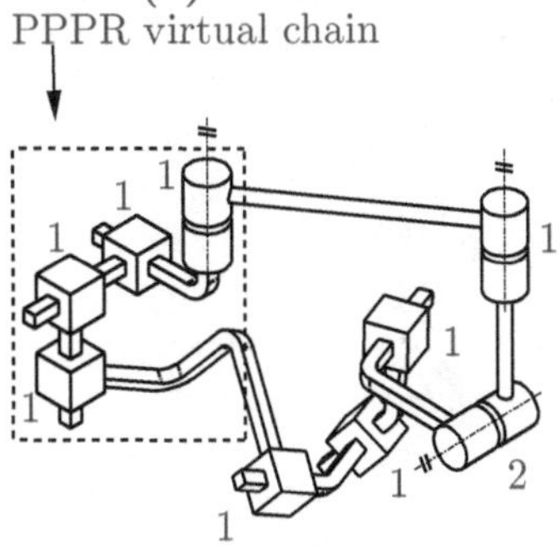

(j) PPPŔR̋V̋.

Fig. 9.4. (*continued*)

and has a 1-ζ_∞-system. Such a single-loop KC is composed of (a) two spatial parallelaxis compositional units or (b) one spatial parallelaxis compositional units and one planar compositional units or can be obtained by inserting one $(R)_L$ joint in a 4-DOF single-loop KC that involves a PPPR virtual chain and has a 2-ζ_∞-system. The 1-ζ_∞-system of the 4-DOF single-loop KCs that involve a PPPR virtual chain and have a 1-ζ_∞-system is the ζ_∞ whose direction is perpendicular to the axes of all the R joints.

Secondly, discard those single-loop KCs in which the twists of all the joints but the virtual chain are linearly dependent. The R̋R̋R̋R̋ŔV̋, R̋R̋R̋ŔR̋V̋, R̋R̋ŔR̋R̋V̋, and ŔŔŔŔR̋V̋ 4-DOF single-loop KC should be discarded since the twists of all the four R̋ joints or the four Ŕ joints are linearly dependent.

In the representation of types of 4-DOF single-loop KCs involving a PPPR virtual chain, R̋ denote R joints whose axes are all parallel to the axis of the R joints of the PPPR virtual chain, and Ŕ denote R joints whose axes are all parallel to a line that is not parallel to the axis of the R joint of the PPPR virtual chain.

It is pointed out that in a 4-DOF single-loop KC, the P joints and the only Ŕ joint can be placed anywhere, and the combination of Ŕ joints can also be placed anywhere. For brevity, we list only the 4-DOF single-loop KCs from which all 4-DOF single-loop KCs can be obtained through the above operations. For example, the R̋R̋PR̋V̋ single-loop KC can be obtained by changing the position of the P joint in the R̋R̋R̋PV̋ single-loop KC [Fig. 9.3b] while the R̋R̋ŔŔR̋V̋ single-loop KC can be obtained by changing the position of the combination of Ŕ joints in the R̋R̋R̋ŔŔV̋ single-loop kinematic chain [Fig. 9.4b]. In a 4-DOF single-loop KC involving a PPPR virtual chain in which there is one Ŕ joint [Figs. 9.4a and 9.4e], the Ŕ joint is inactive.

9.6.2 Step 2b: Generation of Types of Legs

According to Condition (1) for the existence of PPPR= PKCs (Sect. 9.3), the types of legs can be readily obtained from the 4-DOF single-loop KCs obtained in Step 2a, Sect. 9.6.1, by removing the PPPR virtual chain.

The specific geometric conditions which guarantee the satisfaction of Condition (1) for PPPR= PKCs are clearly indicated by the notations that were introduced to represent the types of PPPR= PKCs, PPPR= PMs and their legs, R̋ denote R joints whose axes are all parallel while Ŕ denote R joints within the same leg whose axes are all parallel.

For example, by removing the PPPR virtual chain in an ŔŔR̋R̋R̋V̋ KC [Fig. 9.4b], an ŔŔR̋R̋R̋ [Fig. 9.5a] leg can be obtained. In the ŔŔR̋R̋R̋ leg shown in Fig. 9.5a, the axes of the first two R joints are parallel to each other while the axes of the last three R joints are parallel. This leg has a 1-$\boldsymbol{\zeta}_\infty$-system. The axis of $\boldsymbol{\zeta}^i_{\infty 1}$, a basis of the 1-$\boldsymbol{\zeta}_\infty$-system, is perpendicular to the axes of all the R joints. In the R̋R̋R̋ŔŔ leg shown in Fig. 9.5b, the axes of the first three R joints

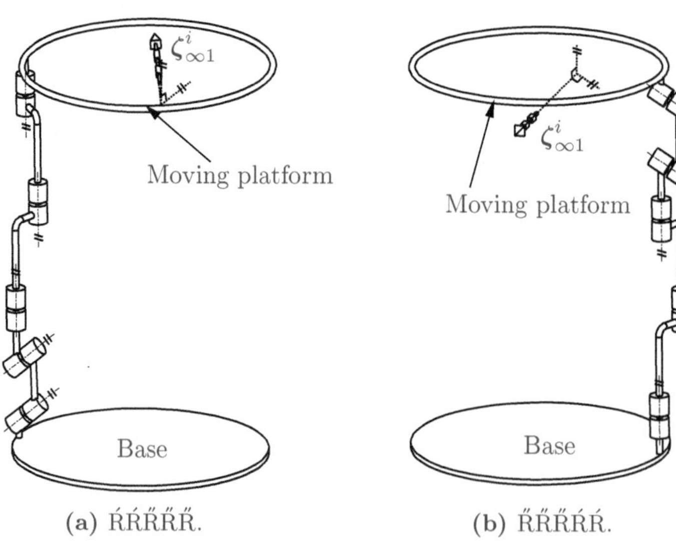

(a) ŔŔR̋R̋R̋. (b) R̋R̋R̋ŔŔ.

Fig. 9.5. Some legs for PPPR= PKCs

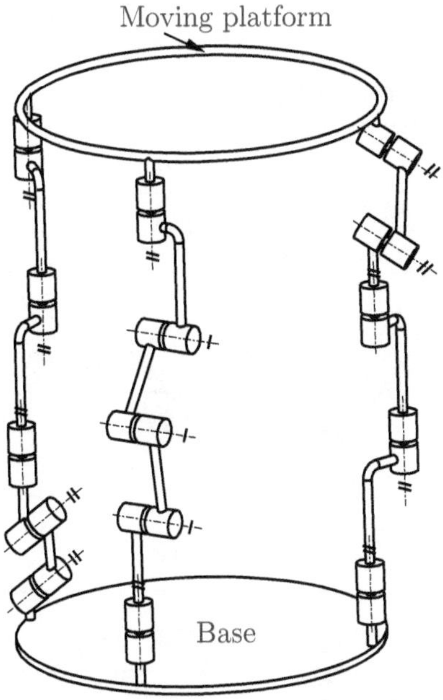

Fig. 9.6. ŔŔR̋R̋R̋-R̋R̋R̋ŔŔ-R̋ŔŔŔR̋ PPPR= PKC

are parallel while the axes of the last two R joints are parallel to each other. This leg has also a 1-ζ_∞-system. The axis of $\zeta^i_{\infty 1}$, a basis of the 1-ζ_∞-system, is perpendicular to the axes of all the R joints.

All the types of legs for PPPR= PKCs obtained are listed in Table 9.1. The legs used to synthesize PPPR= PKCs need to satisfy not only the geometric conditions within a same leg — which can be derived directly from the associated single-loop KCs — but also certain geometric conditions among different legs.

9.7 Step 3: Assembly of Legs

PPPR= PKCs can be generated by assembling a set of legs for PPPR= PKCs shown in Table 9.1 selected according to the combinations of the leg-wrench systems shown in Table 5.3. In assembling PKCs, the following condition should be met: the linear combination of the leg-wrench systems forms the 2-ζ_∞-system (see Condition (2) for PPPR= PKCs in Sect. 9.3).

For a PPPR= PKC in which not all the leg-wrench systems are invariant with respect to the base or the moving platform, the linear combination of the leg-wrench systems usually forms the 2-ζ_∞-system. For a PPPR= PKC in which all the leg-wrench systems are invariant with respect to the base or the moving platform, the base or the moving platform should meet certain conditions to guarantee that the linear combination of the leg-wrench systems forms the 2-ζ_∞-system.

Table 9.1. Legs for PPPR= PKCs

c^i	Class	No.	Type
2	3RP	1–2	PR̋R̋R̋ R̋PR̋R̋
		3–4	R̋R̋PR̋ R̋R̋R̋P
	2R-2P	5-10	permutation of PPR̋R̋
	1R-3P	11–12	PPPR̋ PPR̋P
		13–14	PR̋PP R̋PPP
1	5R	15–16	R̋R̋R̋ŔŔ R̋R̋ŔŔR̋
		17–18	R̋ŔŔR̋R̋ ŔŔR̋R̋R̋
		19–20	R̋R̋ŔŔŔ R̋ŔŔŔR̋
		21	ŔŔŔR̋R̋
	4R-1P	22-41	Permutation of R̋R̋R̋ŔP
		42–43	R̋R̋ŔŔP R̋R̋ŔPŔ
		44–45	R̋R̋PŔŔ R̋PR̋ŔŔ
		46–47	PR̋R̋ŔŔ R̋R̋ŔŔP
		48–49	PR̋ŔŔR̋ R̋ŔŔPR̋
		50–51	R̋ŔPŔR̋ R̋PŔŔR̋
		52–53	ŔŔR̋R̋P ŔŔR̋PR̋
		54–55	ŔŔPR̋R̋ ŔPŔR̋R̋
		56–57	PŔŔR̋R̋ R̋ŔŔŔP
		58–59	R̋ŔŔPŔ R̋ŔPŔŔ
		60–61	R̋PŔŔŔ PR̋ŔŔŔ
		62–63	ŔŔŔR̋P ŔŔŔPR̋
		64–65	ŔŔPŔR̋ ŔPŔŔR̋
		66	PŔŔŔR̋
	3R-2P	67-96	Permutation of R̋R̋ŔPP
		97-112	Permutation of R̋ŔŔPP
	2R-3P	113-132	Permutation of R̋ŔPPP
0	omitted	omitted	omitted

Due to the large number of m-legged PPPR= PKCs, only the 4-legged PPPR= PKCs with identical type of legs are listed in Table 9.2.

9.8 Step 4: Selection of Actuated Joints

9.8.1 t-Components of the Actuation Wrenches

Considering that the order of a screw system is coordinate system free and for simplicity reasons, all the actuation wrenches and wrenches are expressed in a coordinate system with its X-axis parallel to the axis of the rotational degree of

Table 9.2. m-legged PPPR= PKCs with identical type of legs

c^i	Class	No.	Type	N_{OC}		
				m=2	m=3	m=4
2	3R-1P	1–2	m-PR̋R̋R̋ m-R̋PR̋R̋	2	4	6
		3–4	m-R̋R̋PR̋ m-R̋R̋R̋P			
	2R-2P	5-10	m-permutation of PPR̋R̋			
	1R-3P	11–12	m-PPPR̋ m-PPR̋P			
		13–14	m-PR̋PP m-R̋PPP			
1	5R	15–16	m-R̋R̋R̋ŔŔ m-R̋R̋ŔŔR̋	0	1	2
		17–18	m-R̋ŔŔR̋R̋ m-ŔŔR̋R̋R̋			
		19–20	m-R̋R̋ŔŔŔ m-R̋ŔŔŔR̋			
		21	m-ŔŔŔR̋R̋			
	4R-1P	22-41	m-Permutation of R̋R̋R̋ŔP			
		42–43	m-R̋R̋ŔŔP m-R̋R̋ŔPŔ			
		44–45	m-R̋R̋PŔŔ m-R̋PR̋ŔŔ			
		46–47	m-PR̋R̋ŔŔ m-R̋R̋ŔŔP			
		48–49	m-PR̋ŔŔR̋ m-R̋ŔŔPR̋			
		50–51	m-R̋ŔPŔR̋ m-R̋PŔŔR̋			
		52–53	m-ŔŔR̋R̋P m-ŔŔR̋PR̋			
		54-55	m-ŔŔPR̋R̋ m-ŔPŔR̋R̋			
		56–57	m-PŔŔR̋R̋ m-R̋ŔŔŔP			
		58–59	m-R̋ŔŔPŔ m-R̋ŔPŔŔ			
		60–61	m-R̋PŔŔŔ m-PR̋ŔŔŔ			
		62–63	m-ŔŔŔR̋P m-ŔŔŔPR̋			
		64–65	m-ŔŔPŔR̋ m-ŔPŔŔR̋			
		66	m-PŔŔŔR̋			
	3R-2P	67-96	m-Permutation of R̋R̋ŔPP			
		97-112	m-Permutation of R̋ŔŔPP			
	2R-3P	113-132	m-Permutation of R̋ŔPPP			

where N_{OC} denotes the number of overconstraints of a PPPR= PKC

freedom of the moving platform. Let $[\mathbf{0} \quad \mathbf{j}^T]^T$, $[\mathbf{0} \quad \mathbf{k}^T]^T$ denote a basis of $\mathcal{W}$, where $\mathbf{j} = \{1 \quad 0\}^T$ and $\mathbf{k} = \{0 \quad 1\}^T$. The first four scalar components of all the basis wrenches of $\mathcal{W}$ are zero. Thus, the t-component, $\boldsymbol{\zeta}^i_{t\not{\mathcal{D}}j}$, of the $\boldsymbol{\zeta}^i_{\not{\mathcal{D}}j}$ is composed of the first four scalar components of the $\boldsymbol{\zeta}^i_{\not{\mathcal{D}}j}$.

Figure 9.7 shows the actuation wrenches of actuated joints in some legs for PPPR= PMs. The wrench systems of these legs are both a 1-ζ_∞-system (Fig. 9.5). In the $\underline{\acute{\mathrm{R}}}$ŔR̋R̋R̋ leg [Fig. 9.7a], the first R joint is actuated. $\boldsymbol{\zeta}^i_{\not{\mathcal{D}}j}$ can be chosen as any $\boldsymbol{\zeta}_0$ whose axis is parallel to the axes of the last three R joints

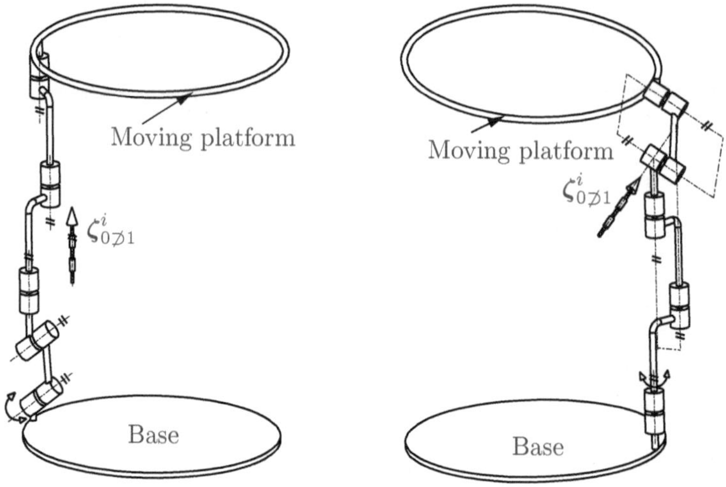

(a) Ŕ̲ŔR̋R̋R̋. (b) R̲̋R̋R̋ŔŔ.

Fig. 9.7. Actuation wrenches of some legs for PPPR= PKCs

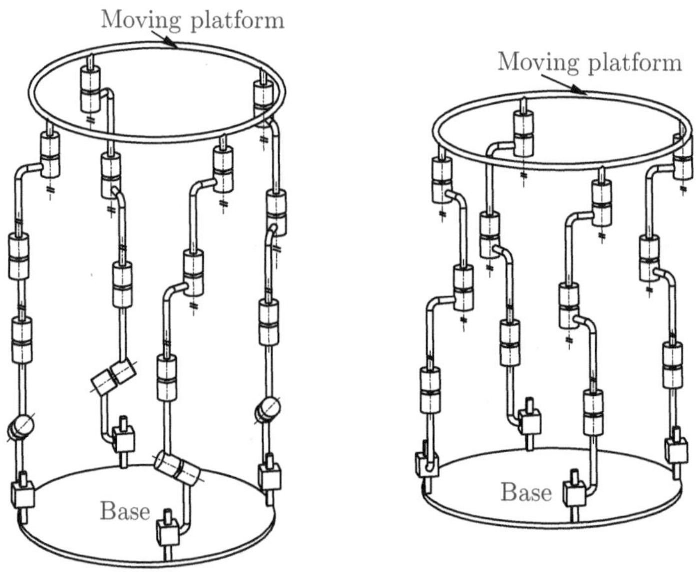

(a) 4-P̲ŔR̋R̋R̋. (b) 4-P̲R̋R̋R̋.

Fig. 9.8. PPPR= PKCs with and without inactive joints

and intersects the axis of the second R joint. In the R̲̋R̋R̋ŔŔ leg [Fig. 9.7b], the first R joint is actuated. $\zeta^i_{\not{0}j}$ can be chosen as any ζ_0 whose axis is the intersection of the plane passing through the axes of two unactuated R̋ joints with the plane passing through the axes of the two Ŕ joints.

9.8.2 Procedure for the Validity Detection of Actuated Joints

The validity detection of actuated joints of PPPR= PMs can be performed using the following steps.

Step 4a. If one or more of the actuated joints of a candidate PPPR= PM are inactive, the set of actuated joints is invalid and the candidate PPPR= PM should be discarded.

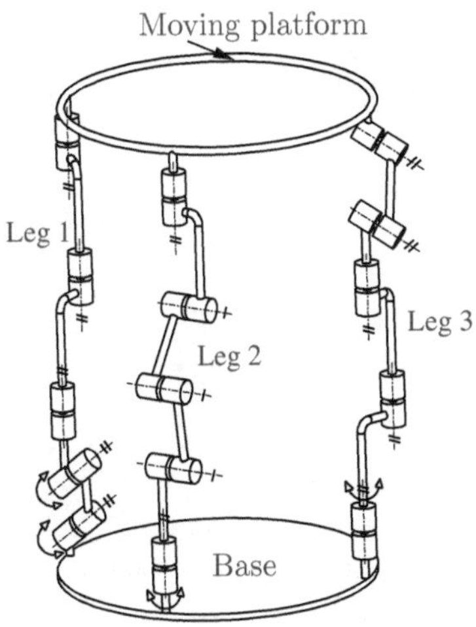

(a) $\underline{\text{ŔŔ}}$R̋R̋R̋-$\underline{\text{R̋}}$R̋ŔŔŔ-$\underline{\text{R̋}}$ŔŔŔR̋.

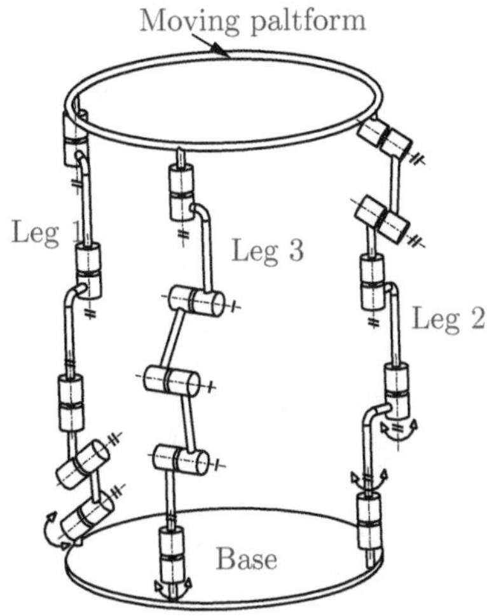

(b) $\underline{\text{Ŕ}}$ŔR̋R̋R̋-$\underline{\text{R̋R̋}}$ŔŔŔ-$\underline{\text{R̋}}$ŔŔŔR̋.

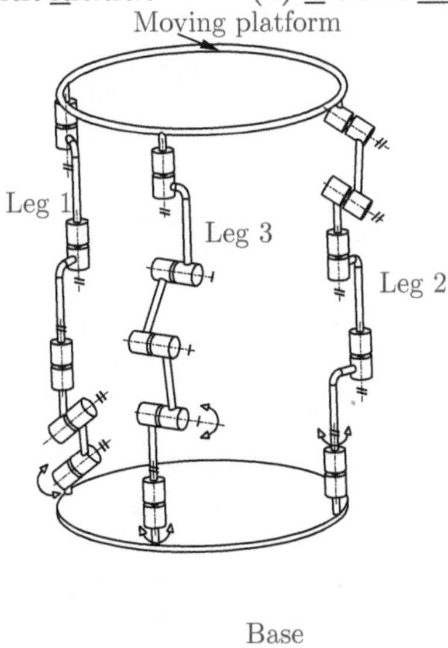

(c) $\underline{\text{Ŕ}}$ŔR̋R̋R̋-$\underline{\text{R̋}}$R̋ŔŔŔ-$\underline{\text{R̋Ŕ}}$ŔŔR̋.

Fig. 9.9. Selection of actuated joints for the ŔŔR̋R̋R̋-R̋R̋ŔŔŔ-R̋ŔŔŔR̋ PPPR= PKC

Step 4b. If the t-components of all the actuation wrenches of actuated joints $\zeta^{i}_{\not{\partial}j}$ are linearly dependent in a general configuration for a candidate PPPR= PM, the set of actuated joints is invalid. In this case, the candidate PPPR= PM should be discarded.

For example, the candidate PPPR= PMs corresponding to the 3-legged ŔŔR̋R̋R̋-R̋R̋R̋ŔŔ-R̋ŔŔŔR̋ PPPR= PKC (Fig. 9.6), are the Ŕ̲Ŕ̲R̋R̋R̋-R̲̋R̋R̋ŔŔ-R̲̋ŔŔŔR̋, Ŕ̲ŔR̋R̋R̋-R̲̋R̲̋R̋ŔŔ-R̲̋ŔŔŔR̋ and Ŕ̲ŔR̋R̋R̋-R̲̋R̋R̋ŔŔ-R̲̋Ŕ̲ŔŔR̋ PPPR= PMs (Fig. 9.9). For the candidate PM shown in Fig. 9.9a, we have

$$\left|\zeta^{1}_{t,\not{\partial}1}\ \zeta^{1}_{t,\not{\partial}2}\ \zeta^{2}_{t,\not{\partial}1}\ \zeta^{3}_{t,\not{\partial}1}\right| = \begin{vmatrix} * & * & * & * \\ 0 & 0 & * & * \\ 0 & 0 & * & * \\ 0 & 0 & * & * \end{vmatrix} = 0.$$

where $*$ denotes an arbitrary number.

According to the validity condition of actuated joints, the set of actuated joints is invalid. The candidate Ŕ̲Ŕ̲R̋R̋R̋-R̲̋R̋R̋ŔŔ-R̲̋ŔŔŔR̋ PPPR= PM [Fig. 9.9(a)] is thus discarded. Similarly, we can prove that both the Ŕ̲ŔR̋R̋R̋-R̲̋R̲̋R̋ŔŔ-R̲̋ŔŔŔR̋ and the Ŕ̲ŔR̋R̋R̋-R̲̋R̋R̋ŔŔ-R̲̋Ŕ̲ŔŔR̋ have a valid set of actuated joints. Therefore, there are only two PPPR= PMs corresponding to the 3-legged ŔŔR̋R̋R̋-R̋R̋R̋ŔŔ-R̋ŔŔŔR̋ PPPR= PKC, i.e., the Ŕ̲ŔR̋R̋R̋-R̲̋R̲̋R̋ŔŔ-R̲̋ŔŔŔR̋ [Fig. 9.9(b)] and Ŕ̲ŔR̋R̋R̋-R̲̋R̋R̋ŔŔ-R̲̋Ŕ̲ŔŔR̋ [Fig. 9.9(c)] PPPR= PMs.

Following the criteria on the selection of actuated joints (Sect. 5.7) and the procedure for the detection of the validity of actuated joints, all the $m(m \geq 2)$-legged PPPR= PMs corresponding to each PPPR= PKC can be generated.

To make the conditions for PPPR= PMs clear in their representation and due to the large number of PPPR= PMs, only 4-legged PPPR= PMs with identical type of legs — including the identical arrangement of actuated joints — which are composed of R and P joints and which satisfy the above criteria are listed in Table 9.3 and also shown in Fig. 9.10. For the 4-R̲̋R̋ŔŔŔ and 4-P̲R̋ŔŔŔ PPPR= PMs, the validity condition of actuated joints is that the axes of the Ŕ joints are not all parallel to a same plane. For the 4-R̲̋ŔŔŔR̋ and 4-P̲ŔŔŔR̋ PPPR= PMs, the validity condition of actuated joints is that not all the axes of the Ŕ joints are perpendicular to the axes of the R̋ joints. Nine of the 11 types of 4-legged PPPR= PMs are new while some special cases of No. 6 and No. 9 PPPR= PMs have been proposed in [126].

The prototype shown in Fig. 1.7b is in fact the input-output partially decoupled case of the P̲R̋R̋R̋-3-P̲R̀R̀R̀R̋ PPPR= PM, which belongs to the 2-1-1-1 family of the PPPR= PMs.

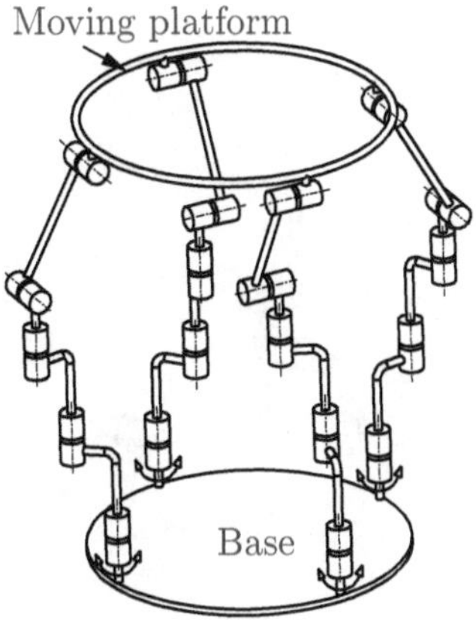

(a) Ŕ̲ŔR̋R̋R̋.

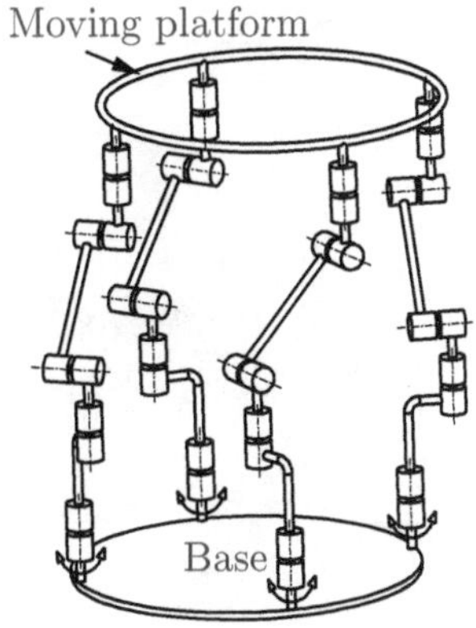

(b) R̲̋R̋R̋ŔŔ.

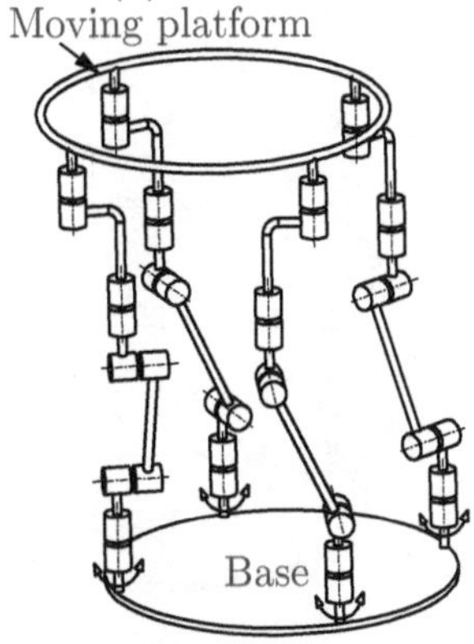

(c) R̲̋R̋R̋ŔŔ.

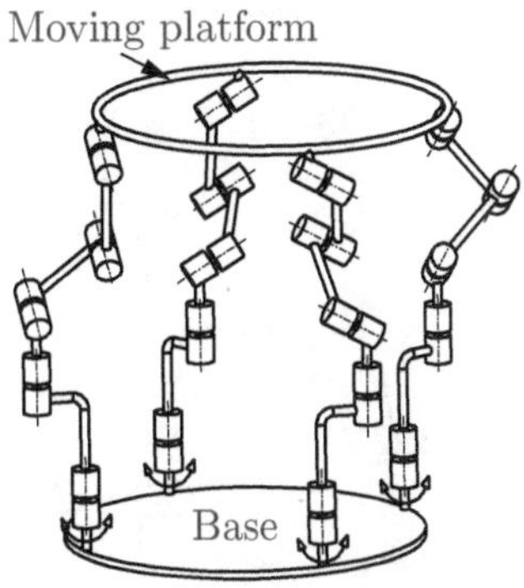

(d) R̲̋R̋R̋ŔŔ.

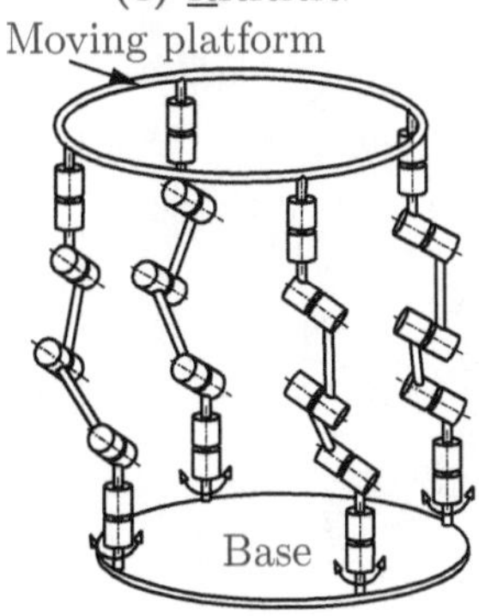

(e) R̲̋R̋R̋ŔŔ.

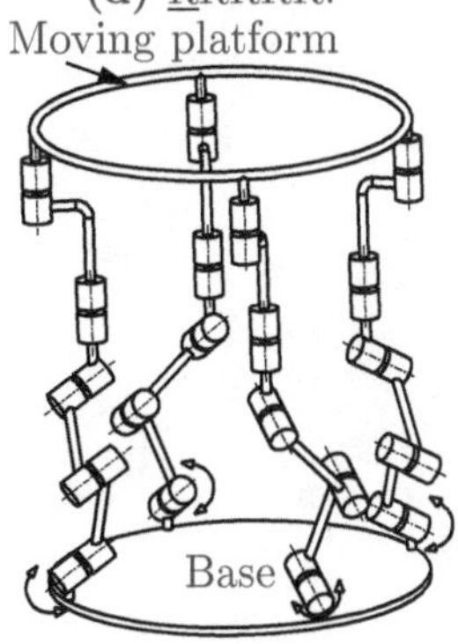

(f) R̲̋R̋R̋ŔŔ.

Fig. 9.10. Eleven 4-legged PPPR= PMs with identical type of legs

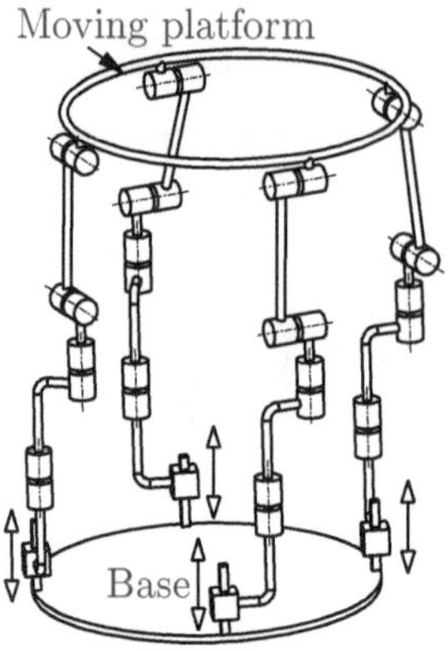

(**g**) R̲̋R̋R̋ŔŔ.

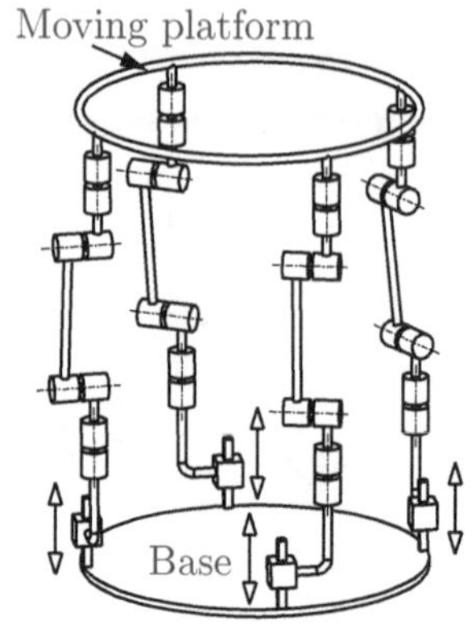

(**h**) R̲̋R̋R̋ŔŔ.

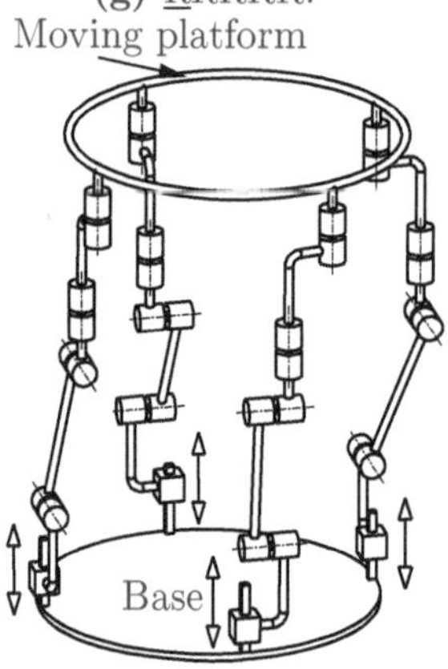

(**i**) R̲̋R̋R̋ŔŔ.

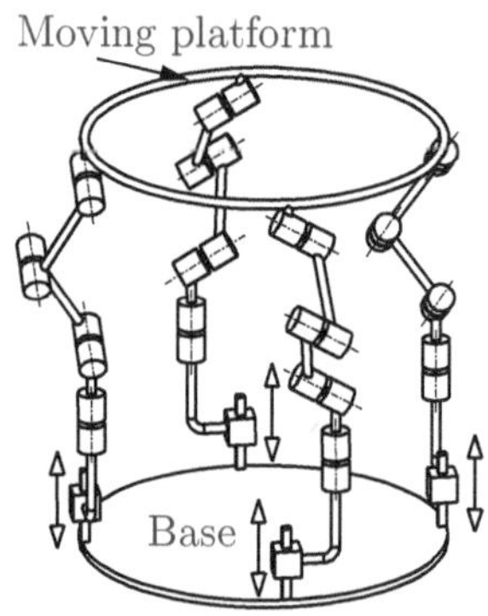

(**j**) R̲̋R̋R̋ŔŔ.

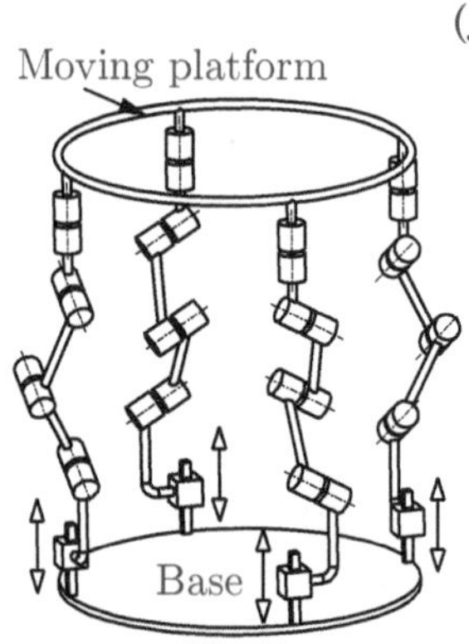

(**k**) R̲̋R̋R̋ŔŔ.

Fig. 9.10. (*continued*)

Table 9.3. Four-legged PPPR= PMs with identical type of legs

c^i	Class	No.	Type
1	5R	1	$4\text{-}\underline{\overset{\prime\prime}{R}}\overset{\prime\prime}{R}\overset{\prime\prime}{R}\overset{\prime}{R}\overset{\prime}{R}$
		2	$4\text{-}\underline{\overset{\prime\prime}{R}}\overset{\prime\prime}{R}\overset{\prime}{R}\overset{\prime}{R}\overset{\prime\prime}{R}$
		3	$4\text{-}\underline{\overset{\prime\prime}{R}}\overset{\prime}{R}\overset{\prime}{R}\overset{\prime\prime}{R}\overset{\prime\prime}{R}$
		4	$4\text{-}\underline{\overset{\prime\prime}{R}}\overset{\prime\prime}{R}\overset{\prime}{R}\overset{\prime}{R}\overset{\prime}{R}$
		5	$4\text{-}\underline{\overset{\prime\prime}{R}}\overset{\prime}{R}\overset{\prime}{R}\overset{\prime}{R}\overset{\prime\prime}{R}$
		6	$4\text{-}\underline{\overset{\prime}{R}}\overset{\prime}{R}\overset{\prime}{R}\overset{\prime\prime}{R}\overset{\prime\prime}{R}$
	4R-1P	7	$4\text{-}\underline{P}\overset{\prime\prime}{R}\overset{\prime\prime}{R}\overset{\prime}{R}\overset{\prime}{R}$
		8	$4\text{-}\underline{P}\overset{\prime\prime}{R}\overset{\prime}{R}\overset{\prime}{R}\overset{\prime\prime}{R}$
		9	$4\text{-}\underline{P}\overset{\prime}{R}\overset{\prime}{R}\overset{\prime\prime}{R}\overset{\prime\prime}{R}$
		10	$4\text{-}\underline{P}\overset{\prime\prime}{R}\overset{\prime}{R}\overset{\prime}{R}\overset{\prime}{R}$
		11	$4\text{-}\underline{P}\overset{\prime}{R}\overset{\prime}{R}\overset{\prime}{R}\overset{\prime\prime}{R}$

9.9 Summary

In this chapter, the type synthesis of PPPR= PMs has been well solved using the general type synthesis approach proposed in Chap. 5. PPPR= PKCs with inactive joints have also been obtained. Both overconstrained and non-overconstrained PPPR= PKCs can be obtained. The validity check of actuated joints of PPPR= PMs has been reduced to the calculation of a 4×4 determinant. Some new PPPR= PMs have also been revealed.

By substituting a combination of an R joint and a P joint with parallel axes, a combination of two R joints with intersecting non-parallel axes and a combination of three R joints with concurrent axes with a C, U, and S joint respectively, PPPR= PMs involving C, U and S joints can be obtained. For a leg in which the axes of the R joints are parallel to two directions, no S joint can exist. For a leg with a $0\text{-}\zeta_\infty$-system, any type of joint may be used. By replacing an R joint with a coaxial H (helical) joint, some PPPR= PMs involving H joints can be obtained. Other PPPR= PMs involving H joints can be obtained starting from the following 4-DOF single-loop KCs involving a PPPR virtual chain and one or more H joints following the above procedure: (1) 4-DOF single-loop KCs obtained from the $\overset{\prime\prime}{R}\overset{\prime\prime}{R}\overset{\prime\prime}{R}\overset{\prime\prime}{R}\overset{\prime\prime}{V}$ [Fig. 9.3a], $\overset{\prime\prime}{R}\overset{\prime\prime}{R}\overset{\prime\prime}{R}\overset{\prime\prime}{R}\overset{\prime}{R}\overset{\prime\prime}{V}$ [Fig. 9.4a], $\overset{\prime\prime}{R}\overset{\prime\prime}{R}\overset{\prime\prime}{R}\overset{\prime}{R}\overset{\prime\prime}{R}\overset{\prime\prime}{V}$ and $\overset{\prime\prime}{R}\overset{\prime\prime}{R}\overset{\prime}{R}\overset{\prime\prime}{R}\overset{\prime\prime}{R}\overset{\prime\prime}{V}$ KC by replacing one or more $\overset{\prime\prime}{R}$ joints each with one coaxial $\overset{\prime\prime}{H}$ joint and (2) 4-DOF single-loop KCs obtained from the $\overset{\prime}{R}\overset{\prime}{R}\overset{\prime}{R}\overset{\prime}{R}\overset{\prime\prime}{R}\overset{\prime\prime}{V}$ KC [Fig. 9.4d] by replacing one or more $\overset{\prime}{R}$ joints each with one coaxial $\overset{\prime}{H}$ joint. Finally, by replacing a P joint with a planar parallelogram or a set of two consecutive P joints with a spatial parallelogram, variations of PPPR= PMs can be obtained.

10. Four-DOF SP= Parallel Mechanisms

Another 4-DOF motion pattern that can find applications in several areas is the one associated with an SP virtual chain. For instance, in laparoscopic robotic surgery, the body is entered through a keyhole. Once the keyhole is established, the robot must be constrained to move around it while maintaining this location fixed. A 4-DOF robot that can extend inside the body would then be behaving as an SP serial KC in which the centre of the S joint is located at the keyhole. The use of such a robot — as opposed to a 6-DOF manipulator — would ensure that the constraint associated with the keyhole is never violated, thereby providing safety. In this chapter, the type synthesis of 4-DOF SP= PMs (also called 3R1T parallel mechanisms) is dealt with using the general approach proposed in Chap. 5. SP= PMs are the parallel counterparts of the 4-DOF SP serial manipulators, which are composed of one S and one P joint. The four steps of the type synthesis of SP= PMs are presented in detail.

10.1 Introduction

Four-DOF SP= PMs are a class of PMs with reduced degrees of freedom. SP= PMs cover a range of applications including motion simulation and laparoscopic surgery [135]. Several SP= PMs have been proposed in the literature ([41, 53, 134, 135]).

It is noted that for a PKC with inactive joints and its kinematic equivalent PKC without inactive joints, the number of over-constraints as well as the reaction forces in the joints are different, although the inactive joints in the PKC make no contribution to the movement of the moving platform [26, 72]. However, no SP= PKC with inactive joints has been proposed yet. Moreover, in [41, 53, 134], the selection of actuated joints [1, 56, 67, 75, 78, 79, 80] has not been dealt with systematically. Therefore, the type synthesis of SP= PMs needs further investigation.

Using the general approach proposed in Chap. 5, the type synthesis of SP= PMs is dealt with in this chapter. The decomposition of the wrench system of

X. Kong and C. Gosselin: Type Syn. of Parallel Mech., STAR 33, pp. 159–172, 2007.
springerlink.com

an SP= PKC is dealt with in Sect. 10.5. The type synthesis of legs for SP= PKCs is performed in Sect. 10.6. In Sect. 10.7, the assembly of legs to generate SP= PKCs is presented while the selection of actuated joints for SP= PMs is discussed in Sect. 10.8.

10.2 Wrench System of an SP= PKC

In any general configuration, the twist system of an SP= PKC or its SP virtual chain is a 3-$\boldsymbol{\xi}_0$-1-$\boldsymbol{\xi}_\infty$-system in which the axes of $\boldsymbol{\xi}_0$ intersect at one point on the base. It can be found without difficulty that the wrench system of an SP= PKC is a 2-$\boldsymbol{\zeta}_0$-system which is composed of all the $\boldsymbol{\zeta}_0$ whose axes pass through the centre of the S joint and are perpendicular to the P joint within the SP virtual chain (Fig. 10.1).

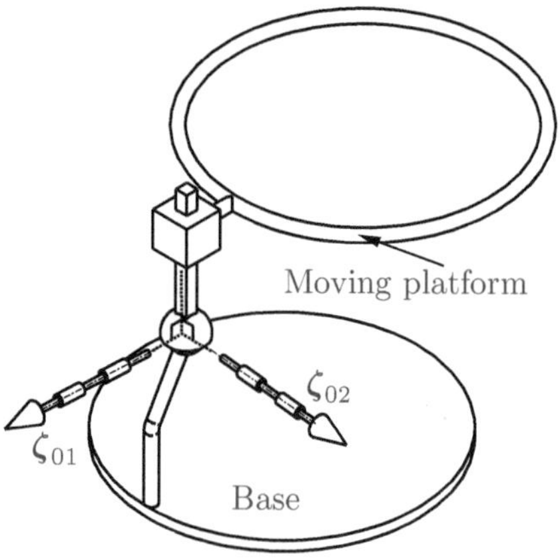

Fig. 10.1. Wrench system of an SP virtual chain

10.3 Conditions for a PKC to Be an SP= PKC

When we connect the base and the moving platform of an SP= PKC by an SP virtual chain, the function of the PKC is not affected. Any of its legs and the SP virtual chain will form a 4-DOF single loop KC. When the order of the leg-wrench system is greater than 0, the single-loop KC formed must be an overconstrained KC.

Based on the concept of SP virtual chain, it follows that a PKC is an SP= PKC if it satisfies the two conditions below:

(1) Each leg of the PKC and a same SP virtual chain form a 4-DOF single-loop KC.
(2) The wrench system of the PKC is the same as that of the SP virtual chain, i.e., a 2-$\boldsymbol{\zeta}_0$-system, in any one general configuration.

10.4 Procedure for the Type Synthesis of SP= PMs

A procedure can thus be proposed for the type synthesis of SP= PKCs as follows:

Step 1. Decomposition of the wrench system of 4-DOF SP= PKCs.
Step 2. Type synthesis of legs for SP= PKCs.
Here, a leg for SP= PKCs refers to a leg satisfying Condition (1) for SP= PKCs.
Step 3. Assembly of legs for SP= PKCs.
SP= PKCs can be generated by taking two or more legs for SP= PKCs, obtained in Step 1, such that Condition (2) for SP= PKCs is satisfied.
Step 4. Selection of actuated joints.
SP= PMs can be generated by selecting actuated joints in different ways for each SP= PKC (Sect. 10.8), obtained in Step 3.

The procedure proposed above will be elaborated upon in the following sections.

10.5 Step 1: Decomposition of the Wrench System of SP= PKCs

The decomposition of the wrench system of an m-legged SP= PKC consists in finding all the leg-wrench systems and all the combinations of m leg-wrench systems for the SP= PKC and a specified number of overconstraints Δ. Step 1 can be performed using (2.24) and (2.28).

10.5.1 Determination of the Leg-Wrench Systems

As the wrench system of a PKC is the linear combination of those of all its legs in a general configuration, any leg-wrench system in a SP= PKC is a sub-system of its wrench system. Following the procedure proposed in Sect. 2.1.3, all the leg-wrench systems for an SP= PKC can be identified without difficulty.

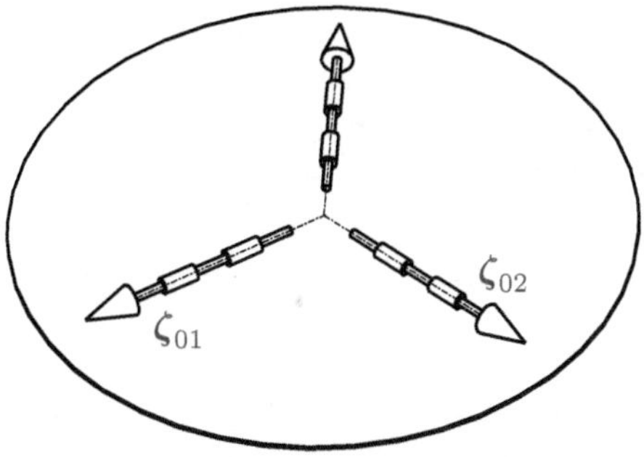

Fig. 10.2. Leg-wrench system of an SP= PKC

In any general configuration, the wrench system of an SP= PKC is the same as that of its SP virtual chain, i.e., a 2-ζ_0-system, which is composed of all the ζ_0 whose axes pass through the centre of the S joint and are perpendicular to the P joint within the SP virtual chain. It then follows that any leg-wrench system with order $c^i > 0$ of an SP= PKC is either a 2-ζ_0-system or a 1-ζ_0-system in a general configuration (Fig. 10.2).

10.5.2 Determination of the Combinations of Leg-Wrench Systems

For the SP motion pattern considered, all the wrenches in the wrench system are of the same pitch. The leg-wrench systems are $c^i(0 \le c^i \le 2)$-systems of the 0 pitch. The combination of leg-wrench systems can be simply represented by the combination of the orders, c^i, of leg-wrench systems.

The combinations of the orders, c^i, of leg-wrench systems have been determined by solving (2.28) and are shown in Table 5.3.

10.6 Step 2: Type Synthesis of Legs

In this section, the type synthesis of legs for SP= PKCs will be performed using the two substeps described in Sect. 5.5.

10.6.1 Step 2a: Type Synthesis of 4-DOF Single-Loop KCs That Involve an SP Virtual Chain and Have a Specified Leg-Wrench System

The type synthesis of 4-DOF single-loop KCs that involve an SP virtual chain and have a specified leg-wrench system can be performed as follows.

Firstly, perform the type synthesis of single-loop KCs that involve an SP virtual chain and have a specified leg-wrench system (See Chap. 3). Secondly, discard those single-loop KCs in which the twists of all the joints but the virtual chain are linearly dependent.

Cases with a 2-ζ_0-System

Firstly, perform the type synthesis of 4-DOF single-loop KCs that involve an SP virtual chain and have a 2-ζ_0-system. According to Chap. 3, there are eight joints in a 4-DOF single-loop KC that involves an SP virtual chain and has a 2-ζ_0-system. Such a single-loop KC is formed by inserting one codirectional compositional unit into one spherical compositional unit (Fig. 7.4). The 2-ζ_0-system is composed of all the ζ_0 whose axes pass through the centre of the SP virtual chain and are perpendicular to the directions of the P joints within the codirectional compositional unit.

Secondly, discard those single-loop KCs in which the twists of all the joints but the virtual chain are linearly dependent. This requires that there be only one P joint within each leg or two P joints within the codirectional compositional unit.

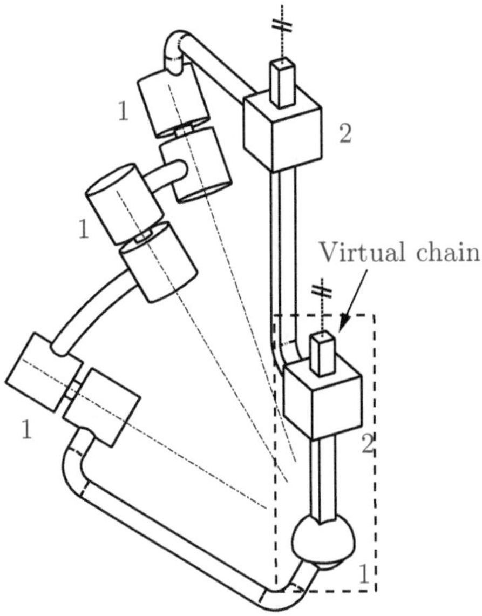

Fig. 10.3. Four-DOF single-loop KCs involving an SP virtual chain ($c^i = 2$)

Cases with a 1-ζ_0-system

Firstly, perform the type synthesis of 4-DOF single-loop KCs that involve an SP virtual chain and have a 1-ζ_0-system. According to Chap. 3, there are nine joints in a 4-DOF single-loop KC that involves an SP virtual chain and has a 1-ζ_0-system. Such a single-loop KC is composed of (a) one spherical compositional unit and one planar compositional unit [Figs. 10.4a–10.4f] or (b) one spherical compositional unit and one planar translational compositional unit [Fig. 10.4g] or can be obtained by inserting one coaxial or codirectional compositional unit into a 4-DOF single-loop KC that involves an SP virtual chain and has a 2-ζ_0-system [Figs. 10.4h–10.4i]. The axis of the basis ζ_0 of the 1-ζ_0-system passes through the centre of the S joint of the SP virtual chain, is perpendicular to the plane associated with the $()_E$ and the P joint within the $()_L$ and intersects the axis of the R joint within the $()_L$.

Secondly, discard those single-loop KCs in which the twists of all the joints but the virtual chain are linearly dependent. This requires that there be only one R joint within the coaxial compositional unit or one P joint within the codirectional compositional unit.

In the representation of the 4-DOF single-loop KCs involving an SP virtual chain and having a c^i-ζ_0-system, $\hat{\mathrm{R}}$ denote R joints whose axes intersect at the centre of the S joint within the SP virtual chain, $(\mathrm{XXX})_E^{|}$ (X represents an R or a P joint) denotes an equivalent planar joint formed by three successive joints whose associated plane is parallel to the direction of the P joint within the SP virtual chain, $\mathrm{P}^{|}$ denotes a P joint whose direction is parallel to the direction of

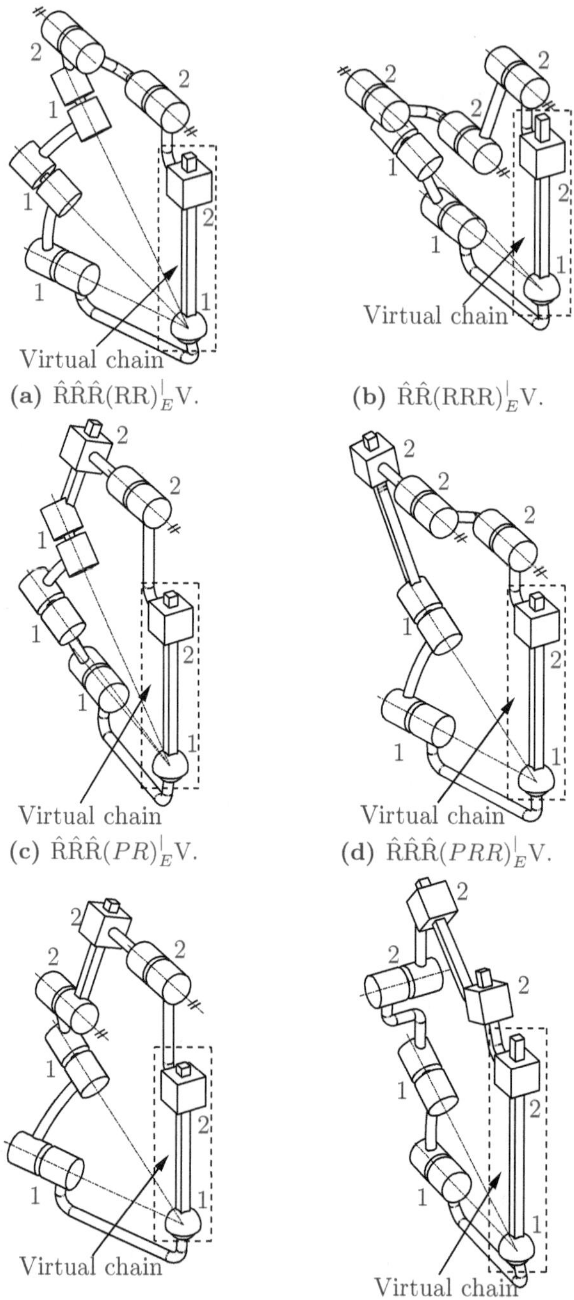

(a) $\hat{R}\hat{R}\hat{R}(RR)^{|}_{E}V$. **(b)** $\hat{R}\hat{R}(RRR)^{|}_{E}V$.

(c) $\hat{R}\hat{R}\hat{R}(PR)^{|}_{E}V$. **(d)** $\hat{R}\hat{R}\hat{R}(PRR)^{|}_{E}V$.

(e) $\hat{R}\hat{R}(RPR)^{|}_{E}V$. **(f)** $\hat{R}\hat{R}(RPP)^{|}_{E}V$.

Fig. 10.4. Four-DOF single-loop KCs involving an SP virtual chain ($c^i = 1$)

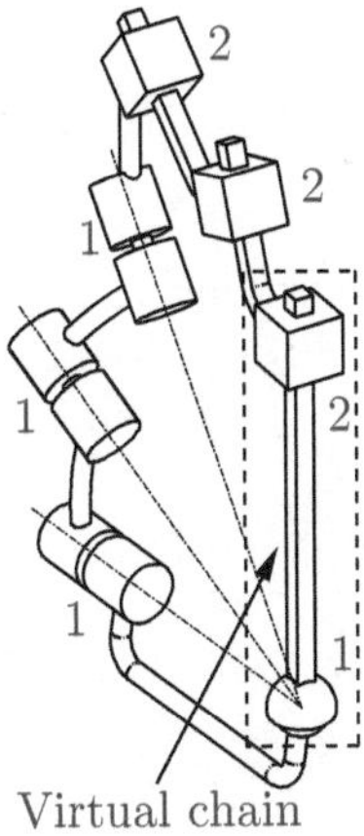

(g) $\hat{R}\hat{R}\hat{R}(PP)^{|}_{E}V$.

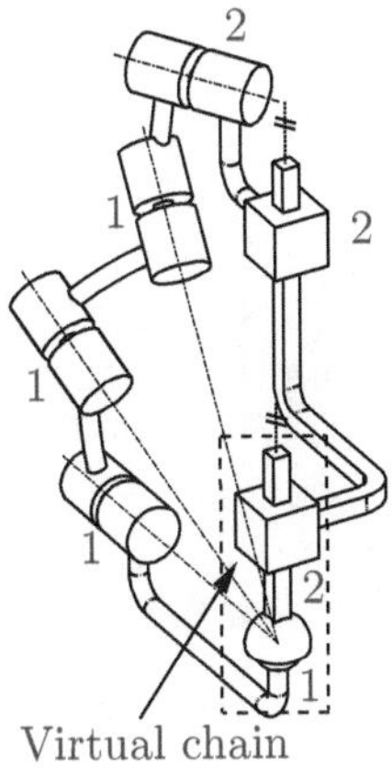

(h) $\hat{R}\hat{R}\hat{R}RP^{|}V$.

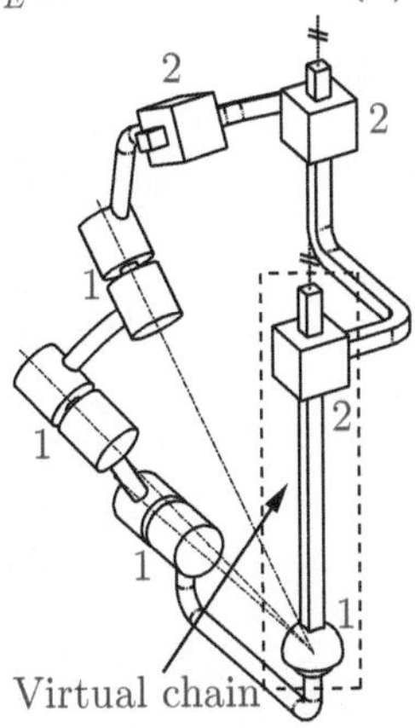

(i) $\hat{R}\hat{R}\hat{R}PP^{|}V$.

Fig. 10.4. (*continued*)

the P joint within the SP virtual chain, and $()_L$ denotes a coaxial or codirectional compositional unit.

It is pointed out that the $(R)_L$ and the $(P)_L$ joints can be placed anywhere in the single-loop KC. For brevity, we list only the 4-DOF single-loop KCs from which all 4-DOF single-loop KCs can be obtained through the above operations. For example, the $\hat{R}\hat{R}\hat{R}P^{|}RV$ single-loop KC can be obtained by changing the position of the P joint in the $\hat{R}\hat{R}\hat{R}RP^{|}V$ single-loop KC [Fig. 10.4h].

10.6.2 Step 2b: Generation of Types of Legs

According to Condition (1) for the existence of SP= PKCs, the types of legs can be readily obtained from the 4-DOF single-loop KCs obtained in Step 2a, Sect. 10.6.1, by removing the SP virtual chain.

The specific geometric conditions which guarantee the satisfaction of Condition (1) for SP= PKCs are clearly indicated by the notations that were introduced to represent the types of SP= PKCs, SP= PMs and their legs.

For example, by removing the virtual chain in a $\hat{\mathrm{R}}\hat{\mathrm{R}}\hat{\mathrm{R}}(\mathrm{RR})^{|}_{E}$V KC [Fig. 10.4a], an $\hat{\mathrm{R}}\hat{\mathrm{R}}\hat{\mathrm{R}}(\mathrm{RR})^{|}_{E}$ leg [Fig.10.5b] can be obtained. Such a leg has a 1-ζ_0-system. The axis of the basis ζ_0 of the 1-ζ_0-system passes through the common point of the axes of three $\hat{\mathrm{R}}$ joints and is parallel to the axes of the R joints within $(\mathrm{RR})^{|}_{E}$.

All the types of legs for SP= PKCs obtained are listed in Table 10.1. The legs used to synthesize SP= PKCs need to satisfy not only the geometric conditions within a same leg, but also the geometric conditions among different legs.

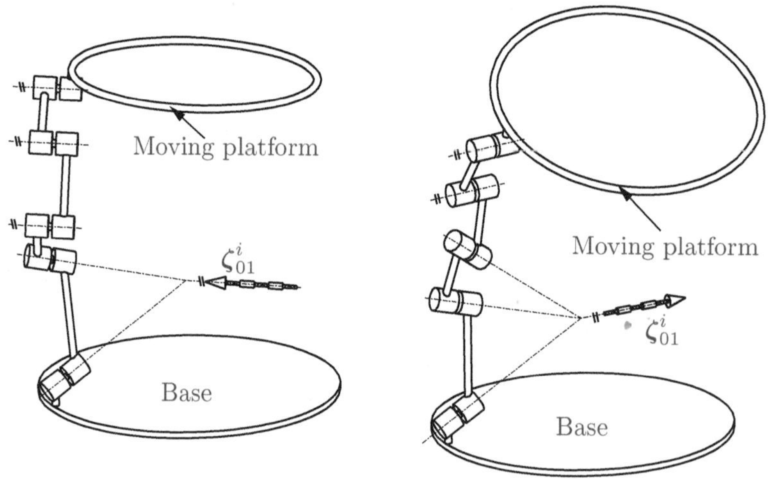

(a) $\hat{\mathrm{R}}\hat{\mathrm{R}}(\mathrm{RRR})^{|}_{E}$ leg. (b) $\hat{\mathrm{R}}\hat{\mathrm{R}}\hat{\mathrm{R}}(\mathrm{RR})^{|}_{E}$ leg.

Fig. 10.5. Some legs for SP= PKCs

10.7 Step 3: Assembly of Legs

SP= PKCs can be generated by assembling a set of legs for SP= PKCs shown in Table 10.1 selected according to the combinations of the leg-wrench systems shown in Table 5.3. In assembling SP= PKCs, the following condition should be met: the linear combination of the leg-wrench systems forms a 2-ζ_0-system (see Condition (2) for SP= PKCs in Sect. 10.3).

For SP= PKCs with a leg having a 2-ζ_0-system, Condition (2) is automatically guaranteed.

For SP= PKCs without a leg having a 2-ζ_0-system, the axis of the basis wrench of any leg with a 1-ζ_0-system is perpendicular to a same line and thus parallel to one plane. In order to guarantee that the linear combination of all the leg-wrench systems forms a 2-ζ_0-system, these SP= PKCs must be such that the

Table 10.1. Legs for SP= PKCs

c^i	Class	No.	Type
2	3R1P	1	$\hat{R}\hat{R}\hat{R}P^{\vert}$
1	5R	2	$\hat{R}\hat{R}(RRR)^{\vert}_{E}$
		3	$\hat{R}\hat{R}\hat{R}(RR)^{\vert}_{E}$
	4R1P	4	$\hat{R}\hat{R}(RRP)^{\vert}_{E}$
		5	$\hat{R}\hat{R}(RPR)^{\vert}_{E}$
		6	$\hat{R}\hat{R}(PRR)^{\vert}_{E}$
		7	$\hat{R}\hat{R}\hat{R}(RP)^{\vert}_{E}$
		8	$\hat{R}\hat{R}\hat{R}(PR)^{\vert}_{E}$
		9-13	Permutation of $\hat{R}\hat{R}\hat{R}RP^{\vert}$
	3R2P	14	$\hat{R}\hat{R}(RPP)^{\vert}_{E}$
		15	$\hat{R}\hat{R}(PRP)^{\vert}_{E}$
		16	$\hat{R}\hat{R}(PPR)^{\vert}_{E}$
		17	$\hat{R}\hat{R}\hat{R}(PP)^{\vert}_{E}$
		18-22	Permutation of $\hat{R}\hat{R}\hat{R}PP^{\vert}$
0	omitted	omitted	omitted

planes of relative motion associated with $()_E$ of all the legs having a 1-ζ_0-system are not all parallel.

For instance, we obtain a 2-legged 2-$\hat{R}\hat{R}\hat{R}(RR)^{\vert}_{E}$ SP= PKC [Fig. 10.6b] by assembling the two $\hat{R}\hat{R}\hat{R}(RR)^{\vert}_{E}$ legs [Fig. 10.5b] selected in Step 3b. In the SP= PKC, the axes of the R joints on the moving platform must not be parallel to each other.

In addition to the types of SP= PKCs proposed in the literature, new types, i.e., the types with a leg having either a 2-ζ_0-system or an inactive joint, are also obtained. Due to the large number of SP= PKCs, only the 4-legged SP= PKCs with legs of the same type are listed in Table 10.2. Among these 4-legged SP= PKCs, there are 12 types without inactive joints (No. 1–8 and 14–17).

10.8 Step 4: Selection of Actuated Joints

10.8.1 t-Component of the Actuation Wrenches

Considering that the order of a screw system is coordinate frame free and for simplicity reasons, all the actuation wrenches and wrenches are expressed in a coordinate system with its origin at the centre of the wrench system of the SP= PKC and the Z-axis perpendicular to all the wrenches within the wrench system of the SP= PKC.

Let $[\mathbf{i} \;\; \mathbf{0}]^T$ and $[\mathbf{j} \;\; \mathbf{0}]^T$ denote a basis of $\mathcal{W}$. Here, $\mathbf{i}$ and $\mathbf{j}$ denote respectively the unit vectors along the X- and Y-axes. The last four scalar components of

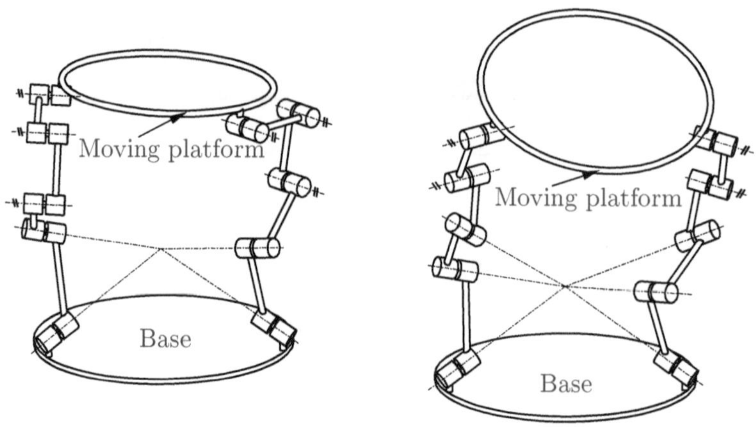

(a) $2\text{-}\hat{R}\hat{R}(RRR)^{|}_{E}$. (b) $2\text{-}\hat{R}\hat{R}\hat{R}(RR)^{|}_{E}$.

Fig. 10.6. Some SP= PKCs

Table 10.2. Four-legged SP= PKCs

c^i	Class	No.	Type
2	3R1P	1	$4\text{-}\hat{R}\hat{R}\hat{R}P^{|}$
1	5R	2	$4\text{-}\hat{R}\hat{R}(RRR)^{|}_{E}$
		3	$4\text{-}\hat{R}\hat{R}\hat{R}(RR)^{|}_{E}$
	4R1P	4	$4\text{-}\hat{R}\hat{R}(RRP)^{|}_{E}$
		5	$4\text{-}\hat{R}\hat{R}(RPR)^{|}_{E}$
		6	$4\text{-}\hat{R}\hat{R}(PRR)^{|}_{E}$
		7	$4\text{-}\hat{R}\hat{R}\hat{R}(RP)^{|}_{E}$
		8	$4\text{-}\hat{R}\hat{R}\hat{R}(PR)^{|}_{E}$
		9-13	4-Permutation of $\hat{R}\hat{R}\hat{R}RP^{|}$
	3R2P	14	$4\text{-}\hat{R}\hat{R}(RPP)^{|}_{E}$
		15	$4\text{-}\hat{R}\hat{R}(PRP)^{|}_{E}$
		16	$4\text{-}\hat{R}\hat{R}(PPR)^{|}_{E}$
		17	$4\text{-}\hat{R}\hat{R}\hat{R}(PP)^{|}_{E}$
		18-22	4-Permutation of $\hat{R}\hat{R}\hat{R}PP^{|}$

all the basis wrenches of $\mathcal{W}$ are zero. Thus, the t-component, $\boldsymbol{\zeta}^i_{t\not{D}j}$, of the $\boldsymbol{\zeta}^i_{\not{D}j}$ is composed of the last four scalar components of the $\boldsymbol{\zeta}^i_{\not{D}j}$.

In the $\underline{\hat{R}}\hat{R}(RRR)^{|}_{E}$ leg [Fig. 10.7a], the first R joint is actuated. The actuation wrench is a $\boldsymbol{\zeta}_0$ whose axis intersects the axis of the second $\hat{R}$ joint, is parallel to the axes of the R joints in $(RRR)^{|}_{E}$ and does not intersect the axis of the actuated joint. In the $\underline{\hat{R}}\hat{R}\hat{R}(RR)^{|}_{E}$ leg [Fig. 10.7b], the first R joint is actuated. The actuation wrench is any $\boldsymbol{\zeta}_0$ whose axis is the intersection of the plane passing

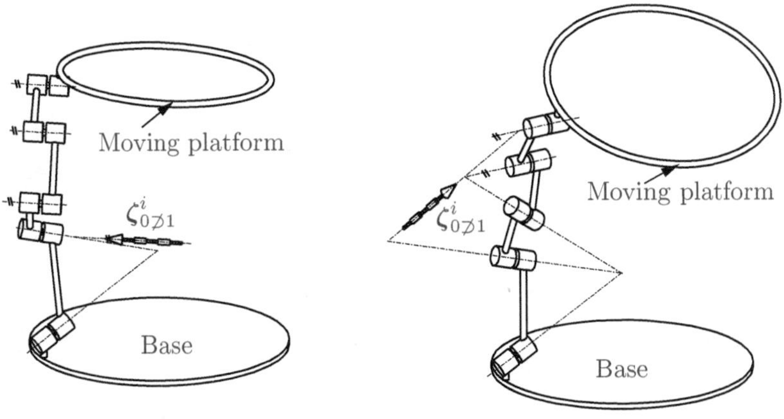

(a) $\underline{\hat{\mathrm{R}}}\hat{\mathrm{R}}(\mathrm{RRR})^{|}_{E}$ leg. **(b)** $\underline{\hat{\mathrm{R}}}\hat{\mathrm{R}}\hat{\mathrm{R}}(\mathrm{RR})^{|}_{E}$ leg.

Fig. 10.7. Actuation wrenches of some legs for SP= PKCs

through the axes of two unactuated $\hat{\mathrm{R}}$ joints and the plane passing through the axes of the two R joints in $(\mathrm{RR})^{|}_{E}$.

10.8.2 Procedure for the Validity Detection of Actuated Joints

The validity detection of actuated joints of SP= PMs can thus be performed using the following steps.

Step 4a. If one or more of the actuated joints of a candidate SP= PM are inactive, the set of actuated joints is invalid and the candidate SP= PM should be discarded.

Step 4b. If the t-components of the actuation wrenches of actuated joints $\zeta^i_{\not{D}j}$ are linearly dependent in a general configuration for a candidate SP= PM, the set of actuated joints is invalid. In this case, the candidate SP= PM should be discarded.

One candidate SP= PM corresponding to the 2-$\hat{\mathrm{R}}\hat{\mathrm{R}}(\mathrm{RRR})^{|}_{E}$ SP= PKC [Fig. 10.6a] is the 2-$\underline{\hat{\mathrm{R}}\hat{\mathrm{R}}}(\mathrm{RRR})^{|}_{E}$ SP= PM [Fig. 10.8a]. The axes of all the actuation wrenches pass through the centre of the common point of the $\hat{\mathrm{R}}$ joints and are perpendicular to the Z-axis. We have

$$\left| \zeta^1_{t,\not{D}1}\ \zeta^1_{t,\not{D}2}\ \zeta^2_{t,\not{D}1}\ \zeta^2_{t,\not{D}2} \right| = \begin{vmatrix} 0\ 0\ 0\ 0 \\ *\ *\ *\ * \\ *\ *\ *\ * \\ *\ *\ *\ * \end{vmatrix} = 0.$$

where $*$ denotes an arbitrary number.

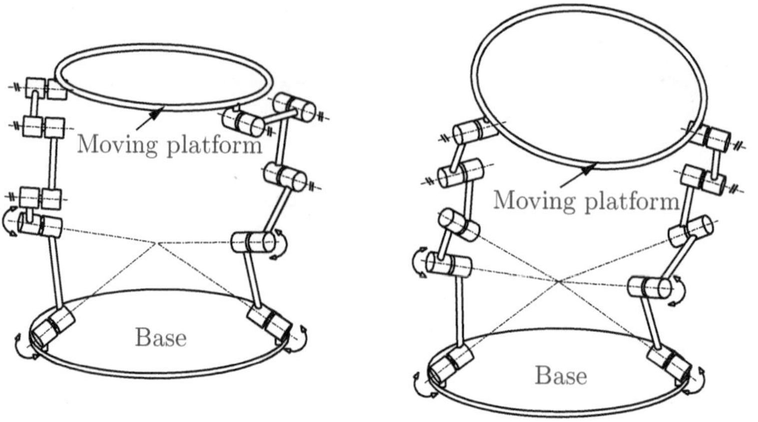

(a) 2-$\underline{\hat{R}\hat{R}}(RRR)\big|_E$ candidate SP= PM.

(b) 2-$\underline{\hat{R}\hat{R}}\hat{R}(RR)\big|_E$ candidate SP= PM.

Fig. 10.8. Selection of actuated joints for some SP= PKCs

Table 10.3. Four-legged SP= PMs

c^i	Class	No.	Type
1	5R	1	4-$\hat{R}\hat{R}(\underline{R}RR)\big|_E$
		2	4-$\underline{\hat{R}}\hat{R}\hat{R}(RR)\big|_E$
		3	4-$\hat{R}\underline{\hat{R}}\hat{R}(RR)\big|_E$
		4	4-$\hat{R}\hat{R}\underline{\hat{R}}(RR)\big|_E$
	4R1P	5	4-$\hat{R}\hat{R}(\underline{R}RP)\big|_E$
		6	4-$\hat{R}\hat{R}(\underline{R}PR)\big|_E$
		7	4-$\hat{R}\hat{R}(\underline{P}RR)\big|_E$
		8	4-$\underline{\hat{R}}\hat{R}\hat{R}(RP)\big|_E$
		9	4-$\hat{R}\underline{\hat{R}}\hat{R}(RP)\big|_E$
		10	4-$\hat{R}\hat{R}\underline{\hat{R}}(RP)\big|_E$
		11	4-$\underline{\hat{R}}\hat{R}\hat{R}(PR)\big|_E$
		12	4-$\hat{R}\underline{\hat{R}}\hat{R}(PR)\big|_E$
		13	4-$\hat{R}\hat{R}\underline{\hat{R}}(PR)\big|_E$
	3R2P	14	4-$\hat{R}\hat{R}(\underline{R}PP)\big|_E$
		15	4-$\hat{R}\hat{R}(\underline{P}RP)\big|_E$
		16	4-$\hat{R}\hat{R}(\underline{P}PR)\big|_E$
		17	4-$\underline{\hat{R}}\hat{R}\hat{R}(PP)\big|_E$
		18	4-$\hat{R}\underline{\hat{R}}\hat{R}(PP)\big|_E$
		19	4-$\hat{R}\hat{R}\underline{\hat{R}}(PP)\big|_E$

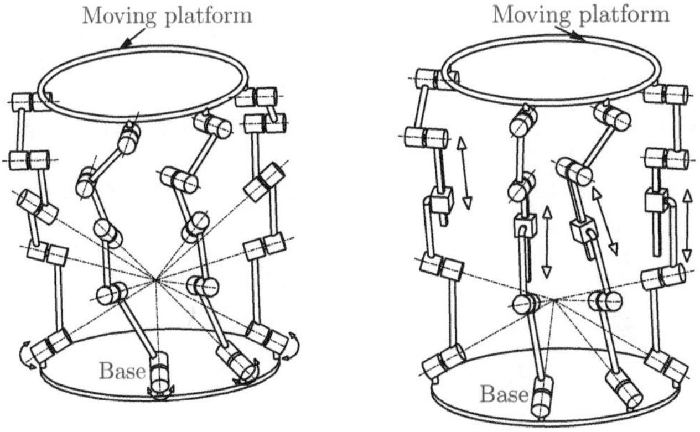

(a) 4-$\underline{\hat{\mathrm{R}}}\hat{\mathrm{R}}\hat{\mathrm{R}}(\mathrm{RR})^{|}_{E}$. **(b)** 4-$\hat{\mathrm{R}}\hat{\mathrm{R}}(\underline{P}RR)^{|}_{E}$.

Fig. 10.9. Four-legged SP= PMs with identical type of legs

According to the validity condition of actuated joints, the set of actuated joints is invalid. The candidate 2-$\underline{\hat{\mathrm{R}}}\hat{\mathrm{R}}(\mathrm{RRR})^{|}_{E}$ SP= PM is thus discarded.

One candidate SP= PM corresponding to the 2-$\hat{\mathrm{R}}\hat{\mathrm{R}}\hat{\mathrm{R}}(\mathrm{RR})^{|}_{E}$ SP= PKC [Fig. 10.6b] is 2-$\underline{\hat{\mathrm{R}}}\underline{\hat{\mathrm{R}}}\hat{\mathrm{R}}(\mathrm{RR})^{|}_{E}$ SP= PM [Fig. 10.8b]. The axis of the actuation wrench of an actuated joint is along the intersection of the plane determined by the axes of the two R joints within $(RR)^{|}_{E}$ and the plane determined by the axes of the $\hat{\mathrm{R}}$ joints except for the actuated joint considered [Fig. 10.7b]. Using the validity condition of actuated joints, it can be proved that the set of actuated joints is valid.

Following the criteria on the selection of actuated joints (Sect. 5.7) and the procedure for the detection of the validity of actuated joints, all the $m(m \geq 2)$-legged SP= PMs corresponding to each SP= PKC can be generated. Due to the large number of SP= PMs, only 4-legged SP= PMs with all legs of the same type satisfying the above criteria are shown in Table 10.3. Figure 10.9 shows two types of 4-legged SP= PMs without unactuated P joints.

10.9 Summary

In this chapter, the type synthesis of SP= PMs has been thoroughly solved using the general type synthesis approach proposed in Chap. 5. SP= PKCs with inactive joints as well as SP= PKCs without inactive joints have been obtained. The validity condition of actuated joints of SP= PMs has been reduced to the calculation of a 4×4 determinant. The SP= PKCs obtained include some new SP= PKCs as well as all the known SP= PKCs. Some new SP= PMs have also been proposed.

Variations of the SP= PKCs can be obtained by (a) substituting a combination of one R joint and one P joint with parallel axes, a combination of two R joints with intersecting axes and a combination of three R joints with concurrent axes with a C joint or its RH, PH, HH form, a U, and an S joint respectively, (b) replacing a P joint within $()_{E}^{|}$ with a planar parallelogram whose plane of motion is parallel to the planes of relative motion associated with $()_{E}^{|}$, and/or (c) replacing an inactive R or P joint with an inactive H (helical) joint.

It is noted that if a combination of an R joint and a P joint with parallel axes, in which one joint is inactive, is replaced with a C joint in the form of RH or HH, there will be no inactive joint in the RH or HH combination.

11. Five-DOF US= Parallel Mechanisms

In this chapter and the following two chapters, the type synthesis of 5-DOF PMs is addressed. Each of the three chapters will focus on a different motion pattern. The use of motion patterns provides a systematic classification of 5-DOF PMs and alleviates the confusion that is often generated in the literature when 5-DOF PMs are described. In this chapter, the type synthesis of 5-DOF US= PMs is dealt with using the general approach proposed in Chap. 5. US= PMs are the parallel counterparts of the 5-DOF US serial manipulators, which are composed of one U and one S joint. The four steps of the type synthesis of US= PMs are presented in detail.

11.1 Introduction

Five-DOF US= PMs are the parallel counterparts of the 5-DOF US serial manipulators. The moving platform of a US= PM can rotate arbitrarily about a point moving along a spherical surface. The first US= PM was proposed in [119].

It is noted that for a PKC with inactive joints and its kinematic equivalent PKC without inactive joints, the number of over-constraints as well as the reaction forces in the joints are different, although the inactive joints in the PKC make no contribution to the movement of the moving platform [26, 72]. However, no US= PKC with inactive joints has been proposed yet. Moreover, in [119], the selection of actuated joints [1, 56, 67, 75, 78, 79, 80] has not been dealt with systematically. Therefore, the type synthesis of US= PMs needs further investigation.

Using the general approach proposed in Chap. 5, the type synthesis of US= PMs is dealt with in this chapter. The decomposition of the wrench system of a US= PKC is dealt with in Sect. 11.5. The type synthesis of legs for US= PKCs is performed in Sect. 11.6. In Sect. 11.7, the assembly of legs to generate US= PKCs is dealt with. The selection of actuated joints for US= PMs is discussed in Sect. 11.8.

X. Kong and C. Gosselin: Type Syn. of Parallel Mech., STAR 33, pp. 173–183, 2007.
springerlink.com

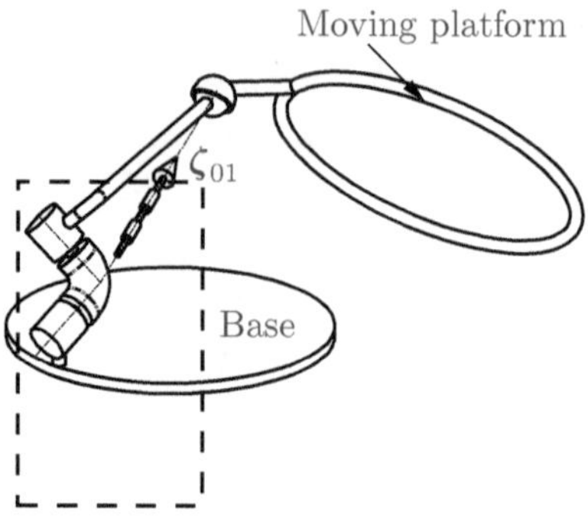

Fig. 11.1. Wrench system of a US= PKC

11.2 Wrench System of a US= PKC

In any general configuration, the twist system of a US= PKC or its US virtual chain is a 2-$\boldsymbol{\xi}_\infty$-3-$\boldsymbol{\xi}_0$-system. In the above twist system, the axes of the basis $\boldsymbol{\xi}_0$ pass through the centre of the S joint, and the axes of the basis $\boldsymbol{\xi}_\infty$ are perpendicular to the line connecting the centres of the S and U joints. Since the virtual power developed by any $\boldsymbol{\zeta}_0$ whose axis passes through both the centre of the U joint and the centre of the S joint along any twist within the 2-$\boldsymbol{\xi}_\infty$-3-$\boldsymbol{\xi}_0$-system is 0, its wrench system is a 1-$\boldsymbol{\zeta}_0$-system (Fig. 11.1). The axis of the basis $\boldsymbol{\zeta}_{01}$ of the 1-$\boldsymbol{\zeta}_0$-system passes through the centres of the S joint and the U joint.

11.3 Conditions for a PKC to Be a US= PKC

When we connect the base and the moving platform of a US= PKC by an appropriate US virtual chain, the function of the PKC is not affected. Any of its legs and the US virtual chain will form a 5-DOF single loop KC. When the order of the leg-wrench system is greater than 0, the single-loop KC formed must be an overconstrained KC.

Based on the concept of US virtual chain, if follows that a PKC is a US= PKC if it satisfies the two conditions below:

(1) Each leg of the PKC and a same US virtual chain form a 5-DOF single-loop KC.
(2) The wrench system of the PKC is the same as that of the US virtual chain, i.e., a 1-$\boldsymbol{\zeta}_0$-system, in any one general configuration.

11.4 Procedure for the Type Synthesis of US= PMs

A procedure can thus be proposed for the type synthesis of US= PKCs as follows:

Step 1. Decomposition of the wrench system of 5-DOF US= PKCs.
Step 2. Type synthesis of legs for US= PKCs.
Here, a leg for US= PKCs refers to a leg satisfying Condition (1) for US= PKCs.

Step 3. Assembly of legs for US= PKCs.
US= PKCs can be generated by taking two or more legs for US= PKCs, obtained in Step 1, such that Condition (2) for US= PKCs is satisfied.
Step 4. Selection of actuated joints.
US= PMs can be generated by selecting actuated joints in different ways for each US= PKC (Sect. 11.8), obtained in Step 3.

The procedure proposed above will be elaborated upon in the following sections.

11.5 Step 1: Decomposition of the Wrench System of US= PKCs

The decomposition of the wrench system of an m-legged US= PKC consists in finding all the leg-wrench systems and all the combinations of m leg-wrench systems for the US= PKC and a specified number of overconstraints Δ. Step 1 can be performed using (2.24) and (2.28).

11.5.1 Determination of the Leg-Wrench Systems

As the wrench system of a PKC is the linear combination of those of all its legs in a general configuration, any leg-wrench system in a US= PKC is a sub-system of its wrench system. Following the procedure proposed in Sect. 2.1.3, all the leg-wrench systems for a US= PKC can be identified without difficulty.

In any general configuration, the wrench system of a US= PKC is the same as that of its US virtual chain, i.e., a 1-$\boldsymbol{\zeta}_0$-system, in which the axis of the basis $\boldsymbol{\zeta}_0$ passes through the centre of the S joint and the centre of the U joint US virtual chain. It then follows that any leg-wrench system with order $c^i > 0$ of a US= PKC is still a 1-$\boldsymbol{\zeta}_0$-system in a general configuration (Fig. 11.1).

11.5.2 Determination of the Combinations of Leg-Wrench Systems

For the US motion pattern considered, all the wrenches in the wrench system are of the same pitch. The leg-wrench systems are $c^i(0 \leq c^i \leq 1)$-systems of the 0 pitch. The combination of leg-wrench systems can be simply represented by the combination of the orders, c^i, of leg-wrench systems.

The combinations of the orders, c^i, of leg-wrench systems have been determined by solving (2.28) and are shown in Table 5.4.

11.6 Step 2: Type Synthesis of Legs

In this section, the type synthesis of legs for US= PKCs will be performed using the two substeps described in Sect. 5.5.

11.6.1 Step 2a: Type Synthesis of 5-DOF Single-Loop KCs That Involve a US Virtual Chain and Have a Specified Leg-Wrench System

The type synthesis of 5-DOF single-loop KCs that involve a US virtual chain and have a specified leg-wrench system can be performed as follows.

Firstly, perform the type synthesis of single-loop KCs that involve a US virtual chain and have a specified leg-wrench system (See Chap. 3). Secondly, discard those single-loop KCs in which the twists of all the joints but the virtual chain are linearly dependent.

Cases with a 1-ζ_0-System

Firstly, perform the type synthesis of 5-DOF single-loop KCs that involve a US virtual chain and have a 1-ζ_0-system. According to Chap. 3, there are ten joints in a 5-DOF single-loop KC that involves a US virtual chain and has a 1-ζ_0-system. Such a single-loop KC is composed of two spherical compositional units (Fig. 11.2). The axis of the basis ζ_0 of the 1-ζ_0-system passes through the centres of the S joint and the U joint of the US virtual chain.

Secondly, discard those single-loop KCs in which the twists of all the joints but the virtual chain are linearly dependent. This requires that, each spherical compositional unit contains at most three R joints whose axes pass through the centre of the U (or S) joint within the virtual chain except the U (or S) joint itself.

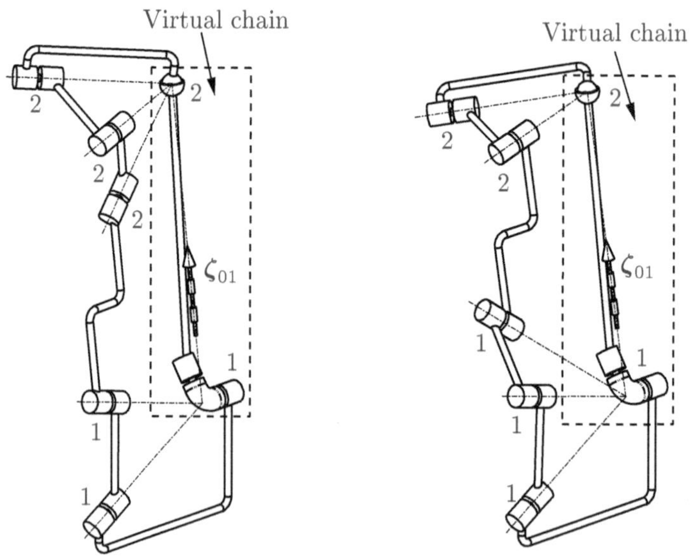

(a) ŘŘR̂R̂R̂V. (b) ŘŘŘR̂R̂V.

Fig. 11.2. Five-DOF single-loop KCs involving a US virtual chain ($c_i = 1$)

In the representation of the 5-DOF single-loop KCs involving a US virtual chain and having a c^i-$\boldsymbol{\zeta}_0$-system, Ř denote R joints whose axes intersect at the centre of the U joint within the US virtual chain, R̂ denote R joints whose axes intersect at the centre of the S joint within the US virtual chain.

$c^i = 0$

Theoretically, any six R and P joints whose twists are linearly independent together with a US virtual chain form a 5-DOF single-loop KC.

It is noted that starting from each of the above 5-DOF single-loop KCs involving a US virtual chain and having a 1-$\boldsymbol{\zeta}_0$-system, one can obtain a class of 5-DOF single-loop KCs involving a US virtual chain and having a 0-$\boldsymbol{\zeta}$-system by the application of an $(\mathrm{R})_L$ coaxial compositional unit or a $(\mathrm{P})_L$ codirectional compositional unit. It is pointed out that the $(\mathrm{R})_L$ or the $(\mathrm{P})_L$ joint is inactive and can thus be placed anywhere in the single-loop KC. For example, the Ř$(\mathrm{R})_L$ŘR̂R̂R̂V single-loop KC can be obtained by changing the position of the $(\mathrm{R})_L$ joint in the ŘŘR̂R̂R̂$(\mathrm{R})_L$V single-loop KC.

11.6.2 Step 2b: Generation of Types of Legs

According to Condition (1) for the existence of US= PKCs (Sect. 11.3), the types of legs can be readily obtained from the 5-DOF single-loop KCs obtained in Step 2a, Sect. 11.6.1, by removing the US virtual chain.

The specific geometric conditions that guarantee the satisfaction of Condition (1) for US= PKCs are clearly indicated by the notations that were introduced to represent the types of US= PKCs, US= PMs and their legs: R̂ denote R joints within a US= PKC, US= PM or a leg whose axes intersect at one common point, and Ř denote R joints within a US= PKC, US= PM or a leg whose axes intersect at another common point. $()_L$ denotes a coaxial or codirectional compositional unit.

For example, by removing the virtual chain in an ŘŘR̂R̂R̂V KC [Fig. 11.2a], an ŘŘR̂R̂R̂ leg [Fig. 11.3a] can be obtained. The leg-wrench system of the ŘŘR̂R̂R̂ leg [Fig. 11.3a] is a 1-$\boldsymbol{\zeta}_0$-system. The basis wrench can be selected as $\boldsymbol{\zeta}_0$ whose axis passes through the intersection of the axes of the Ř joints and the intersection of the axes of the R̂ joints.

All the legs for US= PMs obtained are listed in Table 11.1. For legs with $c^i = 0$, we are interested in two classes:

1. Legs with simple structures: RUS, PUS and UPS legs[119].
2. Legs obtained by adding an R joint or a P joint to each of the legs with $c^i = 1$. For example, the RŘŘŘR̂R̂ leg.

All the types of legs for US= PKCs obtained are listed in Table 11.1. The legs used to synthesize US= PKCs need to satisfy not only the geometric conditions within a same leg, but also the geometric conditions among different legs.

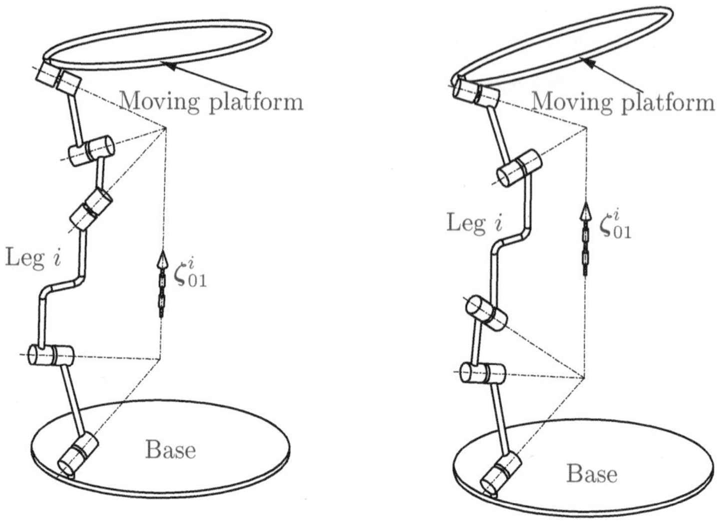

(a) ŘŘR̂R̂R̂. (b) ŘŘŘR̂R̂.

Fig. 11.3. Some legs for US= PKCs

Table 11.1. Legs for US= PKCs

c^i	No.	Type
1	1	ŘŘR̂R̂R̂
	2	ŘŘŘR̂R̂
0	omitted	omitted

11.7 Step 3: Assembly of Legs

US= PKCs can be generated by assembling a set of legs for US= PKCs shown in Table 11.1 selected according to the combinations of the leg-wrench systems shown in Table 5.4. In assembling US= PKCs, the following condition should be met: the linear combination of the leg-wrench systems forms a 1-ζ_0-system (see Condition (2) for US= PKCs in Sect. 11.3).

Since US= PKCs have at least one leg with a 1-ζ_0-system, condition (2) is automatically guaranteed. For example, a 3-legged 3-ŘŘR̂R̂R̂ US= PKC [Fig. 11.4a] can be obtained by assembling three ŘŘR̂R̂R̂ legs [Fig. 11.3a]. The wrench system of the PKC, which is the linear combination of the leg-wrench systems, is still a 1-ζ_0-system. The basis wrench ζ_{01} of the wrench system of the PKC is the same as the basis wrench ζ_{01}^{i} of any of its leg-wrench systems. Similarly, by assembling three ŘŘŘR̂R̂ legs [Fig. 11.3b], a 3-legged 3-ŘŘŘR̂R̂ US= PKC [Fig. 11.4b] can be obtained. The wrench system of the PKC, which

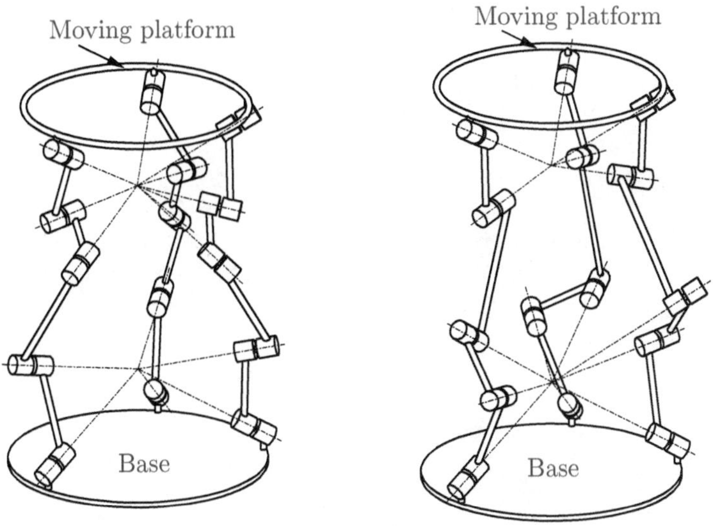

(a) 3-ŘŘR̂R̂R̂. (b) 3-ŘŘŘR̂R̂.

Fig. 11.4. Some US= PKCs

is the linear combination of the leg-wrench systems, is also a 1-$\boldsymbol{\zeta}_0$-system. The basis wrench $\boldsymbol{\zeta}_{01}$ of the wrench system of the PKC is the same as the basis wrench $\boldsymbol{\zeta}_{01}^i$ of any of its leg-wrench systems.

In addition to the types of US= PKCs proposed in the literature, new types, i.e., the types with at least two legs each having a 1-$\boldsymbol{\zeta}_0$-system or the types with at least one leg having an inactive joint, are also obtained.

In comparison with PKCs generating other motion patterns, the number of types of US= PKCs is small. For brevity, only the 5-legged US= PKCs in which each leg has a 1-$\boldsymbol{\zeta}_0$-system are listed in Table 11.2. The number of overconstraints of each of the 5-legged US= PKCs shown in Table 11.2 is 4. Among the six types of US= PKCs, there are two types (Nos. 1 and 6) in which all five legs are of the same type.

Table 11.2. Five-legged US= PKCs

No.	Type
1	5-ŘŘR̂R̂R̂
2	4-ŘŘR̂R̂R̂-ŘŘŘR̂R̂
3	3-ŘŘR̂R̂R̂-2-ŘŘŘR̂R̂
4	2-ŘŘR̂R̂R̂-3-ŘŘŘR̂R̂
5	ŘŘR̂R̂R̂-4-ŘŘŘR̂R̂
6	5-ŘŘŘR̂R̂

11.8 Step 4: Selection of Actuated Joints

11.8.1 t-Component of the Actuation Wrenches

Considering that the order of a screw system is coordinate system free and for simplicity reasons, all the actuation wrenches and wrenches are expressed in a coordinate system with its X-axis coaxial with the axis of the wrench ζ_{01} of the US= PM. The basis wrench of the US= PM can be expressed as $\zeta_{01} = \{1\ \mathbf{0}_{1\times5}\}^T$. The last five scalar components of the basis wrench of $\mathcal{W}$ are zero. Thus, the t-component, $\zeta^i_{t\not{D}j}$, of the $\zeta^i_{\not{D}j}$ is composed of the last five scalar components of the $\zeta^i_{\not{D}j}$.

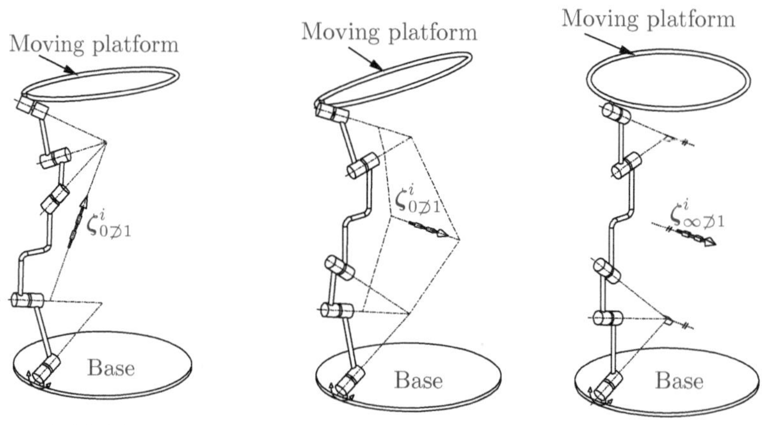

(a) $\underline{\check{R}}\check{R}\hat{R}\hat{R}\hat{R}$ leg.

(b) $\underline{\check{R}}\check{R}\check{R}\hat{R}\hat{R}$ leg.

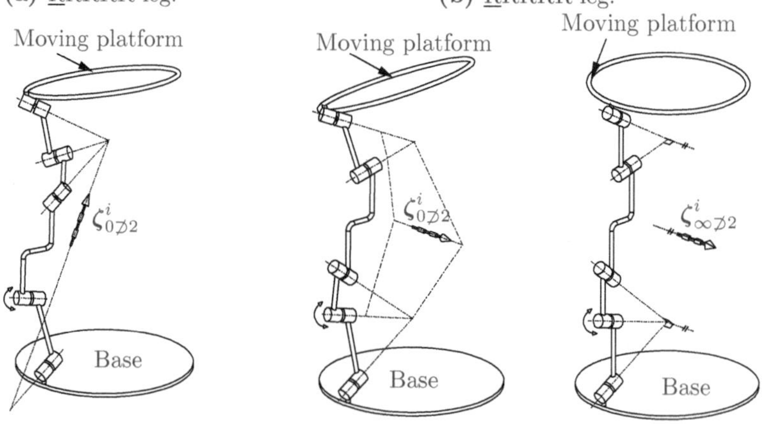

(c) $\check{R}\underline{\check{R}}\hat{R}\hat{R}\hat{R}$ leg.

(d) $\check{R}\underline{\check{R}}\check{R}\hat{R}\hat{R}$ leg.

Fig. 11.5. Actuation wrenches of some legs for US= PKCs

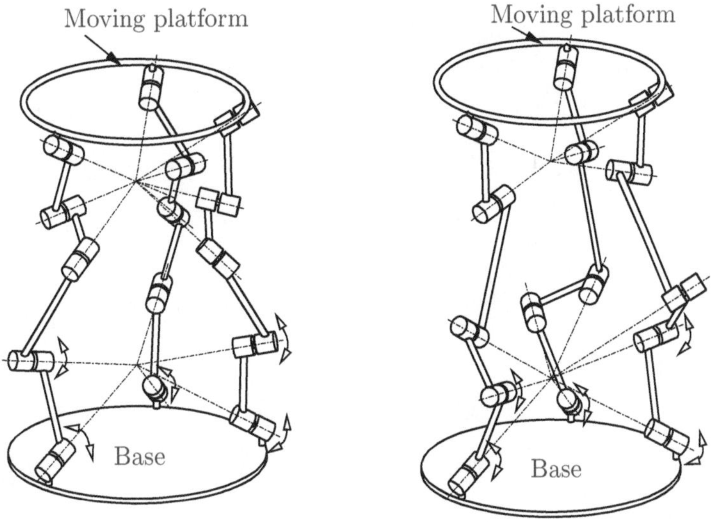

(a) $2\text{-}\underline{\check{R}\check{R}}\hat{R}\hat{R}\hat{R}\text{-}\underline{\check{R}}\check{R}\hat{R}\hat{R}\hat{R}$ candidate US= PM (invalid).

(b) $2\text{-}\underline{\check{R}\check{R}}\check{R}\hat{R}\hat{R}\text{-}\underline{\check{R}}\check{R}\check{R}\hat{R}\hat{R}$ candidate US= PM (valid).

Fig. 11.6. Selection of actuated joints for some US= PKCs

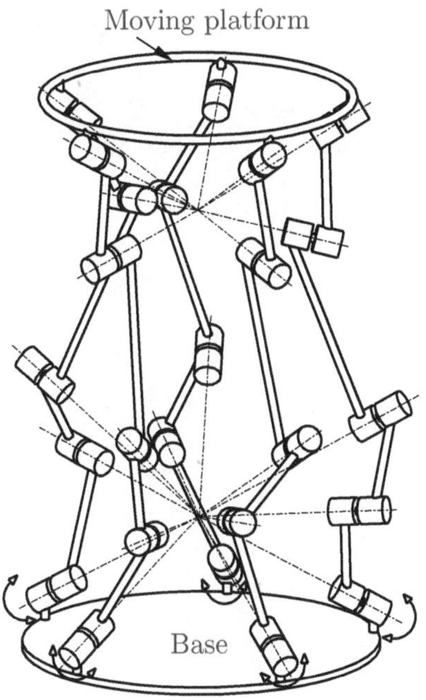

Fig. 11.7. A $5\text{-}\underline{\check{R}}\check{R}\check{R}\hat{R}\hat{R}$ US= PM

Figure 11.5 shows the actuation wrenches of actuated joints in some legs for US= PMs. The wrench systems of these legs are both 1-ζ_0-systems (Fig. 11.3). In the $\underline{\check{\mathrm{R}}}\check{\mathrm{R}}\hat{\mathrm{R}}\hat{\mathrm{R}}\hat{\mathrm{R}}$ leg [Fig. 11.5a], the first R joint is actuated. $\zeta^{i}_{\not{D}1}$ can be chosen as any ζ_0 whose axis passes through the intersection of the axes of the $\hat{\mathrm{R}}$ joints and intersects the axis of the unactuated $\check{\mathrm{R}}$ joint and does not intersect the axis of the actuated $\check{\mathrm{R}}$ joint. In the $\underline{\check{\mathrm{R}}}\check{\mathrm{R}}\check{\mathrm{R}}\hat{\mathrm{R}}\hat{\mathrm{R}}$ leg [Fig. 11.5b], the first R joint is actuated. $\zeta^{i}_{\not{D}1}$ can be chosen as (a) any ζ_0 whose axis is the intersection of the plane defined by the axes of two unactuated $\check{\mathrm{R}}$ joints with the plane defined by the axes of the two $\hat{\mathrm{R}}$ joints if the above two planes are not parallel to each other or (b) any ζ_∞ whose axis is perpendicular to the above two planes if they are parallel to each other.

11.8.2 Procedure for the Validity Detection of Actuated Joints

The validity detection of actuated joints of US= PMs can thus be performed using the following steps.

Step 4a. If one or more of the actuated joints of a candidate US= PM are inactive, the set of actuated joints is invalid and the candidate US= PM should be discarded.

Step 4b. If the t-components of the actuation wrenches of actuated joints $\zeta^{i}_{\not{D}j}$ are linearly dependent in a general configuration for a candidate US= PM, the set of actuated joints is invalid. In this case, the candidate US= PM should be discarded.

For example, the candidate 3-legged US= PM corresponding to the 3-$\check{\mathrm{R}}\check{\mathrm{R}}\hat{\mathrm{R}}\hat{\mathrm{R}}\hat{\mathrm{R}}$ US= PKC [Fig. 11.4a] is the 2-$\underline{\check{\mathrm{R}}}\underline{\check{\mathrm{R}}}\hat{\mathrm{R}}\hat{\mathrm{R}}\hat{\mathrm{R}}$-$\underline{\check{\mathrm{R}}}\check{\mathrm{R}}\hat{\mathrm{R}}\hat{\mathrm{R}}\hat{\mathrm{R}}$ US= PM [Fig. 11.6a]. All the axes of the actuation wrenches pass through the intersection of the axes of the $\hat{\mathrm{R}}$ joints [Fig. 11.5a and Fig. 11.5c]. We have

$$\left|\zeta^{1}_{t,\not{D}1}\ \zeta^{1}_{t,\not{D}2}\ \zeta^{2}_{t,\not{D}1}\ \zeta^{2}_{t,\not{D}2}\ \zeta^{3}_{t,\not{D}1}\right| = \begin{vmatrix} * & * & * & * & * \\ * & * & * & * & * \\ 0 & 0 & 0 & 0 & 0 \\ 0 & 0 & 0 & 0 & 0 \\ 0 & 0 & 0 & 0 & 0 \end{vmatrix} = 0.$$

where $*$ denotes an arbitrary number.

According to the validity condition of actuated joints, the set of actuated joints in invalid. Thus, the candidate 2-$\underline{\check{\mathrm{R}}}\underline{\check{\mathrm{R}}}\hat{\mathrm{R}}\hat{\mathrm{R}}\hat{\mathrm{R}}$-$\underline{\check{\mathrm{R}}}\check{\mathrm{R}}\hat{\mathrm{R}}\hat{\mathrm{R}}\hat{\mathrm{R}}$ US= PM should be discarded.

The candidate 3-legged US= PM corresponding to the 3-$\check{\mathrm{R}}\check{\mathrm{R}}\check{\mathrm{R}}\hat{\mathrm{R}}\hat{\mathrm{R}}$ US= PKC [Fig. 11.4b] is the 2-$\underline{\check{\mathrm{R}}}\underline{\check{\mathrm{R}}}\check{\mathrm{R}}\hat{\mathrm{R}}\hat{\mathrm{R}}$-$\underline{\check{\mathrm{R}}}\check{\mathrm{R}}\check{\mathrm{R}}\hat{\mathrm{R}}\hat{\mathrm{R}}$ US= PM [Fig. 11.6b]. All the actuation wrenches are shown in Fig. 11.5b and Fig. 11.5d. Following the procedure for the validity detection of the actuated joints, it can be proved that the set of actuated joints is valid.

Following the criteria on the selection of actuated joints (Sect. 5.7) and the procedure for the detection of the validity of actuated joints, all the $m(m \geq 2)$-legged US= PMs corresponding to each US= PKC can be generated. It is found that there is only one type of 5-legged US= PM, $\underline{\check{R}}\check{R}\check{R}\hat{R}\hat{R}$ US= PM (Fig. 11.7), with identical type (including the identical arrangement of actuated joints) of legs which are composed of R and P joints and which satisfy the above criteria.

11.9 Summary

In this chapter, the type synthesis of US= PMs has been thoroughly solved using the general type synthesis approach proposed in Chap. 5. US= PKCs with inactive joints as well as US= PKCs without inactive joints have been obtained. The validity condition of actuated joints of US= PMs has been reduced to the calculation of a 5×5 determinant. The US= PKCs obtained include some new US= PKCs as well as all the known US= PKCs. Some new US= PMs have also been proposed.

Variations of the US= PKCs can be obtained by (a) substituting a combination of one R joint and one P joint with parallel axes, a combination of two R joints with intersecting axes and a combination of three R joints with concurrent axes with a C joint or its RH, PH, HH form, U, and S joint respectively, and/or (b) replacing an inactive R or P joint with an inactive H joint.

12. Five-DOF PPPU= Parallel Mechanisms

In this chapter, the type synthesis of 5-DOF PPPU= PMs (also called 3T2R PMs in the literature) is dealt with using the general approach proposed in Chap. 5. A PPPU= PM is the parallel counterpart of the serial PPPU robot and generates 5-DOF PPPU motion. It covers a wide range of applications including, among others, 5-axis machine tools in which an axis-symmetric tool must be positioned and oriented in order to perform the machining of complex surfaces. The four steps of the type synthesis of PPPU= PMs are presented in detail.

12.1 Introduction

The end-effector of a PPPU robot generates PPPU motion (also called 3T2R motion) in which the moving platform can rotate about a U joint whose centre can translate arbitrarily.

In addition to the US= PMs, PPPU= PMs are also 5-DOF PMs of great application potential. PPPU= PMs are the parallel counterparts of the PPPU serial robots, which can be found, for instance, in 5-axis machine tools. In a PPPU= PM, the moving platform generates PPPU motions which are controlled by five actuated joints distributed in different legs.

Fruitful results on the type synthesis of PPPU= PMs have been published since 2001 [31, 35, 55, 62]. In fact, most of the work reported on the systematic type synthesis of PPPU= PMs [31, 35, 55, 62] deals mainly with the systematic type synthesis of PPPU= PKCs. In other words, the selection of actuated joints for 5-DOF PMs has not been dealt with systematically. In [31], the term actuation singularity is used to refer to the situation in which the set of actuated joints is invalid for a PM.

Using the general approach to the type synthesis of PMs proposed in Chap. 5, the type synthesis of PPPU= PMs is dealt with in this chapter. Four steps for the type synthesis of PPPU= PMs, i.e., (1) the decomposition of wrench systems of PPPU= PKCs, (2) the type synthesis of legs for PPPU= PKCs, (3) the assembly of legs to generate PPPU= PKCs, and (4) the selection of actuated joints for PPPU= PMs, are dealt with in Sects. 12.5, 12.6, 12.7, and 12.8, respectively.

X. Kong and C. Gosselin: Type Syn. of Parallel Mech., STAR 33, pp. 185–198, 2007.
springerlink.com

12.2 Wrench System of a PPPU= PKC

In any general configuration, the twist system of a PPPU= PKC or its PPPU virtual chain is a 3-$\boldsymbol{\xi}_\infty$-2-$\boldsymbol{\xi}_0$-system. Since the virtual power developed by any $\boldsymbol{\zeta}_\infty$ whose direction is perpendicular to the axes of the R joints within the PPPU virtual chain along any twist within the 3-$\boldsymbol{\xi}_\infty$-2-$\boldsymbol{\xi}_0$-system is 0, the wrench system of the PPPU= PKC or its virtual chain is thus a 1-$\boldsymbol{\zeta}_\infty$-system (Fig. 12.1) in which the direction of the basis $\boldsymbol{\zeta}_{\infty 1}$ is perpendicular to the axes of the R joints within the PPPU virtual chain.

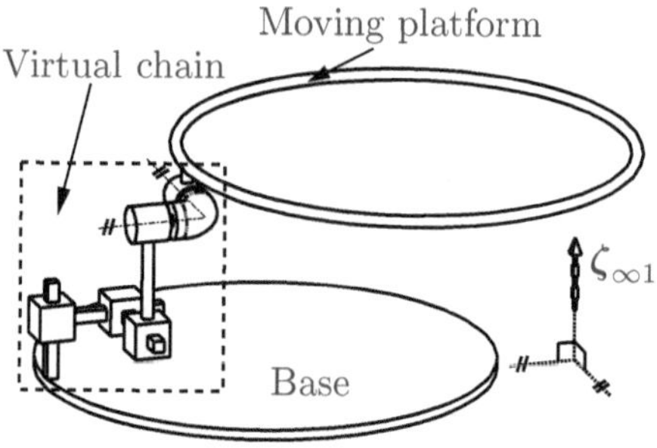

Fig. 12.1. Wrench system of a PPPU= PKC

12.3 Conditions for a PKC to Be a PPPU= PKC

When we connect the base and the moving platform of a PPPU= PKC by an appropriate PPPU virtual chain, the function of the PKC is not affected. Any of its legs and the PPPU virtual chain will form a 5-DOF single loop KC. When the order of the leg-wrench system is greater than 0, the single-loop KC formed must be an overconstrained KC.

Based on the concept of PPPU virtual chain, it follows that a PKC is a PPPU= PKC if it satisfies the two conditions below:

(1) Each leg of the PKC and a same PPPU virtual chain form a 5-DOF single-loop KC.
(2) The wrench system of the PKC is the same as that of the PPPU virtual chain, i.e., a 1-$\boldsymbol{\zeta}_\infty$-system, in any one general configuration.

12.4 Procedure for the Type Synthesis of PPPU= PMs

A general procedure can thus be proposed for the type synthesis of PPPU= PKCs as follows:

Step 1. Decomposition of the wrench system of 5-DOF PPPU= PKCs.
Step 2. Type synthesis of legs for PPPU= PKCs.
Here, a leg for PPPU= PKCs refers to a leg satisfying Condition (1) for PPPU= PKCs.

Step 3. Assembly of legs for PPPU= PKCs.

PPPU= PKCs can be generated by assembling two or more legs for PPPU= PKCs, obtained in Step 1, such that Condition (2) for PPPU= PKCs is satisfied.

Step 4. Selection of actuated joints.

PPPU= PMs can be generated by selecting actuated joints in different ways for each PPPU= PKC (Sect. 3.1), obtained in Step 3.

The procedure proposed above will be elaborated upon in the following sections.

12.5 Step 1: Decomposition of the Wrench System of PPPU= PKCs

The decomposition of the wrench system of an m-legged PPPU= PKC consists in finding all the leg-wrench systems and all the combinations of m leg-wrench systems for the PPPU= PKC and a specified number of overconstraints Δ. Step 1 can be performed using (2.24) and (2.28).

12.5.1 Determination of the Leg-Wrench Systems

As the wrench system of a PKC is the linear combination of the leg-wrench systems in a general configuration, any leg-wrench system in a V= PKC is a sub-system of its wrench system. Following the procedure proposed in Sect. 2.1.3, all the leg-wrench systems for a V= PKC can be identified without difficulty.

In any general configuration, the wrench system of a PPPU= PKC is the same as that of its PPPU virtual chain, i.e., a 1-$\boldsymbol{\zeta}_\infty$-system, the direction of whose basis $\boldsymbol{\zeta}_\infty$ is perpendicular to the axes of the R joints within the PPPU virtual chain (Fig. 12.1). It then follows that any leg-wrench system with order $c^i > 0$ of a PPPU= PKC is a 1-$\boldsymbol{\zeta}_\infty$-system in a general configuration (Fig. 12.1).

12.5.2 Determination of the Combinations of Leg-Wrench Systems

For the PPPU motion pattern considered, all the wrenches in the wrench systems are of the same pitch. The leg-wrench systems are $c^i(0 \leq c^i \leq 1)$-systems of ∞ pitch. The combination of leg-wrench systems can be simply represented by the combination of the orders, c^i, of leg-wrench systems.

The combinations of the orders, c^i, of leg-wrench systems have been determined by solving (2.28) and are shown in Table 5.4.

12.6 Step 2: Type Synthesis of Legs

In this section, the type synthesis of legs for PPPU= PKCs will be performed following the two substeps given in Sect. 5.5.

12.6.1 Step 2a: Type Synthesis of 5-DOF Single-Loop KCs That Involve a PPPU Virtual Chain and Have a Specified Leg-Wrench System

The type synthesis of 5-DOF single-loop KCs that involve a PPPU virtual chain and have a specified leg-wrench system can be performed as follows.

Firstly, perform the type synthesis of single-loop KCs that involve a PPPU virtual chain and have a specified leg-wrench system (See Chap. 3). Secondly, discard those single-loop KCs in which the twists of all the joints but the PPPU virtual chain are linearly dependent.

Cases with a 1-ζ_∞-System

Firstly, perform the type synthesis of 5-DOF single-loop KCs that involve a PPPU virtual chain and have a 1-ζ_∞-system. According to Chap. 3, there are ten joints, including at least two R joints, in a 5-DOF single-loop KC that involves a PPPU virtual chain and has a 1-ζ_∞-system. Such a single-loop KC is composed of (a) two spatial parallelaxis compositional units [Fig. 12.2c, 12.2d, 12.2i, 12.2g, and 12.2h] or (b) one spatial parallelaxis compositional unit and one planar compositional unit [Fig. 12.2a, 12.2b, 12.2e, and 12.2f]. The 1-ζ_∞-system of the 5-DOF single-loop KCs that involve a PPPU virtual chain and have a 1-ζ_∞-system is the ζ_∞ whose direction is perpendicular to the axes of all the R joints.

Secondly, discard those single-loop KCs in which the twists of all the joints but the virtual chain are linearly dependent. The R̋R̄R̄R̄R̄V̋ [Fig. 12.2a], R̋R̋R̋R̋R̄V̋ [Fig. 12.2b], PR̄R̄R̄R̄V̋ [Fig. 12.2e], and PR̋R̋R̋R̋V̋ [Fig. 12.2f] 5-DOF single-loop KCs should be discarded since the twists of the four R̄ joints or the four R̋ joints are linearly dependent.

In the representation of types of 5-DOF single-loop KCs involving a PPPU virtual chain, R̋ denote R joints whose axes are parallel to the axis of the first R joint of the PPPU virtual chain, and R̄ denote R joints whose axes are parallel to the axis of the second R joint of the PPPU virtual chain.

It is pointed out that in a 5-DOF single-loop KC, the P joints can be placed anywhere. For brevity, we list only the 5-DOF single-loop KCs from which all 5-DOF single-loop KCs can be obtained through the above operations.

$c^i = 0$

Theoretically, any six R and P joints whose twists are linearly independent together with a PPPU virtual chain form a 5-DOF single-loop KC.

It is noted that starting from each of the above 5-DOF single-loop KCs involving a PPPU virtual chain and having a 1-ζ_∞-system, one can obtain a class of 5-DOF single-loop KCs involving a PPPU virtual chain and having a 0-ζ-system by the application of an $(R)_L$ coaxial compositional unit. It is pointed out that the $(R)_L$ joint is inactive and can thus be placed anywhere in the single-loop KC. For example, the $(R)_L$R̋R̄R̄R̄R̄V̋ single-loop KC can be obtained by changing the position of the $(R)_L$ joint in R̋R̄R̄R̄R̄$(R)_L$V̋ single-loop KC.

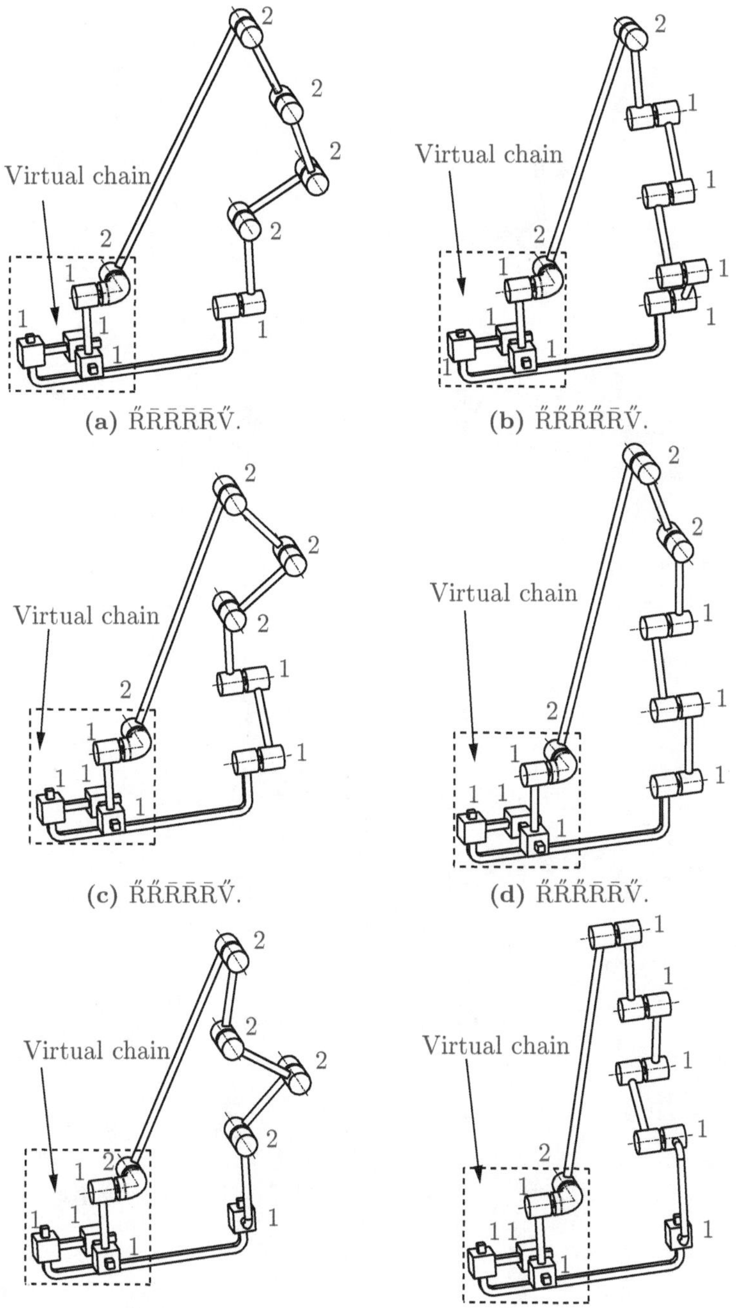

(a) R̋R̄R̄R̄R̄V̋.

(b) R̋R̋R̋R̋R̄V̋.

(c) R̋R̋R̄R̄R̄V̋.

(d) R̋R̋R̋R̄R̄V̋.

(e) PR̄R̄R̄R̄V̋.

(f) PR̋R̋R̋R̋V̋.

Fig. 12.2. Some 5-DOF single-loop KCs involving a PPPU virtual chain ($c^i = 1$)

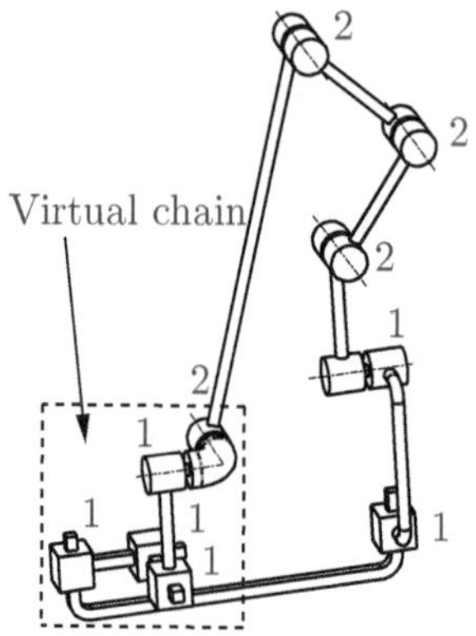

(g) $P\overset{\prime\prime}{R}\bar{R}\bar{R}\bar{R}\overset{\prime\prime}{V}$.

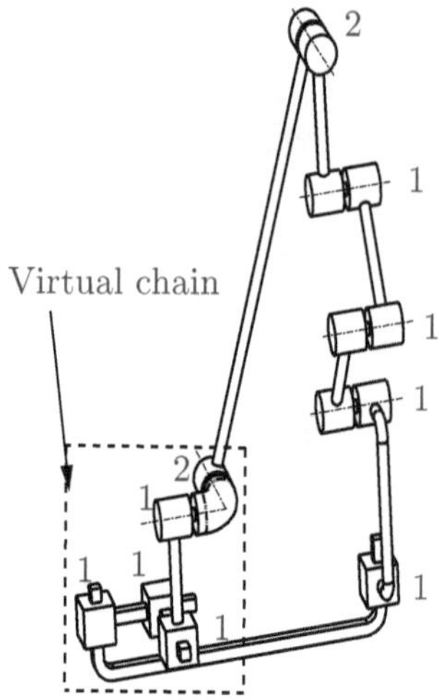

(h) $P\overset{\prime\prime}{R}\overset{\prime\prime}{R}\overset{\prime\prime}{R}\bar{R}\overset{\prime\prime}{V}$.

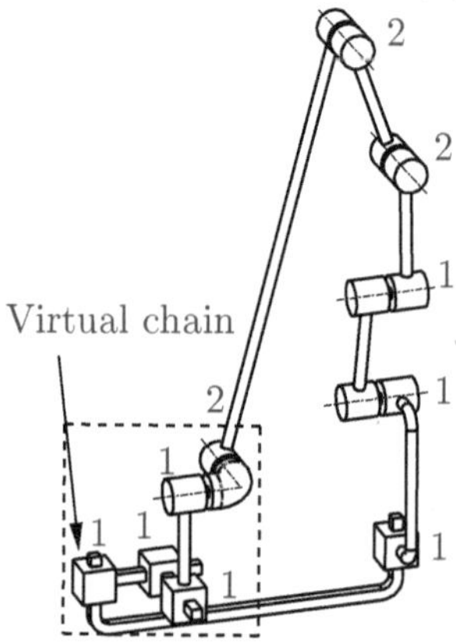

(i) $P\overset{\prime\prime}{R}\overset{\prime\prime}{R}\bar{R}\bar{R}\overset{\prime\prime}{V}$.

Fig. 12.2. (*continued*)

12.6.2 Step 2b: Generation of Types of Legs

According to Condition (1) for the existence of PPPU= PKCs (Sect. 12.3), the types of legs can be readily obtained from the 5-DOF single-loop KCs obtained in Step 2a, Sect. 12.6.1, by removing the PPPU virtual chain.

The specific geometric conditions which guarantee the satisfaction of Condition (1) for PPPU= PKCs are clearly indicated by the notations that were introduced to represent the types of PPPU= PKCs, PPPU= PMs and their legs, $\overset{\prime\prime}{R}$ denote R joints whose axes are all parallel, $\bar{R}$ denote R joints whose axes are also all parallel and not parallel to the axes of the $\overset{\prime\prime}{R}$ joints.

Table 12.1. Legs for PPPU= PKCs

c^i	Class	No.	Type
1	5R	1	$\overset{\prime\prime}{\mathrm{R}}\overset{\prime\prime}{\mathrm{R}}\bar{\mathrm{R}}\bar{\mathrm{R}}\bar{\mathrm{R}}$
		2	$\overset{\prime\prime}{\mathrm{R}}\overset{\prime\prime}{\mathrm{R}}\overset{\prime\prime}{\mathrm{R}}\bar{\mathrm{R}}\bar{\mathrm{R}}$
	4R1P	3	$\overset{\prime\prime}{\mathrm{R}}\overset{\prime\prime}{\mathrm{R}}\overset{\prime\prime}{\mathrm{R}}\bar{\mathrm{R}}\mathrm{P}$
		4	$\overset{\prime\prime}{\mathrm{R}}\overset{\prime\prime}{\mathrm{R}}\overset{\prime\prime}{\mathrm{R}}\mathrm{P}\bar{\mathrm{R}}$
		5	$\overset{\prime\prime}{\mathrm{R}}\overset{\prime\prime}{\mathrm{R}}\mathrm{P}\overset{\prime\prime}{\mathrm{R}}\bar{\mathrm{R}}$
		6	$\overset{\prime\prime}{\mathrm{R}}\mathrm{P}\overset{\prime\prime}{\mathrm{R}}\overset{\prime\prime}{\mathrm{R}}\bar{\mathrm{R}}$
		7	$\mathrm{P}\overset{\prime\prime}{\mathrm{R}}\overset{\prime\prime}{\mathrm{R}}\overset{\prime\prime}{\mathrm{R}}\bar{\mathrm{R}}$
	3R2P	8	$\overset{\prime\prime}{\mathrm{R}}\overset{\prime\prime}{\mathrm{R}}\bar{\mathrm{R}}\mathrm{P}\mathrm{P}$
		9	$\overset{\prime\prime}{\mathrm{R}}\overset{\prime\prime}{\mathrm{R}}\mathrm{P}\bar{\mathrm{R}}\mathrm{P}$
		10	$\overset{\prime\prime}{\mathrm{R}}\mathrm{P}\overset{\prime\prime}{\mathrm{R}}\bar{\mathrm{R}}\mathrm{P}$
		11	$\mathrm{P}\overset{\prime\prime}{\mathrm{R}}\overset{\prime\prime}{\mathrm{R}}\bar{\mathrm{R}}\mathrm{P}$
		12	$\overset{\prime\prime}{\mathrm{R}}\overset{\prime\prime}{\mathrm{R}}\mathrm{P}\mathrm{P}\bar{\mathrm{R}}$
		13	$\overset{\prime\prime}{\mathrm{R}}\mathrm{P}\overset{\prime\prime}{\mathrm{R}}\mathrm{P}\bar{\mathrm{R}}$
		14	$\mathrm{P}\overset{\prime\prime}{\mathrm{R}}\overset{\prime\prime}{\mathrm{R}}\mathrm{P}\bar{\mathrm{R}}$
		15	$\overset{\prime\prime}{\mathrm{R}}\mathrm{P}\mathrm{P}\overset{\prime\prime}{\mathrm{R}}\bar{\mathrm{R}}$
		16	$\mathrm{P}\overset{\prime\prime}{\mathrm{R}}\mathrm{P}\overset{\prime\prime}{\mathrm{R}}\bar{\mathrm{R}}$
		17	$\mathrm{P}\mathrm{P}\overset{\prime\prime}{\mathrm{R}}\overset{\prime\prime}{\mathrm{R}}\bar{\mathrm{R}}$
	2R3P	18	$\overset{\prime\prime}{\mathrm{R}}\bar{\mathrm{R}}\mathrm{P}\mathrm{P}\mathrm{P}$
		19	$\overset{\prime\prime}{\mathrm{R}}\mathrm{P}\bar{\mathrm{R}}\mathrm{P}\mathrm{P}$
		20	$\mathrm{P}\overset{\prime\prime}{\mathrm{R}}\bar{\mathrm{R}}\mathrm{P}\mathrm{P}$
		21	$\overset{\prime\prime}{\mathrm{R}}\mathrm{P}\mathrm{P}\bar{\mathrm{R}}\mathrm{P}$
		22	$\mathrm{P}\overset{\prime\prime}{\mathrm{R}}\mathrm{P}\bar{\mathrm{R}}\mathrm{P}$
		23	$\mathrm{P}\mathrm{P}\overset{\prime\prime}{\mathrm{R}}\bar{\mathrm{R}}\mathrm{P}$
		24	$\overset{\prime\prime}{\mathrm{R}}\mathrm{P}\mathrm{P}\mathrm{P}\bar{\mathrm{R}}$
		25	$\mathrm{P}\overset{\prime\prime}{\mathrm{R}}\mathrm{P}\mathrm{P}\bar{\mathrm{R}}$
		26	$\mathrm{P}\mathrm{P}\overset{\prime\prime}{\mathrm{R}}\mathrm{P}\bar{\mathrm{R}}$
		27	$\mathrm{P}\mathrm{P}\mathrm{P}\overset{\prime\prime}{\mathrm{R}}\bar{\mathrm{R}}$
0	omitted	omitted	omitted

For example, by removing the PPPU virtual chain in an $\overset{\prime\prime}{\mathrm{R}}\overset{\prime\prime}{\mathrm{R}}\overset{\prime\prime}{\mathrm{R}}\bar{\mathrm{R}}\bar{\mathrm{R}}\overset{\prime\prime}{\mathrm{V}}$ KC [Fig. 12.2d], an $\overset{\prime\prime}{\mathrm{R}}\overset{\prime\prime}{\mathrm{R}}\overset{\prime\prime}{\mathrm{R}}\bar{\mathrm{R}}\bar{\mathrm{R}}$ [Fig. 12.3b] leg can be obtained. In the $\overset{\prime\prime}{\mathrm{R}}\overset{\prime\prime}{\mathrm{R}}\overset{\prime\prime}{\mathrm{R}}\bar{\mathrm{R}}\bar{\mathrm{R}}$ leg shown in Fig. 12.3b, the axes of the first three R joints are parallel while the axes of the last two R joints are parallel. This leg has a 1-$\boldsymbol{\zeta}_\infty$-system. The axis of $\boldsymbol{\zeta}^i_{\infty 1}$, a basis of the 1-$\boldsymbol{\zeta}_\infty$-system, is perpendicular to the axes of all the R joints. In the PR̋R̋R̄R̄ leg shown in Fig. 12.3c, the axes of the first two R joints are parallel to each other while the axes of the last two R joints are also parallel to each other. This leg also has a 1-$\boldsymbol{\zeta}_\infty$-system. The axis of $\boldsymbol{\zeta}^i_{\infty 1}$, a basis of the 1-$\boldsymbol{\zeta}_\infty$-system, is perpendicular to the axes of all the R joints.

All the types of legs for PPPU= PKCs obtained are listed in Table 12.1. The legs used to synthesize PPPU= PKCs need to satisfy not only the geometric conditions within a same leg — which can be derived directly from the associated single-loop KCs — but also certain geometric conditions among different legs.

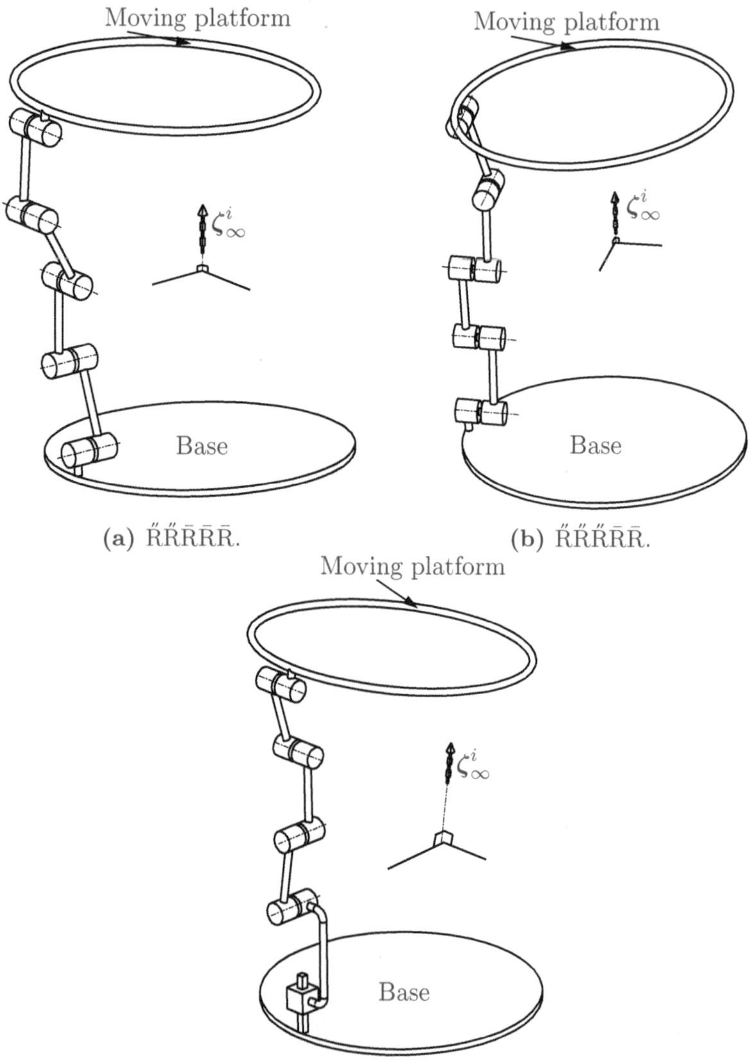

(a) R̋R̋R̄R̄R̄. **(b)** R̋R̋R̋R̄R̄. **(c)** PR̋R̋R̄R̄.

Fig. 12.3. Some legs for PPPU= PKCs

12.7 Step 3: Assembly of Legs

PPPU= PKCs can be generated by assembling a set of legs for PPPU= PKCs shown in Table 12.1 selected according to the combinations of the leg-wrench systems shown in Table 5.4. In assembling PKCs, the following condition should be met: the linear combination of the leg-wrench systems forms a 1-$\boldsymbol{\zeta}_\infty$-system (see Condition (2) for PPPU= PKCs in Sect. 12.3). This simply requires that the number of legs with a 1-$\boldsymbol{\zeta}_\infty$-system is greater than or equal to 1. It is noted that there is no constraint singularity for PPPU= PKCs.

Figure 12.4 shows two PPPU= PKCs. The 3-R̋R̋R̄R̄R̄ PPPU= PKC shown in Fig. 12.4a is composed of three R̋R̋R̄R̄R̄ legs shown in Fig. 12.3a. The 3-R̋R̋R̋R̄R̄ PPPU= PKC shown in Fig. 12.4b is composed of three R̋R̋R̋R̄R̄ legs shown in Fig. 12.3b.

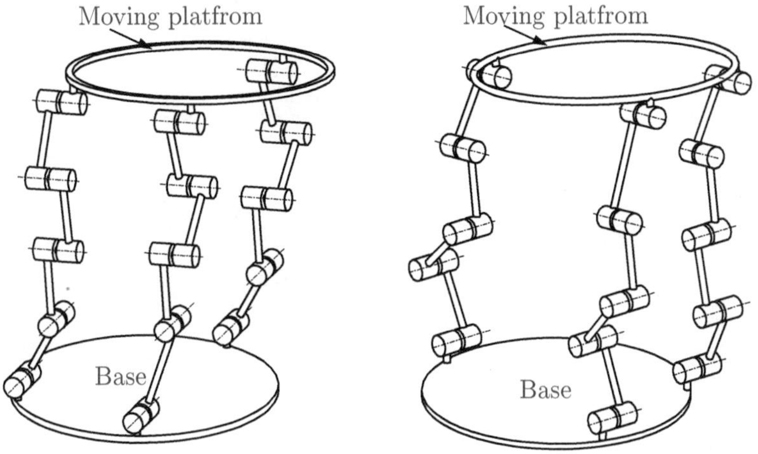

(a) 3-R̋R̋R̄R̄R̄. (b) 3-R̋R̋R̋R̄R̄.

Fig. 12.4. Some PPPU= PKC

Due to the large number of m-legged PPPU= PKCs, only the m-legged PPPU= PKCs with identical type of legs are listed in Table 12.2.

12.8 Step 4: Selection of Actuated Joints

12.8.1 t-Components of the Actuation Wrenches

Considering that the order of a screw system is coordinate system free and for simplicity reasons, all the actuation wrenches and wrenches are expressed in a coordinate system with its Z-axis perpendicular to all the axes of R joints within the PM. Let $\{0 \quad 0 \quad 0 \quad 0 \quad 0 \quad 1\}^T$ denote the basis wrench of $\mathcal{W}$. The first five scalar components of the basis wrench of $\mathcal{W}$ are zero. Thus, the t-component, $\boldsymbol{\zeta}^i_{t\not\mathcal{D}j}$, of the $\boldsymbol{\zeta}^i_{\not\mathcal{D}j}$ is composed of the first five scalar components of the $\boldsymbol{\zeta}^i_{\not\mathcal{D}j}$.

Table 12.2. m-legged PPPU= PKCs with identical type of legs

c^i	Class	No.	Type
1	5R	1	m-$\overset{\prime\prime}{\mathrm{R}}\overset{\prime\prime}{\mathrm{R}}\bar{\mathrm{R}}\bar{\mathrm{R}}\bar{\mathrm{R}}$
		2	m-$\overset{\prime\prime}{\mathrm{R}}\overset{\prime\prime}{\mathrm{R}}\overset{\prime\prime}{\mathrm{R}}\bar{\mathrm{R}}\bar{\mathrm{R}}$
	4R1P	3	m-$\overset{\prime\prime}{\mathrm{R}}\overset{\prime\prime}{\mathrm{R}}\overset{\prime\prime}{\mathrm{R}}\bar{\mathrm{R}}\mathrm{P}$
		4	m-$\overset{\prime\prime}{\mathrm{R}}\overset{\prime\prime}{\mathrm{R}}\overset{\prime\prime}{\mathrm{R}}\mathrm{P}\bar{\mathrm{R}}$
		5	m-$\overset{\prime\prime}{\mathrm{R}}\overset{\prime\prime}{\mathrm{R}}\mathrm{P}\overset{\prime\prime}{\mathrm{R}}\bar{\mathrm{R}}$
		6	m-$\overset{\prime\prime}{\mathrm{R}}\mathrm{P}\overset{\prime\prime}{\mathrm{R}}\overset{\prime\prime}{\mathrm{R}}\bar{\mathrm{R}}$
		7	m-$\mathrm{P}\overset{\prime\prime}{\mathrm{R}}\overset{\prime\prime}{\mathrm{R}}\overset{\prime\prime}{\mathrm{R}}\bar{\mathrm{R}}$
	3R2P	8	m-$\overset{\prime\prime}{\mathrm{R}}\overset{\prime\prime}{\mathrm{R}}\bar{\mathrm{R}}\mathrm{P}\mathrm{P}$
		9	m-$\overset{\prime\prime}{\mathrm{R}}\overset{\prime\prime}{\mathrm{R}}\mathrm{P}\bar{\mathrm{R}}\mathrm{P}$
		10	m-$\overset{\prime\prime}{\mathrm{R}}\mathrm{P}\overset{\prime\prime}{\mathrm{R}}\bar{\mathrm{R}}\mathrm{P}$
		11	m-$\mathrm{P}\overset{\prime\prime}{\mathrm{R}}\overset{\prime\prime}{\mathrm{R}}\bar{\mathrm{R}}\mathrm{P}$
		12	m-$\overset{\prime\prime}{\mathrm{R}}\overset{\prime\prime}{\mathrm{R}}\mathrm{P}\mathrm{P}\bar{\mathrm{R}}$
		13	m-$\overset{\prime\prime}{\mathrm{R}}\mathrm{P}\overset{\prime\prime}{\mathrm{R}}\mathrm{P}\bar{\mathrm{R}}$
		14	m-$\mathrm{P}\overset{\prime\prime}{\mathrm{R}}\overset{\prime\prime}{\mathrm{R}}\mathrm{P}\bar{\mathrm{R}}$
		15	m-$\overset{\prime\prime}{\mathrm{R}}\mathrm{P}\mathrm{P}\overset{\prime\prime}{\mathrm{R}}\bar{\mathrm{R}}$
		16	m-$\mathrm{P}\overset{\prime\prime}{\mathrm{R}}\mathrm{P}\overset{\prime\prime}{\mathrm{R}}\bar{\mathrm{R}}$
		17	m-$\mathrm{P}\mathrm{P}\overset{\prime\prime}{\mathrm{R}}\overset{\prime\prime}{\mathrm{R}}\bar{\mathrm{R}}$
	2R3P	18	m-$\overset{\prime\prime}{\mathrm{R}}\bar{\mathrm{R}}\mathrm{P}\mathrm{P}\mathrm{P}$
		19	m-$\overset{\prime\prime}{\mathrm{R}}\mathrm{P}\bar{\mathrm{R}}\mathrm{P}\mathrm{P}$
		20	m-$\mathrm{P}\overset{\prime\prime}{\mathrm{R}}\bar{\mathrm{R}}\mathrm{P}\mathrm{P}$
		21	m-$\overset{\prime\prime}{\mathrm{R}}\mathrm{P}\mathrm{P}\bar{\mathrm{R}}\mathrm{P}$
		22	m-$\mathrm{P}\overset{\prime\prime}{\mathrm{R}}\mathrm{P}\bar{\mathrm{R}}\mathrm{P}$
		23	m-$\mathrm{P}\mathrm{P}\overset{\prime\prime}{\mathrm{R}}\bar{\mathrm{R}}\mathrm{P}$
		24	m-$\overset{\prime\prime}{\mathrm{R}}\mathrm{P}\mathrm{P}\mathrm{P}\bar{\mathrm{R}}$
		25	m-$\mathrm{P}\overset{\prime\prime}{\mathrm{R}}\mathrm{P}\mathrm{P}\bar{\mathrm{R}}$
		26	m-$\mathrm{P}\mathrm{P}\overset{\prime\prime}{\mathrm{R}}\mathrm{P}\bar{\mathrm{R}}$
		27	m-$\mathrm{P}\mathrm{P}\mathrm{P}\overset{\prime\prime}{\mathrm{R}}\bar{\mathrm{R}}$

Figure 12.5 shows the actuation wrenches of actuated joints in some legs for PPPU= PMs. The wrench systems of these legs are a 1-$\boldsymbol{\zeta}_\infty$-system (Fig. 12.3). In the $\underline{\overset{\prime\prime}{\mathrm{R}}}\overset{\prime\prime}{\mathrm{R}}\bar{\mathrm{R}}\bar{\mathrm{R}}\bar{\mathrm{R}}$ leg [Fig. 12.5a], the first R joint is actuated. $\boldsymbol{\zeta}^i_{\not\supset 1}$ can be chosen as any $\boldsymbol{\zeta}_0$ whose axis intersects the axis of the second R joint and is parallel to the axes of the last three R joints. In the $\underline{\overset{\prime\prime}{\mathrm{R}}}\overset{\prime\prime}{\mathrm{R}}\overset{\prime\prime}{\mathrm{R}}\bar{\mathrm{R}}\bar{\mathrm{R}}$ leg [Fig. 12.5b], the first R joint is actuated. $\boldsymbol{\zeta}^i_{\not\supset 1}$ can be chosen as any $\boldsymbol{\zeta}_0$ whose axis is the intersection of

the plane defined by the axes of two unactuated $\overset{''}{\mathrm{R}}$ joints with the plane defined by the axes of the two $\bar{\mathrm{R}}$ joints. In the $\underline{\mathrm{P}}\overset{''}{\mathrm{R}}\overset{''}{\mathrm{R}}\bar{\mathrm{R}}\bar{\mathrm{R}}$ leg [Fig. 12.5c], the P joint is actuated. $\boldsymbol{\zeta}^i_{\not\partial 1}$ can also be chosen as any $\boldsymbol{\zeta}_0$ whose axis is the intersection of the plane defined by the axes of two unactuated $\overset{''}{\mathrm{R}}$ joints with the plane defined by the axes of the two $\bar{\mathrm{R}}$ joints.

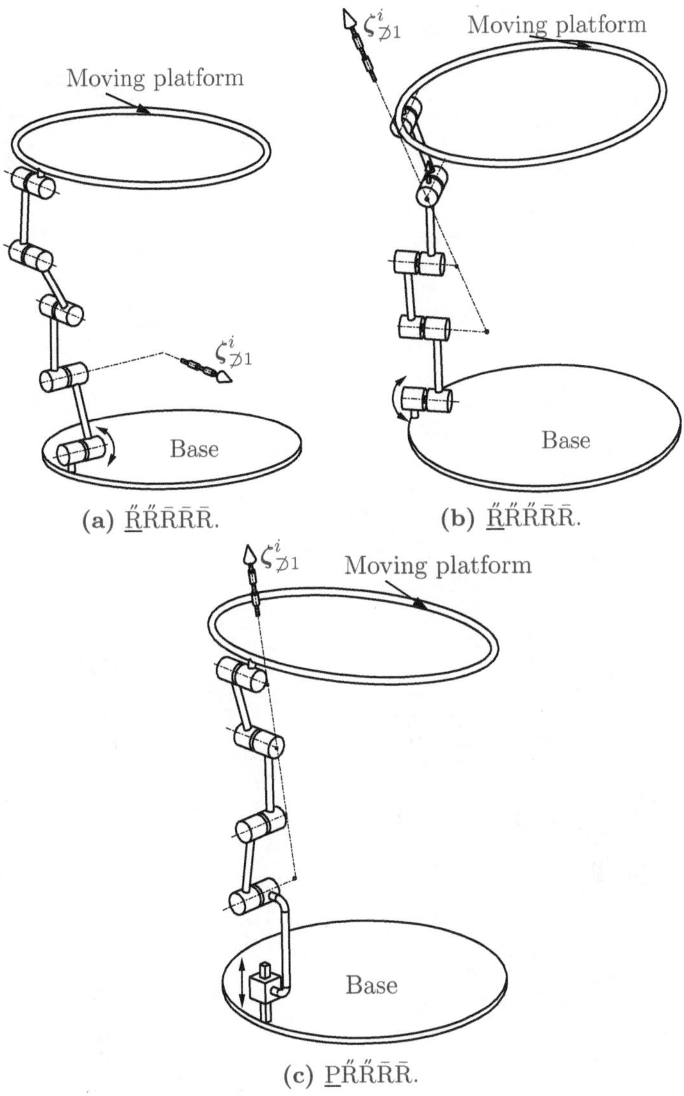

Fig. 12.5. Actuation wrenches of some legs for PPPU= PKCs

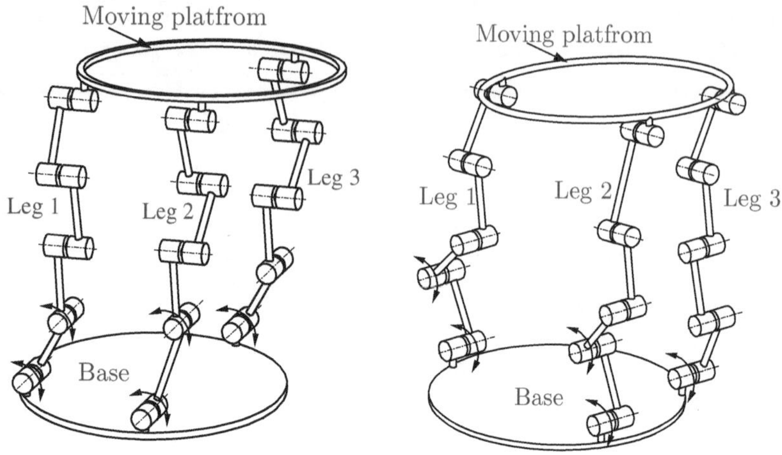

(a) $2\text{-}\underline{\overset{\prime\prime}{R}\overset{\prime\prime}{R}}\bar{R}\bar{R}\bar{R}\text{-}\underline{\overset{\prime\prime}{R}}\overset{\prime\prime}{R}\bar{R}\bar{R}\bar{R}$.
(b) $2\text{-}\underline{\overset{\prime\prime}{R}\overset{\prime\prime}{R}}\overset{\prime\prime}{R}\bar{R}\bar{R}\text{-}\underline{\overset{\prime\prime}{R}}\overset{\prime\prime}{R}\overset{\prime\prime}{R}\bar{R}\bar{R}$.

Fig. 12.6. Selection of actuated joints for some PPPU= PKC

Table 12.3. Five-legged PPPU= PMs with identical type of legs

c^i	Class	No.	Type
1	5R	1	$5\text{-}\underline{\overset{\prime\prime}{R}}\overset{\prime\prime}{R}\overset{\prime\prime}{R}\bar{R}\bar{R}$
	4R1P	2	$5\text{-}\underline{P}\overset{\prime\prime}{R}\overset{\prime\prime}{R}\bar{R}\bar{R}$

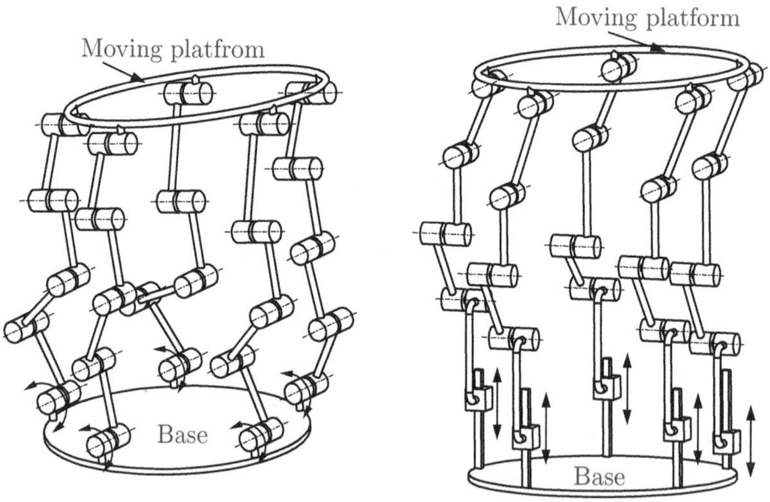

(a) $5\text{-}\underline{\overset{\prime\prime}{R}}\overset{\prime\prime}{R}\overset{\prime\prime}{R}\bar{R}\bar{R}$.
(b) $5\text{-}\underline{P}\overset{\prime\prime}{R}\overset{\prime\prime}{R}\bar{R}\bar{R}$.

Fig. 12.7. Two 5-legged PPPU= PMs with identical type of legs

12.8.2 Procedure for the Validity Detection of Actuated Joints

The validity detection of actuated joints of PPPU= PMs can be performed using the following steps.

Step 4a. If one or more of the actuated joints of a candidate PPPU= PM are inactive, the set of actuated joints is invalid and the candidate PPPU= PM should be discarded.

Step 4b. If the t-components of all the actuation wrenches of actuated joints $\boldsymbol{\zeta}^{i}_{\not\supset j}$ are linearly dependent in a general configuration for a candidate PPPU= PM, the set of actuated joints is invalid. In this case, the candidate PPPU= PM should be discarded.

For example, the candidate PPPU= PM corresponding to the 3-legged 3-R̋R̋R̄R̄R̄ PPPU= PKC [Fig. 12.4a] is the 2-$\underline{\text{R̋R̋}}$R̄R̄R̄-$\underline{\text{R̋}}$R̋R̄R̄R̄ [Fig. 12.6a]. For this candidate PM, the axes of all the actuation wrenches are perpendicular to the Z-axis [Fig. 12.5a]. We have

$$\left| \boldsymbol{\zeta}^{1}_{t,\not\supset 1}\ \boldsymbol{\zeta}^{1}_{t,\not\supset 2}\ \boldsymbol{\zeta}^{2}_{t,\not\supset 1}\ \boldsymbol{\zeta}^{2}_{t,\not\supset 2}\ \boldsymbol{\zeta}^{3}_{t,\not\supset 1} \right| = \begin{vmatrix} * & * & * & * & * \\ * & * & * & * & * \\ 0 & 0 & 0 & 0 & 0 \\ * & * & * & * & * \\ * & * & * & * & * \end{vmatrix} = 0.$$

where $*$ denotes an arbitrary number.

According to the validity condition of actuated joints, the set of actuated joints is invalid. Thus, the 2-$\underline{\text{R̋R̋}}$R̄R̄R̄-$\underline{\text{R̋}}$R̋R̄R̄R̄ [Fig. 12.6a] PM should be discarded.

The candidate PPPU= PM corresponding to the 3-legged 3-R̋R̋R̋R̄R̄ PPPU= PKC [Fig. 12.4b] is the 2-$\underline{\text{R̋R̋}}$R̋R̄R̄-$\underline{\text{R̋}}$R̋R̋R̄R̄ [Fig. 12.6b]. Following the procedure for the validity detection of actuated joints, it is found that the set of the actuated joints is valid.

Following the criteria on the selection of actuated joints (Sect. 5.7) and the procedure for the detection of the validity of actuated joints, all the $m(m \geq 2)$-legged PPPU= PMs corresponding to each PPPU= PKC can be generated.

To make the conditions for PPPU= PMs clear in their representation and due to the large number of PPPU= PMs, only 5-legged PPPU= PMs with identical type (including the identical arrangement of actuated joints) of legs which are composed of R and P joints and which satisfy the above criteria are listed in Table 12.3 and also shown in Fig. 12.7.

12.9 Summary

In this chapter, the type synthesis of PPPU= PMs has been solved using the general type synthesis approach proposed in Chap. 5. PPPU= PKCs with inactive joints have also been obtained. Both overconstrained and non-overconstrained PPPU= PKCs can be obtained. The validity check of actuated joints of PPPU= PMs has been reduced to the calculation of a 5×5 determinant. Some new PPPU= PMs have also been revealed.

By substituting a combination of an R joint and a P joint with parallel axes, a combination of two R joints with intersecting non-parallel axes and a combination of three R joints with concurrent axes with a C, U, and S joint respectively, PPPU= PMs involving C, U and S joints can be obtained. For a leg in which the axes of the R joints are parallel to two directions, no S joint can exist. For a leg with a 0-ζ_∞-system, any type of joints may be used. By replacing an R joint with a coaxial H (helical) joint, some PPPU= PMs involving H joints can be obtained. Other PPPU= PMs involving H joints can be obtained starting from the following 5-DOF single-loop KCs involving a PPPU virtual chain and one or more H joints following the above procedure: (1) 5-DOF single-loop KCs obtained from the $\overset{\prime\prime}{R}\bar{R}\bar{R}\bar{R}\bar{R}\overset{\prime\prime}{V}$ KC [Fig. 12.2a] or $P\bar{R}\bar{R}\bar{R}\bar{R}\overset{\prime\prime}{V}$ [Fig. 12.2e] KCs by replacing one or more $\bar{R}$ joints each with one coaxial $\bar{H}$ joint, and (2) 5-DOF single-loop KCs obtained from the $\overset{\prime\prime}{R}\overset{\prime\prime}{R}\overset{\prime\prime}{R}\overset{\prime\prime}{R}\bar{R}\overset{\prime\prime}{V}$ [Fig. 12.2b] or $P\overset{\prime\prime}{R}\overset{\prime\prime}{R}\overset{\prime\prime}{R}\bar{R}\overset{\prime\prime}{V}$ [Fig. 12.2f] KCs by replacing one or more $\overset{\prime\prime}{R}$ joints each with one coaxial $\overset{\prime\prime}{H}$ joint. Finally, by replacing a P joint with a planar parallelogram or a set of two consecutive P joints with a spatial parallelogram, variations of PPPU= PMs can be obtained.

13. Five-DOF PPS= Parallel Mechanisms

In this chapter, the type synthesis of 5-DOF PPS= PMs (also called 3R2T parallel mechanisms) is dealt with using the general approach proposed in Chap. 5. PPS= PMs are the parallel counterparts of the 5-DOF PPS serial manipulators, which are composed of two P joints and one S joint. The four steps of the type synthesis of PPS= PMs are presented in detail.

13.1 Introduction

The moving platform of a PPS= PM can rotate arbitrarily about a point moving along a plane. The type synthesis of PPS= PMs has been performed in [31, 41, 51, 53].

As pointed out in previous chapters, for a PKC with inactive joints and its kinematic equivalent PKC without inactive joints, the number of over-constraints as well as the reaction forces in the joints are different, although the inactive joints in the PKC make no contribution to the movement of the moving platform [26, 72]. However, no PPS= PKC with inactive joints has been proposed yet. Moreover, in [41, 53], the selection of actuated joints [1, 56, 67, 75, 78, 79, 80] has not been dealt with systematically. Therefore, the type synthesis of PPS= PMs needs further investigation.

Using the general approach proposed in Chap. 5, the type synthesis of PPS= PMs is dealt with in this chapter. The decomposition of the wrench system of a PPS= PKC is dealt with in Sect. 13.5. The type synthesis of legs for PPS= PKCs is performed in Sect. 13.6. In Sect. 13.7, the assembly of legs to generate PPS= PKCs is dealt with while the selection of actuated joints for PPS= PMs is discussed in Sect. 13.8.

13.2 Wrench System of a PPS= PKC

In any general configuration, the twist system of a PPS= PKC or its PPS virtual chain is a 2-$\boldsymbol{\xi}_\infty$-3-$\boldsymbol{\xi}_0$-system. It can be found without difficulty that its wrench system is a 1-$\boldsymbol{\zeta}_0$-system in which the basis is a $\boldsymbol{\zeta}_0$ whose axis passes through the centre of the S joint and is perpendicular to the directions of the P joints.

X. Kong and C. Gosselin: Type Syn. of Parallel Mech., STAR 33, pp. 199–211, 2007.
springerlink.com

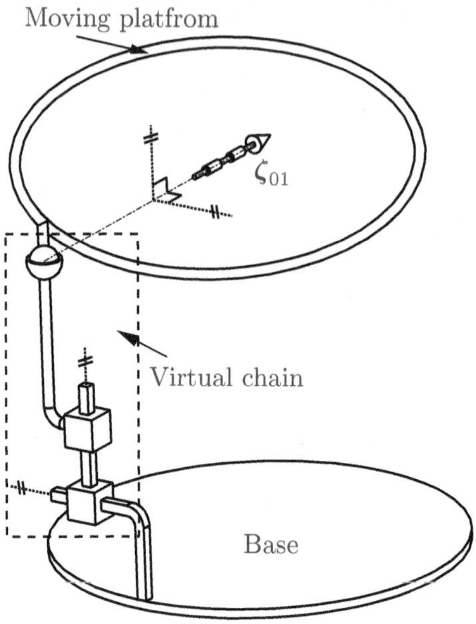

Fig. 13.1. Wrench system of a PPS virtual chain

13.3 Conditions for a PKC to Be a PPS= PKC

When we connect the base and the moving platform of a PPS= PKC by a PPS virtual chain, the function of the PKC is not affected. Any of its legs and the PPS virtual chain will form a 5-DOF single loop KC. When the order of the leg-wrench system is greater than 0, the single-loop KC formed must be an overconstrained KC.

Based on the concept of PPS virtual chain, it follows that a PKC is a PPS= PKC if it satisfies the two conditions below:

(1) Each leg of the PKC and a same PPS virtual chain form a 5-DOF single-loop KC.
(2) The wrench system of the PKC is the same as that of the PPS virtual chain, i.e., a 1-ζ_0-system, in any one general configuration.

13.4 Procedure for the Type Synthesis of PPS= PMs

A procedure can thus be proposed for the type synthesis of PPS= PKCs as follows:

Step 1. Decomposition of the wrench system of 5-DOF PPS= PKCs.
Step 2. Type synthesis of legs for PPS= PKCs.
Here, a leg for PPS= PKCs refers to a leg satisfying Condition (1) for PPS= PKCs.

Step 3. Assembly of legs for PPS= PKCs.
PPS= PKCs can be generated by taking two or more legs for PPS= PKCs, obtained in Step 1, such that Condition (2) for PPS= PKCs is satisfied.

Step 4. Selection of actuated joints.
PPS= PMs can be generated by selecting actuated joints in different ways for each PPS= PKC (Sect. 13.8), obtained in Step 3.

The procedure proposed above will be elaborated upon in the following sections.

13.5 Step 1: Decomposition of the Wrench System of PPS= PKCs

The decomposition of the wrench system of an m-legged PPS= PKC consists in finding all the leg-wrench systems and all the combinations of m leg-wrench systems for the PPS= PKC and a specified number of overconstraints Δ. Step 1 can be performed using (2.24) and (2.28).

13.5.1 Determination of the Leg-Wrench Systems

As the wrench system of a PKC is the linear combination of those of all its legs in a general configuration, any leg-wrench system in a PPS= PKC is a sub-system of its wrench system. Following the procedure proposed in Sect. 2.1.3, all the leg-wrench systems for a PPS= PKC can be identified without difficulty.

In any general configuration, the wrench system of a PPS= PKC is the same as that of its PPS virtual chain, i.e., a 1-ζ_0-system, in which the basis wrench is the ζ_0 whose axis passes through the centre of the S joint and is perpendicular to the directions of the two P joints within the PPS virtual chain. It then follows that any leg-wrench system with order $c^i > 0$ of a PPS= PKC is a 1-ζ_0-system in a general configuration (Fig. 13.1).

13.5.2 Determination of the Combinations of Leg-Wrench Systems

For the PPS motion pattern considered, all the wrenches in the wrench system are of the same pitch. The leg-wrench systems are $c^i (0 \leq c^i \leq 1)$-systems of the 0 pitch. The combination of leg-wrench systems can be simply represented by the combination of the orders, c^i, of leg-wrench systems.

The combinations of the orders, c^i, of leg-wrench systems have been determined by solving (2.28) and are shown in Table 5.4.

13.6 Step 2: Type Synthesis of Legs

In this section, the type synthesis of legs for PPS= PKCs will be performed using the two substeps described in Sect. 5.5.

13.6.1 Step 2a: Type Synthesis of 5-DOF Single-Loop KCs That Involve a PPS Virtual Chain and Have a Specified Leg-Wrench System

The type synthesis of 5-DOF single-loop KCs that involve a PPS virtual chain and have a specified leg-wrench system can be performed as follows.

Firstly, perform the type synthesis of single-loop KCs that involve a PPS virtual chain and have a specified leg-wrench system (See Chap. 3). Secondly, discard those single-loop KCs in which the twists of all the joints but the virtual chain are linearly dependent.

Cases with a 1-ζ_0-System

Firstly, perform the type synthesis of 5-DOF single-loop KCs that involve a PPS virtual chain and have a 1-ζ_0-system. According to Chap. 3, there are ten joints, including at least three R joints, in a 5-DOF single-loop KC that involves a PPS virtual chain and has a 1-ζ_0-system. Such a single-loop KC is composed of (a) one spherical compositional unit and one planar compositional unit [Figs. 13.2a–13.2e] or (b) one spherical compositional unit and one planar translational compositional unit [Fig. 13.2f]. The axis of the basis ζ_0 of the 1-ζ_0-system passes through the centre of the S joint of the PPS virtual chain and is perpendicular to the plane associated with the $()_E$.

Secondly, discard those single-loop KCs in which the twists of all the joints but the virtual chain are linearly dependent.

In the representation of the 5-DOF single-loop KCs involving a PPS virtual chain and having a 1-ζ_0-system, $\hat{\text{R}}$ denote R joints whose axes intersect at the centre of the S joint within the PPS virtual chain, $()^{\parallel}_E$ denotes an equivalent planar joint formed by two or three successive joints whose associate plane of motion is parallel to the directions of the P joints within the PPS virtual chain.

It is pointed out that in a 5-DOF single-loop KC, the P joints within a $()^{\parallel}_E$ can be placed anywhere within the $()^{\parallel}_E$. For brevity, we list only the 5-DOF single-loop KCs from which all 5-DOF single-loop KCs can be obtained through the above operations.

$c^i = 0$

Theoretically, any six R and P joints whose twists are linearly independent together with a PPS virtual chain form a 5-DOF single-loop KC.

It is noted that starting from each of the above 5-DOF single-loop KCs involving a PPS virtual chain and having a 1-ζ_0-system, one can obtain a class of 5-DOF single-loop KCs involving a PPS virtual chain and having a 0-ζ-system by the application of an $(\text{R})_L$ coaxial compositional unit or a $(\text{P})_L$ codirectional compositional unit. It is pointed out that the $(\text{R})_L$ or the $(\text{P})_L$ joint is inactive and can thus be placed anywhere in the single-loop KC. For example, the $(\text{RRR})^{\parallel}_E(\text{R})_L\hat{\text{R}}\hat{\text{R}}\text{V}$ single-loop KC can be obtained by changing the position of the $(\text{R})_L$ joint in the $(\text{RRR})^{\parallel}_E\hat{\text{R}}\hat{\text{R}}(\text{R})_L\text{V}$ single-loop KC.

13.6.2 Step 2b: Generation of Types of Legs

According to Condition (1) for the existence of PPS= PKCs, the types of legs can be readily obtained from the 5-DOF single-loop KCs obtained in Step 2a, Sect. 13.6.1, by removing the PPS virtual chain.

The specific geometric conditions that guarantee the satisfaction of Condition (1) for PPS= PKCs are clearly indicated by the notations that were introduced

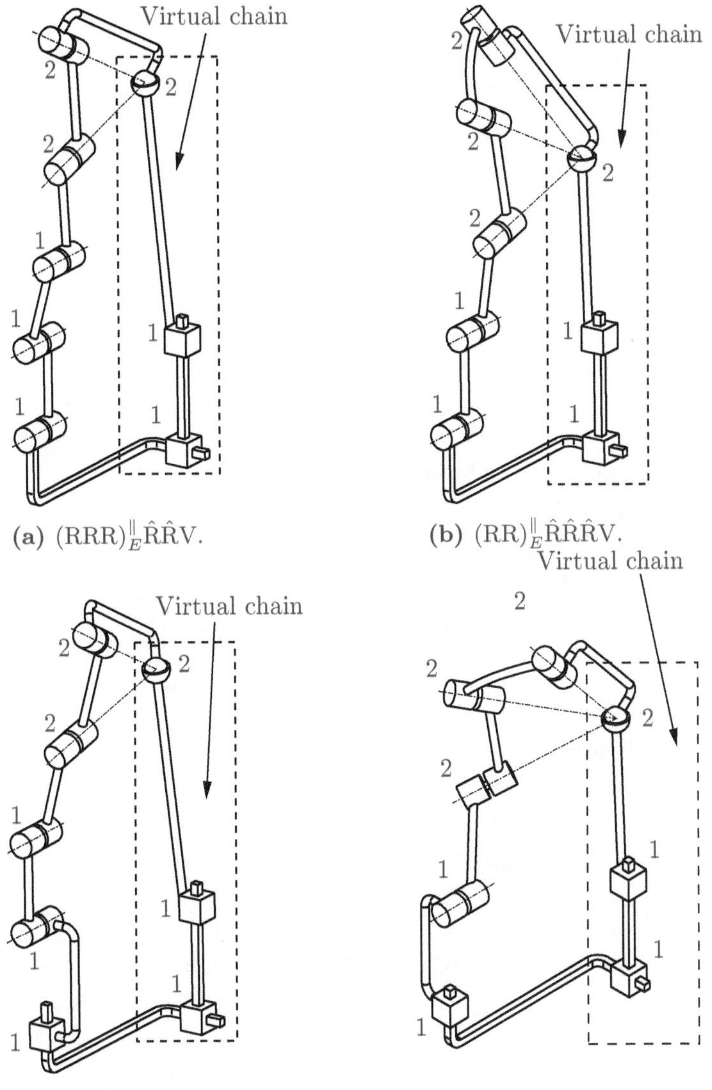

(a) (RRR)$^{\|}_{E}\hat{R}\hat{R}$V. **(b)** (RR)$^{\|}_{E}\hat{R}\hat{R}\hat{R}$V.

(c) (PRR)$^{\|}_{E}\hat{R}\hat{R}$V. **(d)** (PR)$^{\|}_{E}\hat{R}\hat{R}\hat{R}$V.

Fig. 13.2. Five-DOF single-loop KCs involving a PPS virtual chain ($c^i = 1$)

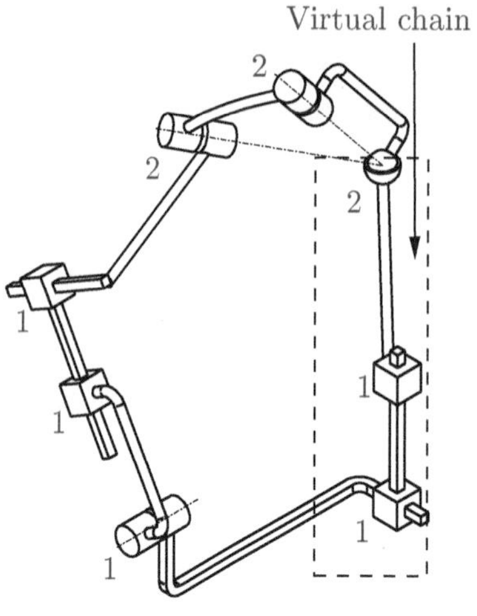

(e) $(\mathrm{RPP})_E^{\parallel}\hat{\mathrm{R}}\hat{\mathrm{R}}\mathrm{V}$.

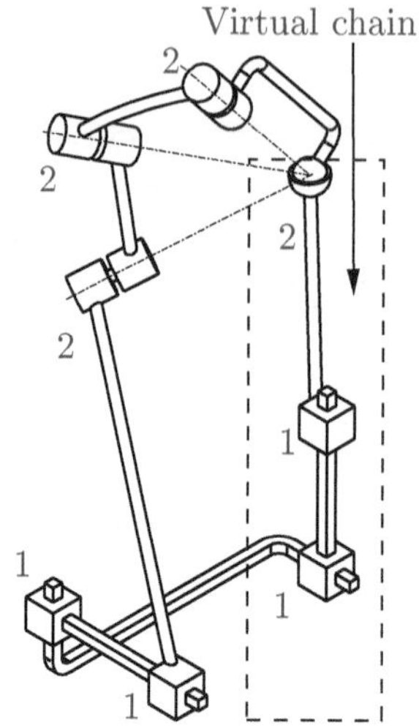

(f) $(\mathrm{PP})_E^{\parallel}\hat{\mathrm{R}}\hat{\mathrm{R}}\hat{\mathrm{R}}\mathrm{V}$.

Fig. 13.2. (*continued*)

to represent the types of PPS= PKCs, PPS= PMs and their legs: $\hat{\mathrm{R}}$ denote R joints within a PPS= PKC, PPS= PM or a leg whose axes intersect at a common point, $()_E^{\parallel}$ denotes an equivalent planar joint formed by two or three successive joints. Within a PPS= PKC or PPS= PM, the planes of motion associate with all the $()_E^{\parallel}$ are parallel.

For example, by removing the virtual chain in an $(\mathrm{RR})_E^{\parallel}\hat{\mathrm{R}}\hat{\mathrm{R}}\hat{\mathrm{R}}\mathrm{V}$ KC [Fig. 13.2b], an $(\mathrm{RR})_E^{\parallel}\hat{\mathrm{R}}\hat{\mathrm{R}}\hat{\mathrm{R}}$ leg [Fig.13.3b] can be obtained. Such a leg has a 1-$\boldsymbol{\zeta}_0$-system. The axis of the basis $\boldsymbol{\zeta}_0$ of the 1-$\boldsymbol{\zeta}_0$-system passes through the common point of the axes of three $\hat{\mathrm{R}}$ joints and is parallel to the axes of the R joints within $(\mathrm{RR})_E^{\parallel}$.

All the types of legs for PPS= PKCs obtained are listed in Table 13.1. The legs used to synthesize PPS= PKCs need to satisfy not only the geometric conditions within a same leg, but also the geometric conditions among different legs.

13.7 Step 3: Assembly of Legs

PPS= PKCs can be generated by assembling a set of legs for PPS= PKCs shown in Table 13.1 selected according to the combinations of the leg-wrench systems shown in Table 5.4. In assembling PPS= PKCs, the following condition should be met: the linear combination of the leg-wrench systems forms a 1-$\boldsymbol{\zeta}_0$-system (see Condition (2) for PPS= PKCs in Sect. 13.3). This simply requires that the number of legs with a 1-$\boldsymbol{\zeta}_o$-system is greater than or equal to 1. It is noted that there is no constraint singularity for PPS= PKCs.

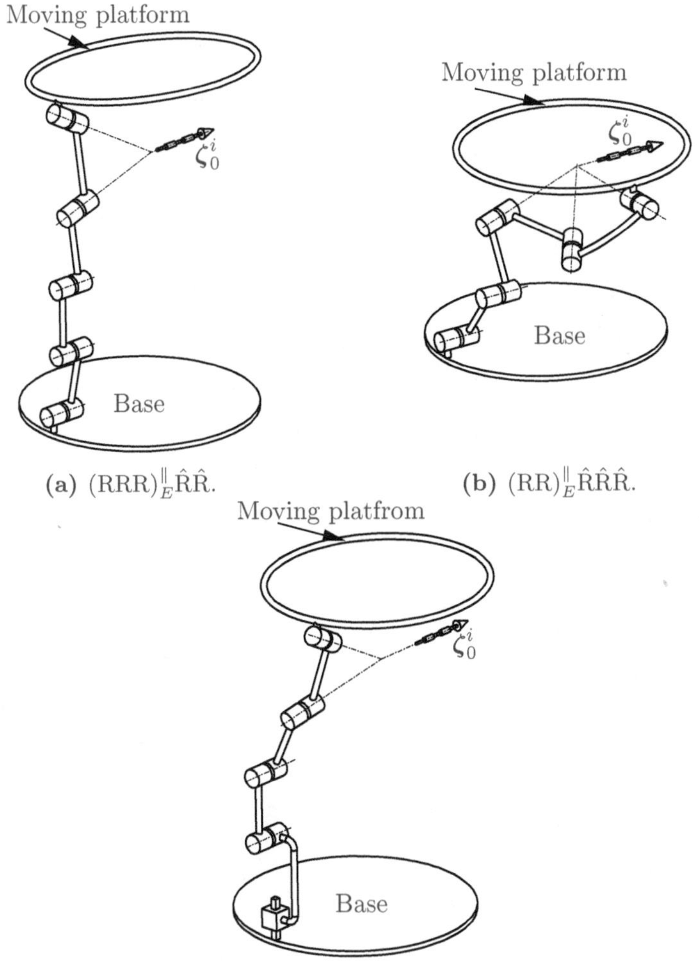

(a) $(\mathrm{RRR})_E^{\|}\hat{\mathrm{R}}\hat{\mathrm{R}}$. **(b)** $(\mathrm{RR})_E^{\|}\hat{\mathrm{R}}\hat{\mathrm{R}}\hat{\mathrm{R}}$. **(c)** $(\mathrm{PRR})_E^{\|}\hat{\mathrm{R}}\hat{\mathrm{R}}$.

Fig. 13.3. Some legs for PPS= PKCs

In addition to the types of PPS= PKCs proposed in the literature, new types, i.e., the types with a leg having an inactive joint, are also obtained. Due to the large number of PPS= PKCs, only the 5-legged PPS= PKCs with legs of the same type are listed in Table 13.2.

Figure 13.4 shows two PPS= PKCs. The 3-$(\mathrm{RRR})_E^{\|}\hat{\mathrm{R}}\hat{\mathrm{R}}$ PPS= PKC shown in Fig. 13.4a is composed of three $(\mathrm{RRR})_E^{\|}\hat{\mathrm{R}}\hat{\mathrm{R}}$ legs shown in Fig. 13.3a. The 3-$(\mathrm{RR})_E^{\|}\hat{\mathrm{R}}\hat{\mathrm{R}}\hat{\mathrm{R}}$ PPS= PKC shown in Fig. 13.4b is composed of three $(\mathrm{RR})_E^{\|}\hat{\mathrm{R}}\hat{\mathrm{R}}\hat{\mathrm{R}}$ legs shown in Fig. 13.3b.

Table 13.1. Legs for PPS= PKCs

c^i	Class	No.	Type
1	5R	1	$(\mathrm{RRR})^{\parallel}_{E}\hat{\mathrm{R}}\hat{\mathrm{R}}$
		2	$(\mathrm{RR})^{\parallel}_{E}\hat{\mathrm{R}}\hat{\mathrm{R}}\hat{\mathrm{R}}$
	4R1P	3	$(\mathrm{RRP})^{\parallel}_{E}\hat{\mathrm{R}}\hat{\mathrm{R}}$
		4	$(\mathrm{RPR})^{\parallel}_{E}\hat{\mathrm{R}}\hat{\mathrm{R}}$
		5	$(\mathrm{PRR})^{\parallel}_{E}\hat{\mathrm{R}}\hat{\mathrm{R}}$
		6	$(\mathrm{RP})^{\parallel}_{E}\hat{\mathrm{R}}\hat{\mathrm{R}}\hat{\mathrm{R}}$
		7	$(\mathrm{PR})^{\parallel}_{E}\hat{\mathrm{R}}\hat{\mathrm{R}}\hat{\mathrm{R}}$
	3R2P	8	$(\mathrm{RPP})^{\parallel}_{E}\hat{\mathrm{R}}\hat{\mathrm{R}}$
		9	$(\mathrm{PRP})^{\parallel}_{E}\hat{\mathrm{R}}\hat{\mathrm{R}}$
		10	$(\mathrm{PPR})^{\parallel}_{E}\hat{\mathrm{R}}\hat{\mathrm{R}}$
		11	$(\mathrm{PP})^{\parallel}_{E}\hat{\mathrm{R}}\hat{\mathrm{R}}\hat{\mathrm{R}}$
0	omitted	omitted	omitted

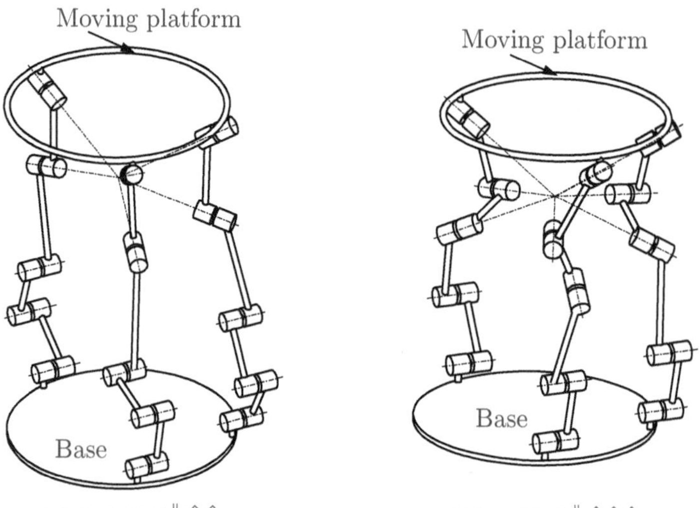

(a) 3-$(\mathrm{RRR})^{\parallel}_{E}\hat{\mathrm{R}}\hat{\mathrm{R}}$. **(b)** 3-$(\mathrm{RR})^{\parallel}_{E}\hat{\mathrm{R}}\hat{\mathrm{R}}\hat{\mathrm{R}}$.

Fig. 13.4. Some PPS= PKCs

13.8 Step 4: Selection of Actuated Joints

13.8.1 t-Component of the Actuation Wrenches

Considering that the order of a screw system is coordinate system free and for simplicity reasons, all the actuation wrenches and wrenches are expressed in a

Table 13.2. Five-legged PPS= PKCs

c^i	Class	No.	Type
1	5R	1	$5\text{-}(\text{RRR})^{\parallel}_{E}\hat{\text{R}}\hat{\text{R}}$
	4R1P	2	$5\text{-}(\text{RR})^{\parallel}_{E}\hat{\text{R}}\hat{\text{R}}\hat{\text{R}}$
		3	$5\text{-}(\text{RRP})^{\parallel}_{E}\hat{\text{R}}\hat{\text{R}}$
		4	$5\text{-}(\text{RPR})^{\parallel}_{E}\hat{\text{R}}\hat{\text{R}}$
		5	$5\text{-}(\text{PRR})^{\parallel}_{E}\hat{\text{R}}\hat{\text{R}}$
		6	$5\text{-}(\text{RP})^{\parallel}_{E}\hat{\text{R}}\hat{\text{R}}\hat{\text{R}}$
		7	$5\text{-}(\text{PR})^{\parallel}_{E}\hat{\text{R}}\hat{\text{R}}\hat{\text{R}}$
	3R2P	8	$5\text{-}(\text{RPP})^{\parallel}_{E}\hat{\text{R}}\hat{\text{R}}$
		9	$5\text{-}(\text{PRP})^{\parallel}_{E}\hat{\text{R}}\hat{\text{R}}$
		10	$5\text{-}(\text{PPR})^{\parallel}_{E}\hat{\text{R}}\hat{\text{R}}$
		11	$5\text{-}(\text{PP})^{\parallel}_{E}\hat{\text{R}}\hat{\text{R}}\hat{\text{R}}$

coordinate system with its origin at the intersection of the $\hat{\text{R}}$ joints and its X-axis perpendicular to the planes of motion associated with $()^{\parallel}_{E}$.

Let $\{1 \quad 0 \quad 0 \quad 0 \quad 0 \quad 0\}^T$ denote the basis wrench of $\mathcal{W}$. The last five scalar components of the basis wrench of $\mathcal{W}$ are zero. Thus, the t-component, $\boldsymbol{\zeta}^i_{t\not{D}j}$, of the $\boldsymbol{\zeta}^i_{\not{D}j}$ is composed of the last five scalar components of the $\boldsymbol{\zeta}^i_{\not{D}j}$.

In the $(\underline{\text{R}}\text{RR})^{\parallel}_{E}\hat{\text{R}}\hat{\text{R}}$ leg [Fig. 13.5a], the first R joint is actuated. The actuation wrench is any $\boldsymbol{\zeta}_0$ whose axis is the intersection of the plane passing through the axes of two $\hat{\text{R}}$ joints and the plane passing through the axes of the two unactuated R joints in $(\text{RRR})^{\parallel}_{E}$. In the $(\underline{\text{R}}\text{R})^{\parallel}_{E}\hat{\text{R}}\hat{\text{R}}\hat{\text{R}}$ leg [Fig. 13.5b], the first R joint is actuated. The actuation wrench is a $\boldsymbol{\zeta}_0$ whose axis passes through the intersection of the axes of the $\hat{\text{R}}$ joints, intersects the axis of the second R joint and is not parallel to the axis of the first R joint.

13.8.2 Procedure for the Validity Detection of Actuated Joints

The validity detection of actuated joints of PPS= PMs can thus be performed using the following steps.

Step 4a. If one or more of the actuated joints of a candidate PPS= PM are inactive, the set of actuated joints is invalid and the candidate PPS= PM should be discarded.

Step 4b. If the *t*-components of the actuation wrenches of actuated joints $\boldsymbol{\zeta}^i_{\not{D}j}$ are linearly dependent in a general configuration for a candidate PPS= PM, the set of actuated joints is invalid. In this case, the candidate PPS= PM should be discarded.

One candidate PPS= PM corresponding to the $3\text{-}(\text{RR})^{\parallel}_{E}\hat{\text{R}}\hat{\text{R}}\hat{\text{R}}$ PPS= PKC [Fig. 13.4b] is the $2\text{-}(\underline{\text{RR}})^{\parallel}_{E}\hat{\text{R}}\hat{\text{R}}\hat{\text{R}}\text{-}(\underline{\text{R}}\text{R})^{\parallel}_{E}\hat{\text{R}}\hat{\text{R}}\hat{\text{R}}$ PPS= PM [Fig. 13.6b]. The axes

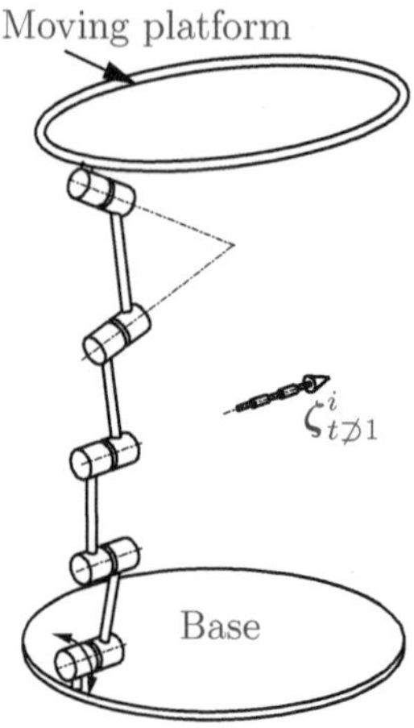

(a) $(\underline{R}RR)^{\parallel}_{E}\hat{R}\hat{R}\hat{R}$.

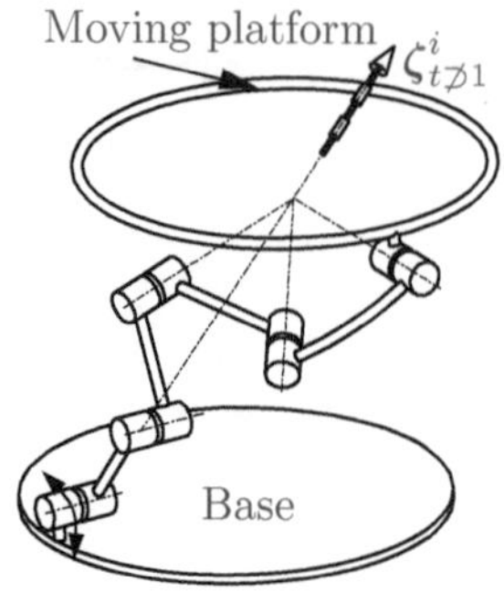

(b) $(\underline{R}R)^{\parallel}_{E}\hat{R}\hat{R}\hat{R}$.

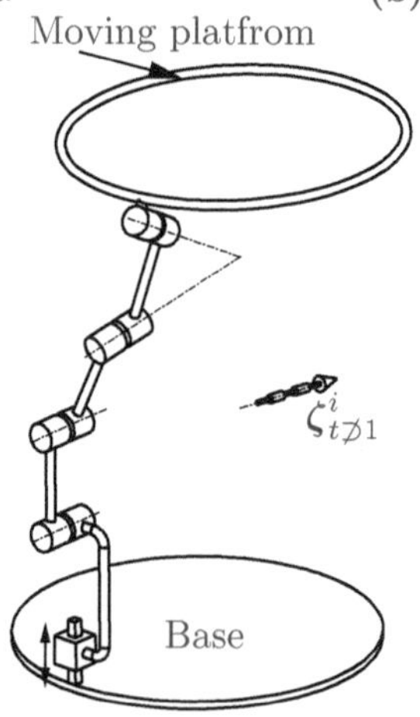

(c) $(\underline{P}RR)^{\parallel}_{E}\hat{R}\hat{R}$.

Fig. 13.5. Actuation wrenches of some legs for PPS= PKCs

of all the actuation wrenches pass through the centre of the common point of the $\hat{R}$ joints. We have

$$\left| \zeta^{1}_{t,\not\supset 1}\ \zeta^{1}_{t,\not\supset 2}\ \zeta^{2}_{t,\not\supset 1}\ \zeta^{2}_{t,\not\supset 2}\ \zeta^{3}_{t,\not\supset 1} \right| = \begin{vmatrix} * & * & * & * & * \\ * & * & * & * & * \\ 0 & 0 & 0 & 0 & 0 \\ 0 & 0 & 0 & 0 & 0 \\ 0 & 0 & 0 & 0 & 0 \end{vmatrix} = 0.$$

where $*$ denotes an arbitrary number.

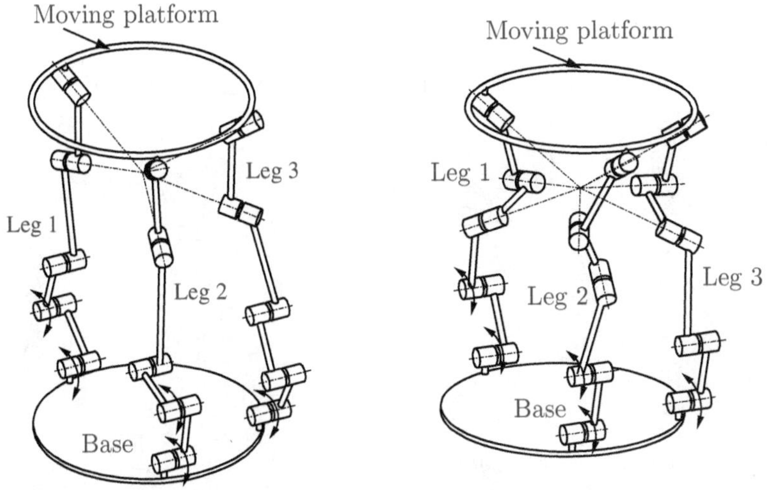

(a) $2\text{-}(\underline{RR}R)^{\parallel}_{E}\hat{R}\hat{R}\text{-}(\underline{R}RR)^{\parallel}_{E}\hat{R}\hat{R}$ candidate PPS= PM.

(b) $2\text{-}(\underline{RR})^{\parallel}_{E}\hat{R}\hat{R}\hat{R}\text{-}(\underline{R}R)^{\parallel}_{E}\hat{R}\hat{R}\hat{R}$ candidate PPS= PM.

Fig. 13.6. Selection of actuated joints for some PPS= PKCs

According to the validity condition of actuated joints, the set of actuated joints is invalid. The candidate $2\text{-}(\underline{RR})^{\parallel}_{E}\hat{R}\hat{R}\hat{R}\text{-}(\underline{R}R)^{\parallel}_{E}\hat{R}\hat{R}\hat{R}$ is thus discarded.

One candidate PPS= PM corresponding to the $3\text{-}(RRR)^{\parallel}_{E}\hat{R}\hat{R}$ PPS= PKC [Fig. 13.4a] is the $2\text{-}(\underline{RR}R)^{\parallel}_{E}\hat{R}\hat{R}\text{-}(\underline{R}RR)^{\parallel}_{E}\hat{R}\hat{R}$ PPS= PM [Fig. 13.6a]. The axis of the actuation wrench of an actuated joint is along the intersection of the plane determined by the axes of the R joints within $(RRR)^{\parallel}_{E}$ except for the considered actuated joint and the plane determined by the axes of the $\hat{R}$ joints [Fig. 13.5a]. Using the validity condition of actuated joints, it can be proved that the set of actuated joints is valid.

Following the criteria on the selection of actuated joints (Sect. 5.7) and the procedure for the detection of the validity of actuated joints, all the $m(m \geq 2)$-legged

Table 13.3. Five-legged PPS= PMs

c^i	Class	No.	Type
1	5R	1	$5\text{-}(\underline{R}RR)^{\parallel}_{E}\hat{R}\hat{R}$
		2	$5\text{-}(R\underline{R}R)^{\parallel}_{E}\hat{R}\hat{R}$
		3	$5\text{-}(RR\underline{R})^{\parallel}_{E}\hat{R}\hat{R}$
	4R1P	4	$5\text{-}(RR\underline{P})^{\parallel}_{E}\hat{R}\hat{R}$
		5	$5\text{-}(R\underline{P}R)^{\parallel}_{E}\hat{R}\hat{R}$
		6	$5\text{-}(\underline{P}RR)^{\parallel}_{E}\hat{R}\hat{R}$

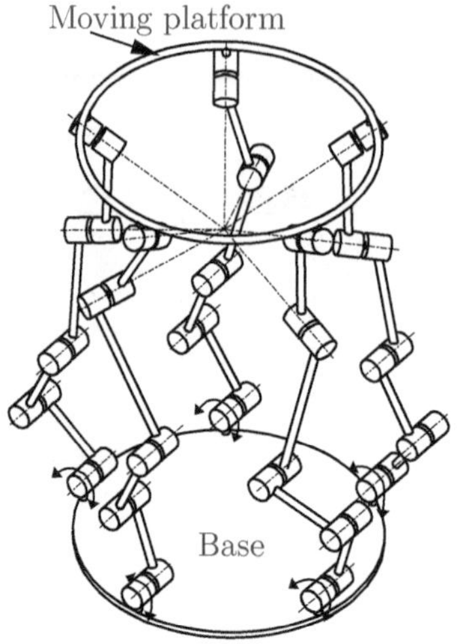

(a) $5\text{-}(\underline{R}RR)^{\parallel}_{E}\hat{R}\hat{R}$.

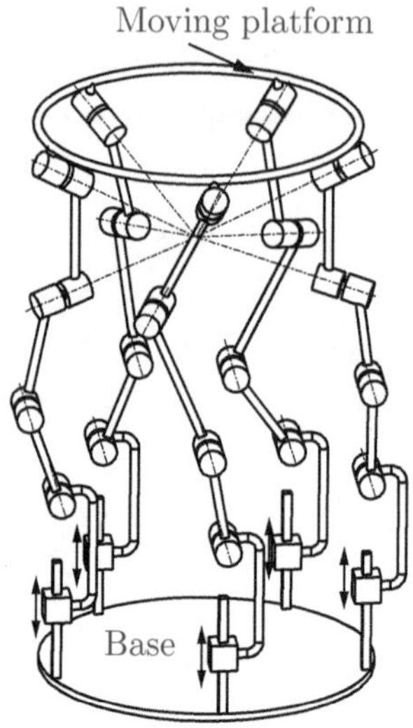

(b) $5\text{-}(\underline{P}RR)^{\parallel}_{E}\hat{R}\hat{R}$.

Fig. 13.7. Five-legged PPS= PMs with identical type of legs

PPS= PMs corresponding to each PPS= PKC can be generated. Due to the large number of PPS= PMs, only 5-legged PPS= PMs with all legs of the same type satisfying the above criteria are shown in Table 13.3. Figure 13.7 shows two types of 5-legged PPS= PMs without unactuated P joints.

13.9 Summary

In this chapter, the type synthesis of PPS= PMs has been thoroughly solved using the general type synthesis approach proposed in Chap. 5. PPS= PKCs with inactive joints as well as PPS= PKCs without inactive joints have been obtained. The validity condition of actuated joints of PPS= PMs has been reduced to the calculation of a 5×5 determinant. The PPS= PKCs obtained include some new PPS= PKCs as well as all the known PPS= PKCs. Some new PPS= PMs have also been proposed.

Variations of PPS= PKCs can be obtained by (a) substituting a combination of one R joint and one P joint with parallel axes, a combination of two R joints with intersecting axes and a combination of three R joints with concurrent axes with a C joint or its RH, PH, HH form, U, and S joint respectively, (b) replacing a P joint within $()^{\parallel}_{E}$ with a planar parallelogram whose plane of motion is parallel to the planes of relative motion associated with $()^{\parallel}_{E}$, and/or (c) replacing an inactive R or P joint with an inactive H joint.

It is noted that if a combination of an inactive R joint and a P joint with parallel axes, is replaced with a C joint in the form of RH or HH, there will be no inactive joint in the RH or HH combination. If a combination of an inactive P joint and an R joint with parallel axes, is replaced with a C joint in the form of PH or HH, there will be no inactive joint in the PH or HH combination.

14. Parallel Mechanisms with a Parallel Virtual Chain

In previous chapters, the type synthesis of PMs with motion patterns described by serial virtual chains was performed. However, some motion patterns cannot be described by serial virtual chains. Such motion patterns may nevertheless have practical relevance. Therefore, in some instances, the type synthesis of PMs with a parallel virtual chain is required.

In this chapter, a method for the type synthesis of PMs with a parallel virtual chain is developed based on the type synthesis of PMs with a serial virtual chain. A general procedure is proposed which consists of four main steps, namely, (1) the decomposition of the number of legs, (2) the type synthesis of sub-PKCs, (3) the combination of sub-PKCs to generate PKCs, and (4) the selection of the actuated joints. These steps will be discussed successively in separate sections.

14.1 Introduction

The type synthesis of PMs with a serial virtual chain presented in previous chapters will be used here as a foundation for the type synthesis of PMs with a parallel virtual chain. A general procedure is proposed for the type synthesis of PMs with a parallel virtual chain in Sect. 14.2. In Sects. 14.3–14.5, the proposed procedure is illustrated by synthesizing three classes of PMs with a parallel virtual chain.

14.2 Procedure for the Type Synthesis of Parallel Mechanisms

Consider an m-legged PM with an m_V-legged parallel virtual chain. Each leg of the m_V-legged parallel virtual chain can be regarded as a serial virtual chain, the i-th of which is called the V_i virtual chain. In the m-legged PM, there must be a sub-PKC, which is composed of m_i legs with a c_i ($c_i > 0$) -$\boldsymbol{\zeta}$-system, which generates the motion pattern of the V_i virtual chain. Let m_0 denote the

X. Kong and C. Gosselin: Type Syn. of Parallel Mech., STAR 33, pp. 213–222, 2007.
springerlink.com

number of legs with a 0-$\boldsymbol{\zeta}$-system, i.e., legs that exert no constraint on the moving platform. We have

$$\sum_{i=1}^{m_V} m_i + m_0 = m. \tag{14.1}$$

It is noted that m_i may be greater than or equal to one. The sub-PKCs associated with the V_i virtual chain can be obtained using the results of the type synthesis of PMs with a serial virtual chain presented in previous chapters, where we have been focusing on PMs with at least two legs. As it will be seen in the following sections, to find the 1-legged sub-PKCs associated with the V_i virtual chain is straightforward.

A general procedure can be proposed for the type synthesis of PMs with a parallel virtual chain as follows:

Step 1. Decompose the number of legs of the PKC, i.e., find the number of legs, m_i, corresponding to each V_i virtual chain. This can be determined using (14.1).

Step 2. For $i = 1$ to m_V, perform the type synthesis of the i-th sub-PKC generating the V_i virtual chain motion.

Step 3. Generate PKCs. PKCs can be generated by assembling a combination of m_V sub-PKCs, each corresponding to one V_i virtual chain, obtained in Step 2 as well as m_0 legs with a 0-$\boldsymbol{\zeta}$-system.

Step 4. Generate PMs by selecting actuated joints in different ways for each PKC (Sect. 3.1), obtained in Step 3.

The validity detection of actuated joints of V= PMs can thus be performed using the following steps.

Step 4a. If one or more of the actuated joints of a possible V= PM are inactive, the set of actuated joints is invalid and the possible V= PM should be discarded.

Step 4b. If in a general configuration, all the actuation wrenches, $\boldsymbol{\zeta}^i_{\mathcal{D}j}$, of the $\mathcal{F}$ actuated joints together with a set of basis wrenches of the wrench system $\mathcal{W}$ of the PKC does not constitute a 6-system for a candidate PM, the set of actuated joints is invalid. In this case, the candidate V= PM should be discarded.

Following the criteria on the selection of actuated joints (Sect. 5.7) and the procedure for the detection of the validity of actuated joints, all the $m(m \geq 2)$-legged V= PMs corresponding to each V= PKC can be generated.

Steps 1 through 4 described above will be illustrated with examples in the following sections.

14.3 Type Synthesis of 3-PPS= PMs

A 3-PPS= PM is a PM which generates the 3-PPS motion. The associated 3-PPS virtual chain (Fig. 14.1) is composed of three PPS chains connected in-parallel.

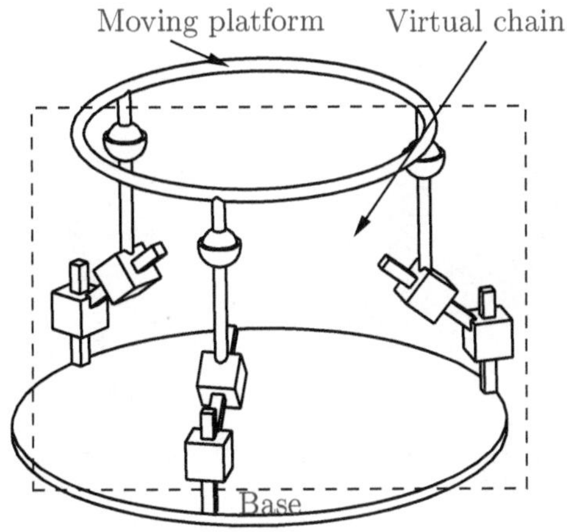

Fig. 14.1. 3-PPS virtual chain

14.3.1 Decomposition of Number of Legs

In this section, we confine ourselves to 3-legged PMs. In this case, we obtain using (14.1) that

$$m_0 = 0 \tag{14.2}$$

$$m_1 = m_2 = m_3 = 1 \tag{14.3}$$

14.3.2 Type Synthesis of Sub-PKCs

Since all the legs of the 3-PPS virtual chain are of the same type, we need only to synthesize sub-PKCs which are composed of one leg that has the same wrench system as the PPS sub-virtual chain.

Following the procedure presented in Chap. 5, we can obtain that the types of legs are $\acute{R}\acute{R}\acute{R}\hat{R}\hat{R}$, $\acute{R}\acute{R}P\hat{R}\hat{R}$, $\acute{R}P\acute{R}\hat{R}\hat{R}$, $P\acute{R}\acute{R}\hat{R}\hat{R}$, $\acute{R}PP\hat{R}\hat{R}$, $P\acute{R}P\hat{R}\hat{R}$, $PP\acute{R}\hat{R}\hat{R}$, $\acute{R}\acute{R}\hat{R}\hat{R}\hat{R}$ (Fig. 14.2a), $\acute{R}P\hat{R}\hat{R}\hat{R}$ (Fig. 14.2b), $P\acute{R}\hat{R}\hat{R}\hat{R}$, and $PP\hat{R}\hat{R}\hat{R}$.

From the above legs, we can obtain a large number of m_i-legged sub-PKCs.

14.3.3 Combination of Sub-PKCs

Selecting three — according to Step 1 — sub-PKCs each corresponding to one leg in the 3-PPS virtual chain and assembling them together, we can obtain many types of 3-PPS= PKCs with identical types of legs or different types of legs.

For brevity, we only show three 3-PPS= PKCs with identical types of legs in Fig. 14.3.

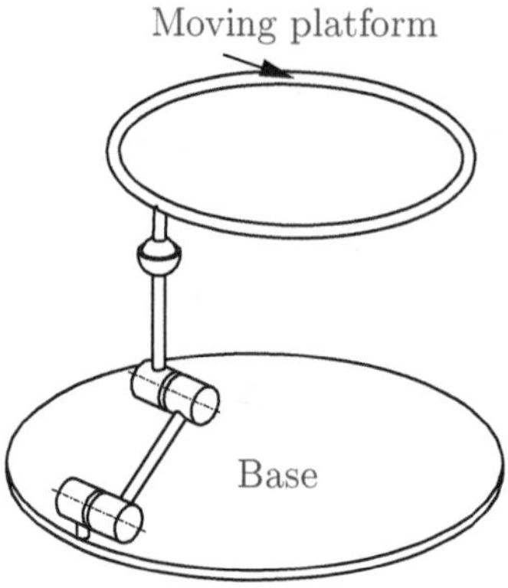

(**a**) $\acute{R}\acute{R}\hat{R}\hat{R}\hat{R}$.

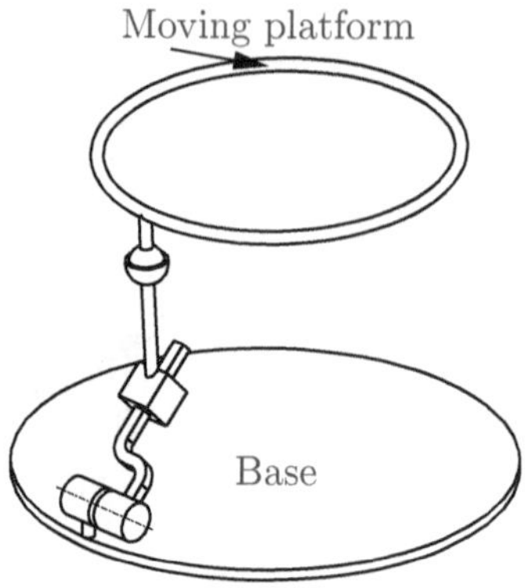

(**b**) $\acute{R}P\hat{R}\hat{R}\hat{R}$.

Fig. 14.2. Some sub-PKCs for 3-PPS= PKCs

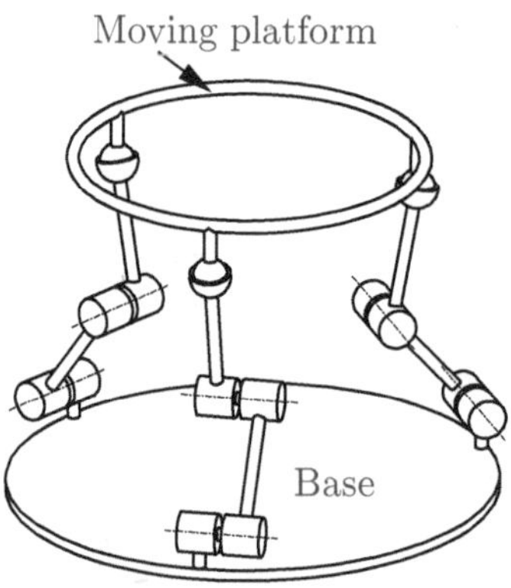

(**a**) 3-$\acute{R}\acute{R}$S.

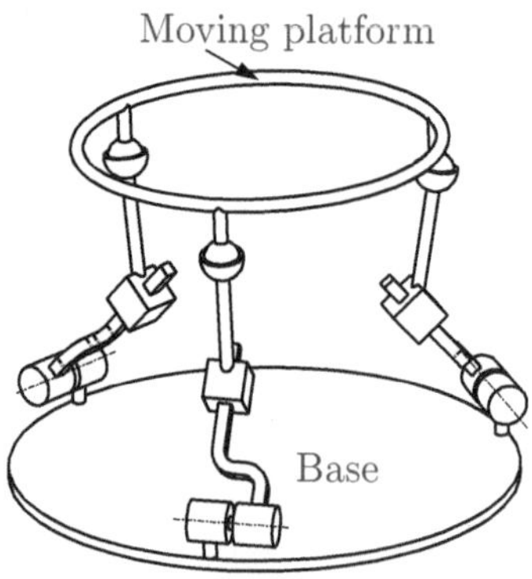

(**b**) 3-$\acute{R}$PS.

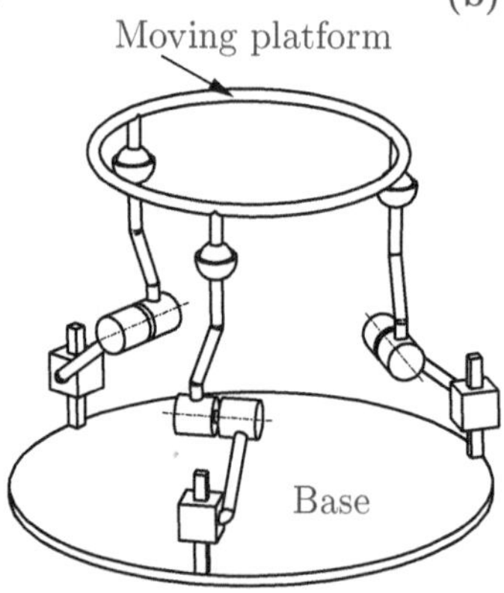

(**c**) 3-P$\acute{R}$S.

Fig. 14.3. Some 3-PPS= PKCs

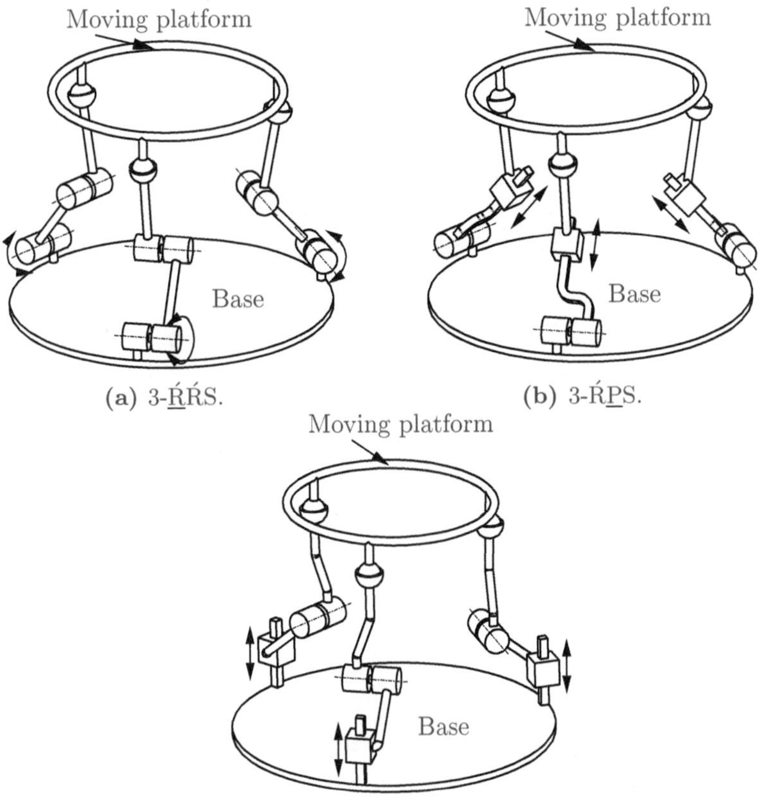

(a) 3-$\underline{\text{R}}$ŔS. (b) 3-Ŕ$\underline{\text{P}}$S. (c) 3-$\underline{\text{P}}$ŔS.

Fig. 14.4. Some 3-PPS= PMs

14.3.4 Selection of Actuated Joints

Following the procedure for the selection of actuated joints, we can obtain all the 3-PPS= PMs corresponding to each 3-PPS= PKC obtained in Step 2. Figure 14.4 shows several 3-PPS= PMs, obtained, respectively, from the 3-PPS= PKCs shown in Fig. 14.3.

14.4 Type Synthesis of 2-PPPU= PMs

A 2-PPPU= PM is a PM which generates the 2-PPPU= PM motion. The associated 2-PPPU virtual chain (Fig. 14.5) is composed of two PPPU chains connected in-parallel.

14.4.1 Decomposition of Number of Legs

In this section, we confine ourselves to 4-legged PMs. In this case, the combinations of m_0 and m_i can be obtained using (14.1) and are listed in Table 14.1.

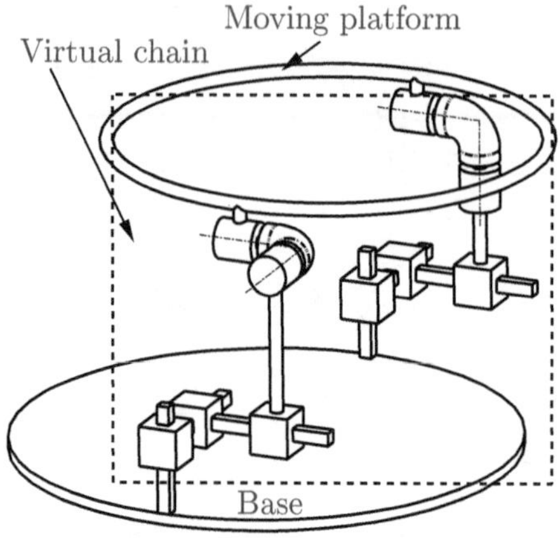

Fig. 14.5. Wrench system of a 2-PPPU= PKC

14.4.2 Type Synthesis of Sub-PKCs

Since all the legs of the 2-PPPU virtual chain are of the same type, we need only to synthesize m_i-legged sub-PKCs associated with the PPPU virtual chain.

Following the procedure proposed in Chap. 5, we can obtain that the types of legs are R̋R̋R̋R̄R̄, R̋R̋R̄R̄R̄, PR̋R̋R̄R̄, R̋PR̋R̄R̄, R̋R̋PR̄R̄, R̋R̋R̋PR̄, R̋R̋R̋R̄P, PR̋R̄R̄R̄, R̋PR̄R̄R̄, R̋R̋PR̄R̄, R̋R̋R̄PR̄, R̋R̋R̄R̄P, PPR̋R̄R̄, PR̋PR̄R̄, R̋PPR̄R̄, R̋R̋R̋PP, PR̋R̋PR̄, R̋PR̋PR̄, R̋R̋PPR̄, PR̋R̋R̄P, R̋PR̋R̄P, R̋R̋PR̄P, and permutation of PPPR̋R̄.

From the above legs, we can obtain a large number of m_i-legged sub-PKCs.

14.4.3 Combination of Sub-PKCs

Selecting a set of m_V sub-PKCs according to step 1 as well as m_0 legs with a 0-ζ-system and assembling them together, we can obtain many types of 2-PPPU= PKCs.

Table 14.1. Combinations of m_i and m_0 for 4-legged 2-PPPU= PKCs

No	m_1	m_2	m_0
1	3	1	0
2	2	2	0
3	2	1	1
4	1	1	2

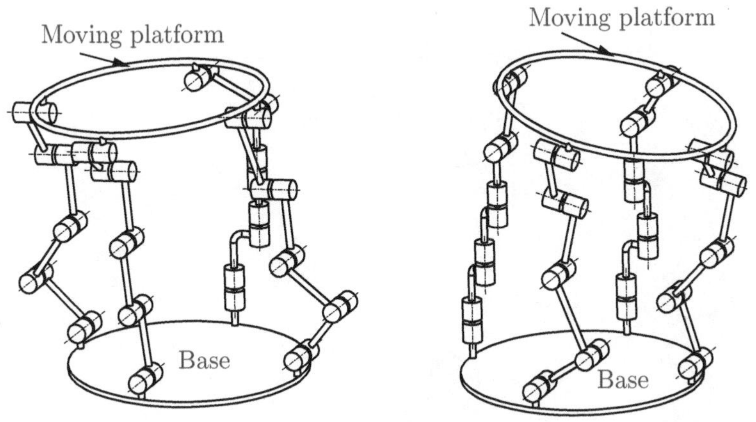

(a) $3\text{-}\overset{\prime\prime}{R}\overset{\prime\prime}{R}\overset{\prime\prime}{R}\bar{R}\bar{R}\_\overset{\prime\prime}{R}\overset{\prime\prime}{R}\overset{\prime\prime}{R}\bar{R}\bar{R}$.
(b) $2\text{-}\overset{\prime\prime}{R}\overset{\prime\prime}{R}\overset{\prime\prime}{R}\bar{R}\bar{R}\_2\text{-}\overset{\prime\prime}{R}\overset{\prime\prime}{R}\overset{\prime\prime}{R}\bar{R}\bar{R}$.

Fig. 14.6. Some 2-PPPU= PKCs

For brevity, we only show in Fig. 14.6 two 2-PPPU= PKCs which correspond respectively to rows No. 1 and 2 of Table 14.1.

14.4.4 Selection of Actuated Joints

Following the procedure for the selection of actuated joints, we can obtain all the 2-PPPU= PMs corresponding to each 2-PPPU= PKC obtained in Step 2. Figure 14.7 shows two 2-PPPU= PMs, which correspond respectively to the 2-PPPU= PKCs shown in Fig. 14.6.

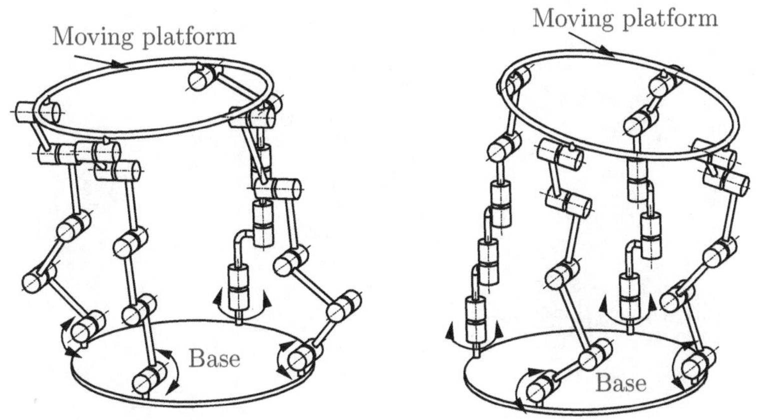

(a) $3\text{-}\underline{\overset{\prime\prime}{R}}\overset{\prime\prime}{R}\overset{\prime\prime}{R}\bar{R}\bar{R}\_\underline{\overset{\prime\prime}{R}}\overset{\prime\prime}{R}\overset{\prime\prime}{R}\bar{R}\bar{R}$.
(b) $2\text{-}\underline{\overset{\prime\prime}{R}}\overset{\prime\prime}{R}\overset{\prime\prime}{R}\bar{R}\bar{R}\_2\text{-}\underline{\overset{\prime\prime}{R}}\overset{\prime\prime}{R}\overset{\prime\prime}{R}\bar{R}\bar{R}$.

Fig. 14.7. Some 2-PPPU= PMs

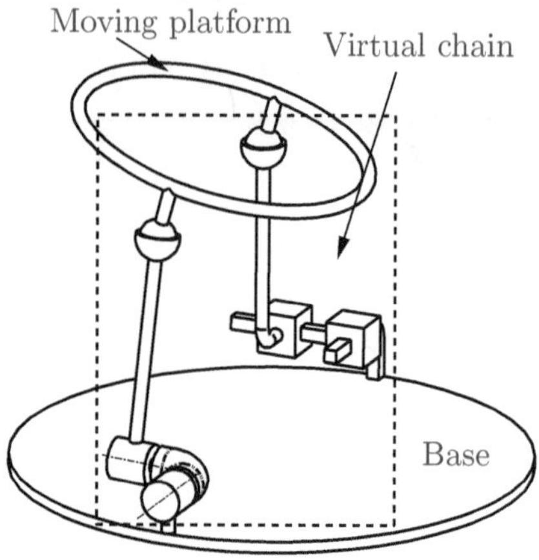

Fig. 14.8. Wrench system of a US-PPS= PKC

14.5 Type Synthesis of US-PPS= PMs

A US-PPS= PM is a PM which generates the US-PPS motion. The associated US-PPS virtual chain (Fig. 14.8) is composed of one US leg and one PPS leg connected in-parallel.

14.5.1 Decomposition of Number of Legs

In this section, we confine ourselves to 4-legged PMs. In this case, the combinations of m_0 and m_i can be obtained using (14.1) and are listed in Table 14.2.

Table 14.2. Combinations of m_i and m_0 for 4-legged US-PPS= PKCs

No	m_1	m_2	m_0
1	3	1	0
2	2	2	0
3	2	1	1
4	1	1	2

14.5.2 Type Synthesis of Sub-PKCs

Since the US-PPS virtual chain has two types of legs, i.e., the US and PPS legs, we need to synthesize m_i-legged sub-PKCs associated either with the US virtual chain or the PPS virtual chain.

Following the procedure proposed in Chap. 5, we can obtain that the types of legs for the US sub-PKCs are ŘŘR̂R̂R̂ and ŘŘŘR̂R̂ and that the types of legs for the PPS sub-PKCs are (RRR)$^{|}_{E}$R̂R̂, (RR)$^{|}_{E}$R̂R̂R̂, (PRR)$^{|}_{E}$R̂R̂, (RPR)$^{|}_{E}$R̂R̂, (RRP)$^{|}_{E}$R̂R̂, (PR)$^{|}_{E}$R̂R̂R̂, (RP)$^{|}_{E}$R̂R̂R̂, (PPR)$^{|}_{E}$R̂R̂, (PRP)$^{|}_{E}$R̂R̂, (RPP)$^{|}_{E}$R̂R̂, and (PP)$^{|}_{E}$R̂R̂R̂.

From the above legs, we can obtain a large number of m_i-legged US= sub-PKCs and PPS= sub-PKCs.

14.5.3 Combination of Sub-PKCs

Selecting a set of m_V sub-PKCs according to step 1 as well as m_0 legs with a 0-$\boldsymbol{\zeta}$-system and assembling them together, we can obtain many types of US-PPS= PKCs.

For brevity, we only show in Fig. 14.9 two US-PPS= PKCs which correspond respectively to rows No. 1 and 2 of Table 14.2.

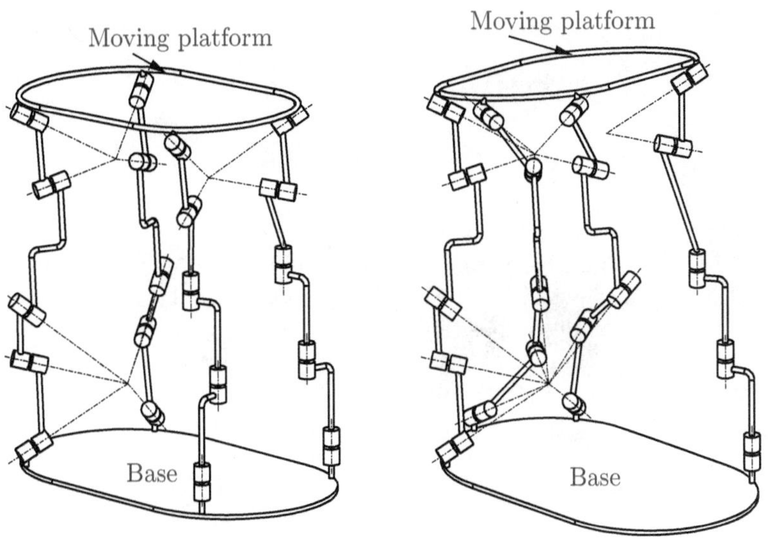

(a) 2-ŘŘŘR̂R̂_2-(RRR)$^{|}_{E}$R̂R̂. **(b)** 3-ŘŘŘR̂R̂_(RRR)$^{|}_{E}$R̂R̂.

Fig. 14.9. Some US-PPS= PKCs

14.5.4 Selection of Actuated Joints

Following the procedure for the selection of actuated joints, we can obtain all the US-PPS= PMs corresponding to each US-PPS= PKC obtained in Step 2. Figure 14.10 shows two US-PPS= PMs, which correspond respectively to the US-PPS= PKCs shown in Fig. 14.9.

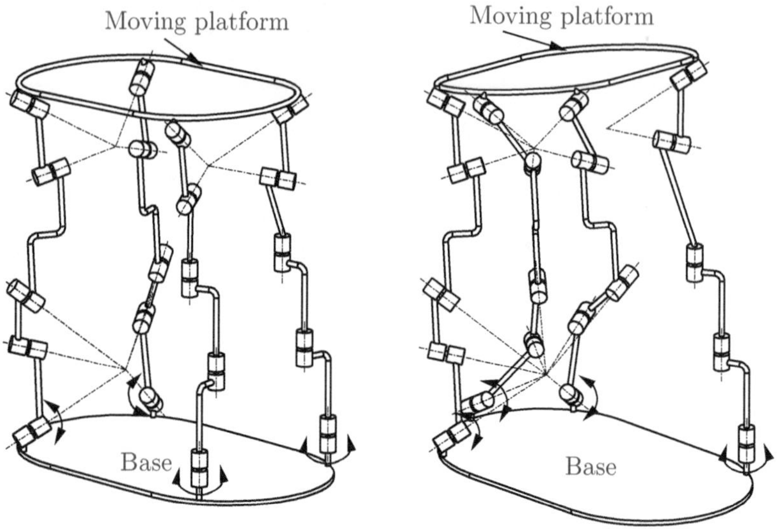

(a) $2\text{-}\underline{\check{R}}\check{R}\check{R}\hat{R}\hat{R}\_2\text{-}(\underline{R}RR)|_{E}\hat{R}\hat{R}$. **(b)** $3\text{-}\underline{\check{R}}\check{R}\check{R}\hat{R}\hat{R}\_(\underline{R}RR)|_{E}\hat{R}\hat{R}$.

Fig. 14.10. Some US-PPS= PMs

14.6 Summary

In this chapter, a method has been proposed for the type synthesis of PMs with a parallel virtual chain based on the type synthesis of PMs with a serial virtual chain. It has been shown that a PM with a parallel virtual chain can be obtained in four steps: (1) the decomposition of number of legs, (2) the type synthesis of sub-PKCs, (3) the combination of sub-PKCs to generate PKCs, and (4) the selection of the actuated joints.

15. Conclusions

知之为知之，不知为不知，是知也。
——孔子(公元前 551-479)

When you know a thing, to hold that you know it, and when you do not know a thing, to allow that you do not know it — this is wisdom.
— Confucius (551–479 BC)

15.1 Major Contributions

This book presented a systematic study of the type synthesis of PMs. The major contributions are highlighted below:

1. Representation and classification of motion patterns.
   With the introduction of the concept of virtual chain, motion patterns can be represented in a rigorous way. A number of motion patterns having practical application potential have been identified.
2. The virtual-chain approach to the type synthesis of PMs.
   Based on the concept of virtual chain and screw theory, PMs can be constructed using several of the seven compositional units via the construction of the multi-DOF single-loop KCs in four steps: (1) the decomposition of the wrench system of a parallel kinematic chain associated with a virtual chain (V= PKC), (2) the type synthesis of legs, (3) the assembly of legs to generate V= PKCs, and (4) the selection of the actuated joints. Once a PM is constructed, part of one or more of its compositional units may be invisible due to the removal of the virtual chain in the process of type synthesis. It is observed that [80]:
   - The virtual-chain approach requires fewer derivations than the approaches proposed in [13, 33, 65, 79].
   - The virtual-chain approach is conceptually simpler and therefore easier to understand than the method proposed in [58].

X. Kong and C. Gosselin: Type Syn. of Parallel Mech., STAR 33, pp. 223–225, 2007.
springerlink.com

- The virtual-chain approach is more general than the approaches presented in [47, 121]. When the latter approaches are used for the type synthesis, some 3-DOF and 4-DOF PMs with 5-DOF legs may not be obtained.

Unlike in most of the existing approaches to the type synthesis, the issue of the selection of actuated joints was systematically investigated here. The application of inactive joints was also emphasized.

3. Types of PMs with great application potential.
   Using the virtual-chain approach, the type synthesis of several PMs with a serial virtual chain was performed. A number of PPP= PMs, S= PMs, PPR= PMs, PPPR= PMs, SP= PMs, US= PMs, PPPU= PMs and PPS= PMs involving R and P joints have been obtained. Based on the results of the type synthesis of PMs with a serial virtual chain, the type synthesis of two classes of PMs generating a parallel virtual chain was also performed. The types of 3-PPS= PMs and 2-PPPRR= PMs involving R and P joints have been obtained.
   The results cover all the PMs generating the same motion pattern that are proposed in the literature as well as some new types. Based on these results, the reader can develop more and more innovative devices based on PMs including parallel manipulators, haptic devices, medical robots, nano-manipulators and micro-manipulators (Appendix A).
4. Mobility analysis of PMs.
   As a by-product of the proposed approach for the type synthesis of PMs, a novel approach has been proposed for the mobility analysis of PMs in Appendix B. The proposed process for the mobility analysis of PMs is the reverse of the process proposed for the type synthesis of PMs. This provides an efficient method for readers to understand PMs that are not covered in this book.

15.2 Future Research

Despite the advances reported in this book, there are still several issues that may deserve more attention in the future due to the complex nature of PMs. Meanwhile, the development of nanotechnology, biotechnology, MEMS technology and other high-technologies also provides new opportunities and challenges for the research on devices based on PMs (Appendix A). The following are issues that deserve further investigation.

1. To determine new motion patterns for practical applications, such as each of the parallel modules of a hybrid machine tool, and then perform the type synthesis of PMs. In addition to PMs generating a specific motion pattern presented in this book, the type synthesis of PMs generating other motion patterns has also been performed and some new PMs, such as the 3-DOF UP= PMs (see [82]), have been obtained. These results have not been presented here because (1) the application of the general procedure has

been well illustrated using the type synthesis of eight class of PMs, and (2) the inclusion of these results would further increase the number of pages. The classification of motion patterns from a theoretical perspective is also an open issue [108].

2. To perform the type synthesis of PMs with multiple operation modes. With the advances in the type synthesis of PMs with a single operation mode, it is natural to develop PMs with multiple operation modes. One of the merits of PMs with multiple operation modes is to reduce the number of actuators needed [86]. Up to now, only a few PMs with multiple operational modes have been proposed [32, 86].
3. To extend the proposed approach for the type synthesis of PMs to the type synthesis of mechanisms with more complicated structures [50]. Some applications do need such manipulators despite their complicated structures [17, 129].
4. To perform the type synthesis of compliant PMs. As compared with the conventional PMs, PMs with flexure joints do not suffer from backlash in the joints and can thus be used for nanomanipulation. Together with the advances in the research on compliant mechanisms [48, 89, 111], the work proposed in this book provides a solid foundation for the type synthesis of compliant PMs.
5. To perform the type synthesis of micro PMs. Taking into consideration the available MEMS fabrication technologies, the results in this book provide a solid starting point for the type synthesis of micromanipulators [60].

In addition, the type synthesis of PMs with simple kinematic characteristics is also of practical importance. The control of such PMs is usually very simple. Combined with the techniques for kinematic analysis, this work is also expected to be used for the type synthesis of input-output decoupled PMs [72], partially decoupled PMs [104], selectively actuated PMs [63], and analytic PMs [83].

A. Design of Devices Based on Parallel Mechanisms

In this appendix, we briefly illustrate the application of PMs in the design of innovative parallel manipulators, haptic devices, medical robots, nano-manipulator and micro-manipulators. Several common issues and specific issues are addressed for the design of devices based on PMs.

A.1 Common Issues

Although the appearance and the function of different devices based on PMs are quite different, there are several common issues to be addressed in their design.

A.1.1 Overall Design Process

The development of a device based on PMs usually follows a typical procedure such as the one illustrated schematically in Fig. A.1.

A.1.2 Selection of Working Mode

In performing the inverse displacement analysis of a PM, any one of its legs can be treated as a serial manipulator. The concept of postures of serial manipulator can also be applied to a leg in a PM. A working mode of a PM is defined as a combination of the postures of all its legs. For a PM having multiple solutions to its inverse displacement analysis, there are multiple working modes. The postures of at least one leg are different in different working modes of a PM. For a given PM, the workspace may vary with the working modes. In practice, the assembly mode which has a larger workspace for a given PM should be selected to perform a required task.

A.1.3 Redundancy in the Task Space

Generally, it is recommended that no redundant DOF in the task space be introduced in the design such that the number of actuators used is the minimum.

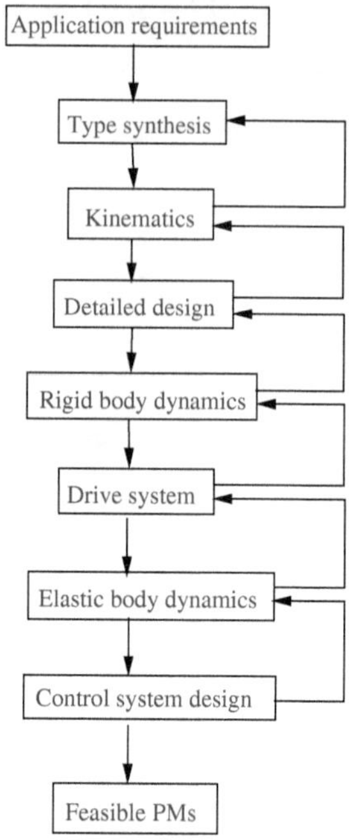

Fig. A.1. Flowchart of the design of devices based on PMs

However, if the structural error of a PM without redundant DOF in the task space exceeds the permitted limit, then a PM with redundant DOF in the task space should be considered.

For example, in order to generate a PPP motion, we can use a PPP= PM (also translational parallel mechanism). If the orientational structural error of a PPP= PM exceeds the permitted limit, then a PM with redundant DOF in the task space, such as a PPPR= PM, PPPU= PM or 6-DOF PM should be used.

A.1.4 Number of Legs

The number of legs in a PM is determined by the arrangement of actuators, the DOF of actuators, the workspace and stiffness requirement.

In most PMs, it is preferable to place all the actuators on the base. If single-DOF actuators are used, then the number of legs is equal to the DOF of the PM. If multi-DOF motors are used, the number of legs is less than the DOF of the PM.

If the objective is to design a PM with a large workspace, the number of legs my be chosen such that it is smaller than the DOF of the PM. Conversely, if the objective is to to increase the stiffness of a PM, the number of legs may be chosen such that it is greater than the DOF of the PM.

A.1.5 Selection of Types of Legs

Usually, it is preferable to use PMs with identical type of legs. However, if one cannot achieve a satisfactory result by using PMs with identical type of legs, one then should consider the application of PMs with different types of legs (see [5] for example).

A.2 Specific Issues

In this section, we will list some specific issues to be considered in the development of parallel manipulators, haptic devices, medical robots, nano-manipulators and micro-manipulators.

A.2.1 Parallel Manipulators

The forward displacement analysis of a parallel manipulator is usually of a highly nonlinear nature. The development of analytic parallel manipulators, especially parallel manipulators with a set of linear input-output equations may help to promote the use of PMs in a given application.

By considering the geometric interpretation of the forward displacement analysis of PPP= PMs, a large number of PPP= PMs with a linear solution to their forward displacement analysis were proposed in [39, 71, 72]. By further revealing the condition for a PPP= PM to have a linear solution to its inverse displacement analysis, a class of PPP= PMs with a set of linear input-output equations were obtained in [73].

Taking into consideration the workspace requirement, one can further propose the preferred designs of PPP= parallel manipulators (also called translational parallel manipulators).

Figure 6.11 shows two preferred designs of PPP= parallel manipulators, which are obtained from the 3-$\underline{\text{P}}\grave{\text{R}}\grave{\text{R}}\grave{\text{R}}$ PPP= PM. The characteristic of the manipulator shown in Fig. 6.11a is that its input-output equations are fully decoupled. Each output of the parallel manipulator is controlled exclusively by one actuator. Such a character greatly simplifies the control of the manipulator. The characteristic of the manipulator shown in Fig. 6.11b is that the range of translation along the direction parallel to the direction of linear actuators is only limited by the stroke of the linear actuators.

Figure 1.7a shows the prototype of the Tripteron which is based on the preferred design shown in Fig. 6.11a. For the 3-$\underline{\text{P}}\grave{\text{R}}\grave{\text{R}}\grave{\text{R}}$ PPP= parallel manipulator, there usually exist two sets of solutions to the joint variables of the unactuated joints for each leg and eight solutions to its inverse displacement analysis for a given position of the moving platform. The PPP= parallel manipulator has thus eight working modes. Different working modes are separated by the inverse singularity of one or more legs. Due to link interferences, the workspaces of the manipulator under different working modes may be different.

If the inverse kinematic singularities at the boundary of the workspace are eliminated by limiting the range of motion of the actuated joints, a PPP= parallel manipulator with a set of linear input-output equations will always remain in the working mode in which it was first assembled.

The prototype of the Tripteron is driven by three DC brushless servo motors having a continuous stall torque of 0.388 N m and a maximum operating speed of 6000 RPM. The control of the prototype consists of a simple PID loop on each DC motor. The control is implemented in Simulink and executed on RT-LAB's real-time platform by Opal-RT. The prototype was tested and trajectories involving accelerations of up to 8 g were produced.[40] A manufacturer of such serial Cartesian robots could undoubtedly improve their performance by adopting the Tripteron concept.

A.2.2 Haptic Devices

Haptic devices are devices that allow a computer to recreate the sense of touch for the users of a Virtual Reality environments. Haptic devices provide controlled force feedback to the fingers of the user so that they feel as though they are touching objects in the virtual landscape. Haptic devices have many applications such as surgeon training, keyhole surgery and entertainment.

High-performance haptic devices should have the following features [101]: low inertia, high stiffness, low friction, backdriveability, near-zero backlash, gravitational counterbalancing, sizeable workspace, multiple DOF, and human matched force capabilities. The haptic device must not distort the reflected forces/torques by its mechanical characteristics. In addition, compactness is highly desirable. It can be seen that PMs are appropriate candidates for haptic devices since they are well-known for their high stiffness and low inertia, which enables large bandwidth transmission of forces [5]. Other desired properties can be achieved through suitable mechanical design and proper control scheme. It is preferable not to use unactuated P joints in a haptic device.

For the purpose of wrist rehabilitation, the moving platform needs to undergo three-DOF rotations about the centre of the wrist. Thus, haptic devices for wrist rehabilitation can be developed based on the class of S= PMs. Figure 1.2g shows the haptic device developed at Laval University [5] based on the $\hat{\text{R}}\hat{\text{R}}\hat{\text{R}}$-$\hat{\text{R}}\hat{\text{R}}\hat{\text{R}}$-$\hat{\text{R}}(\text{RRR})_E\hat{\text{R}}$ S= PM. It is noted that in order to avoid link interference, two types of legs are used in the above S= PM. Two legs are of type $\hat{\text{R}}\hat{\text{R}}\hat{\text{R}}$ (No. 1 in Table 7.1), and the last leg is of type $\hat{\text{R}}(\text{RRR})_E\hat{\text{R}}$ (No. 21 in Table 7.1).

A.2.3 Medical Robots

Medical robots are used to expand the surgeons' capability. Instead of directly handling the instruments, the surgeon teleoperates a robot which performs the actual motion. This eliminates hand tremor, while allowing motion scaling and very high precision. A survey of the design issues in surgical robotics is presented in [113].

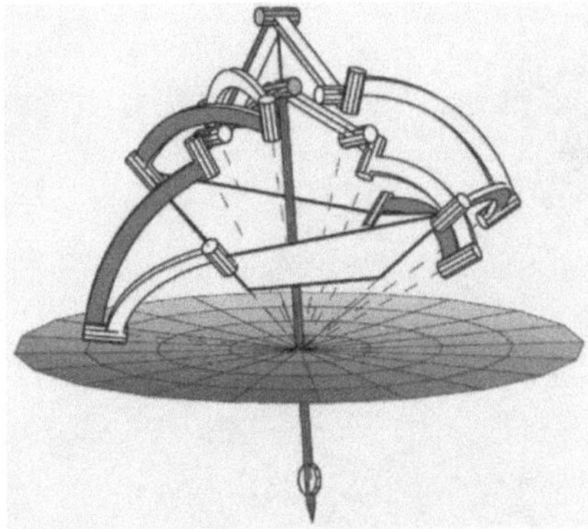

Fig. A.2. $4\text{-}\underline{\hat{R}}\hat{R}\hat{R}(RR)^{|}_{E}SP{=}$ PM (©2005 IEEE, reprinted with permission, from [135])

Minimally invasive surgery is performed through sets of small incisions, rather than the traditional single major incision in the patient's body. This greatly reduces the recovery time of the patient. In order to place the instrument to carry out the surgery, the RCM (remote centre-of-motion) is a central design feature of any robot for keyhole surgery.

Of particular interest is laparascopic surgery, performed inside the abdomen of the patient. The surgeon operates with long, thin instruments inserted into the abdomen through tiny holes. According to [135], the mobility required for the mechanism carrying the instrument during a minimally invasive operation is four DOF in order to guarantee that the tool pivots about the point at which it enters in the patient's body and translates along its axis (insertion/retraction).

Based on the above motion requirement, the class of SP= PMs (Chap. 10) can be used to carry out the minimally invasive surgery.

Figure A.2 shows the $4\text{-}\underline{\hat{R}}\hat{R}\hat{R}(RR)^{|}_{E}$ SP= PM proposed in [135] by taking into consideration the workspace requirement and the RCM consideration.

A.2.4 Nano-manipulators

Nano-manipulators (also called Flexure parallel manipulators in [103]) refer here to manipulators with (sub)nanometer precision. Such precision can be achieved by replacing conventional joints of a PM by flexure joints. The key issue in the replacement is to ensure that the stiffness of the manipulator is high.

Nano-manipulators belong to compliant mechanisms, which gain some or all of their motion from the elastic deformation of flexible segments. The advantages of compliant mechanisms over traditional rigid-body mechanisms are that they are more resistant to wear due to the elimination of rubbing parts and that backlash and clearances are eliminated.

Figure A.3 shows a plastic model of the nano-manipulator proposed in [120]. Figure 1.3b shows a prototype of the nano-manipulator proposed in [103]. Unlike in the manipulator proposed in [120], in which each flexure joint is very wide,

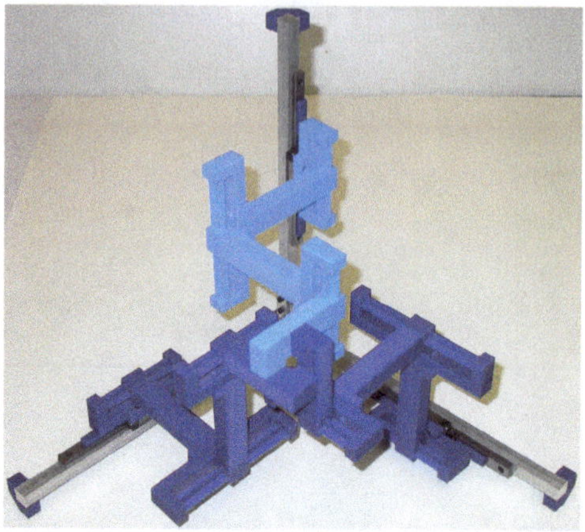

Fig. A.3. 3-$\underline{\text{P}}\grave{\text{R}}\grave{\text{R}}\grave{\text{R}}$ PPP= PM

three groups of legs are arranged in parallel thus greatly reducing the width of each flexure joint. It was pointed out in [103] that the error of the nano-manipulator is proved — using optical laser sensors — to be less than 3% and can be ignored in a motion range of 20 μm. Also, from the experiments, the resolution and the repeatability of the prototype are determined as 0.2 μm and less than 0.4 μm, respectively, within a motion range of 10 μm.

A.2.5 Micro-manipulators

Using the available MEMS fabrication technologies, micro-manipulators can be developed starting from the PMs proposed in this book. Both rigid-body and compliant micro-manipulators can be fabricated [60].

As pointed out in [60], the placement and geometry of the legs are governed by the limitations in the surfacing or bulk micro-machining process. For the current surfacing micro-machining process, the axis of any R joint should be either parallel or perpendicular to a plane, and the directions of P joints must be parallel to the same plane. For some surfacing micro-machining processes, one link can be layered over another. For other surfacing micro-machining processes, one link cannot be layered over another. In addition, it is preferable to use linear actuated joints located on the base and not to use unactuated P joints.

Now, let us look briefly at the micro-manipulators corresponding to the 3-$\underline{\grave{\text{R}}}\grave{\text{R}}\text{P}\grave{\text{R}}$ PPP= PM (See No. 3 of Table 6.5).

Firstly, arrange the R and P joints such that the axis of any R joint is either parallel or perpendicular to a plane, and the directions of P joints must be parallel to the same plane (Fig. A.4a). Here and in Fig. A.5, the base of the PM is not shown for brevity. Secondly, replace an unactuated P joint with a planar parallelogram (Fig. A.4b) for each leg. Thirdly, in order to use a linear actuator to drive the R joint on the base in the micro-manipulator, connect a

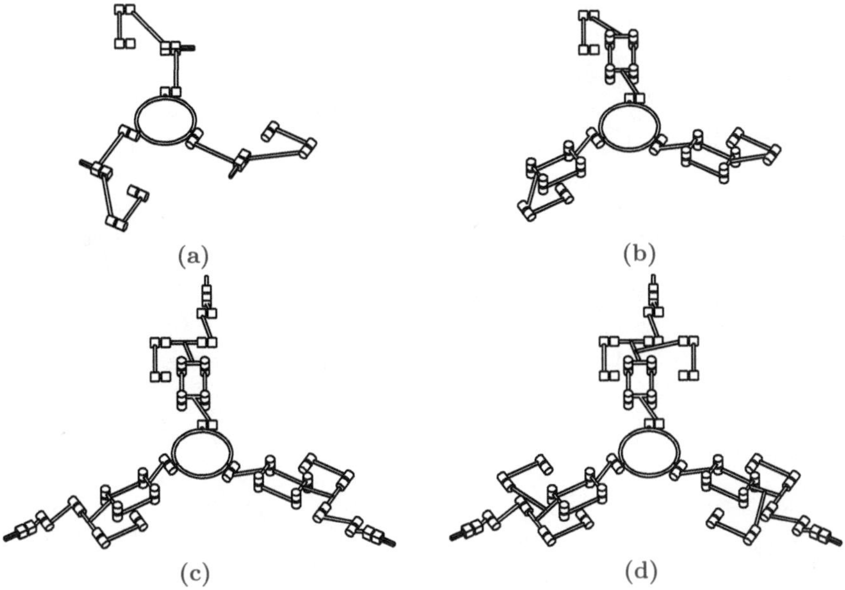

Fig. A.4. Design of a micro-manipulator based on the 3-$\underline{\grave{R}}\grave{R}P\grave{R}$ PPP= PM

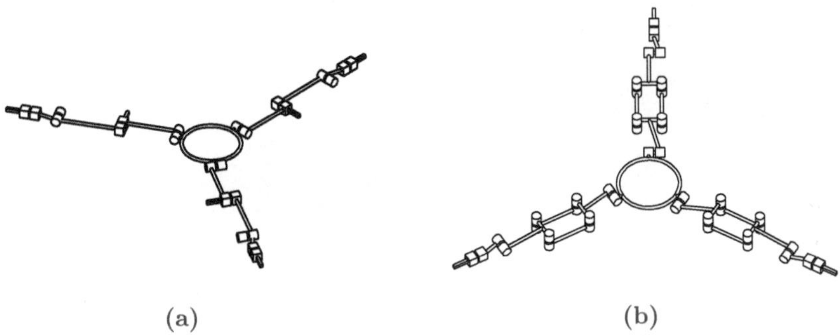

Fig. A.5. Design of a micro-manipulator based on the 3-$\underline{P}\grave{R}P\grave{R}$ PPP= PM

serial KC, which is composed of two links connected by an R joint, to the first link by an R joint and to the base by a P joint. Then, the serial KC, the first link and the base form a planar slider four-bar linkage (Fig. A.4c). Thus, the rotation of the R joints located on the base can be controlled through the slider by using a linear actuator. Finally, in order to improve the stiffness of the micro parallel manipulator, add another link to connect the first two joints in each leg (Fig. A.4d).

Figure 1.3c shows the micro-manipulator [60] corresponding to the PM shown in Fig. A.4d.

Similarly, we can obtain a PPP= PM (Fig. A.5b) from which a micro-manipulator can be fabricated using the surfacing micro-machining process. The

PPP= PM is obtained from the $3\text{-}\underline{P}\grave{R}P\grave{R}$ PPP= PM as follows: Firstly, arrange the R and P joints such that the axis of any R joint is either parallel or perpendicular to a plane, and the directions of P joints must be parallel to the same plane (Fig. A.5a). Secondly, replace an unactuated P joint with a planar parallelogram (Fig. A.5b). The PM has fewer links than the one shown in Fig. A.4d[1].

A.3 Summary

In this appendix, several examples were used to illustrate the application of the results presented in this book to the design of devices based on PMs. It is hoped that many innovative devices based on PMs can be developed based on the approach and the results presented in this book.

[1] A systematic study on the type synthesis of micro-manipulators will be published in our future publications.

B. Mobility Analysis of Parallel Mechanisms

Since it is almost impossible to cover all the types of PMs in a single book, it is necessary to provide a systematic approach for the mobility analysis of PMs in order for the reader to understand the PMs not listed in this book. Based on the reverse process of the type synthesis of PMs proposed in this book, a systematic approach for the mobility analysis of PMs [74] is proposed. The proposed approach is also illustrated with several examples. For other recent works on the mobility analysis of PMs, see [22, 23, 36, 47, 49, 69, 72, 91, 106, 107, 121, 130].

B.1 Principle of Full-Cycle Mobility Inspection

B.1.1 Equivalent Serial Kinematic Chain

Similar to the concept of virtual chain in the type synthesis, the concept of equivalent serial KC is introduced for the mobility analysis of PMs.

For a k-legged PKC [Fig. B.1a], which is composed of k legs of an m-legged PKC (Fig. 2.17), let $\mathcal{W}_{[k]}$ denote the wrench system of the k-legged PKC. An *equivalent serial KC* of the k-legged PKC is defined as a serial KC which has the same twist system and wrench system as the k-legged PKC.

For a specified twist system of the moving platform, there is no difficulty in determining a serial KC with the same twist system. For example, if a twist system of a PKC is the 3-$\boldsymbol{\xi}_\infty$-system, a serial KC with the same twist system can be any one PPP serial chain in which the direction of the P joints are not all parallel to one plane. If a twist system of a PKC is a 3-$\boldsymbol{\xi}_\infty$-1-$\boldsymbol{\xi}_0$-system, a serial KC with the same twist system can be any one PPPR serial chain in which the axis of the R joint is parallel to the axis of the $\boldsymbol{\xi}_0$ and the direction of the P joints are not all parallel to one plane. If for a k-legged PKC there exists a leg i which satisfies $\mathcal{W}^i = \mathcal{W}_{[k]}$, leg i can be directly selected as the equivalent serial KC.

If the equivalent serial KC is not a leg of the k-legged PKC, we can connect the moving platform and the base of the k-legged PKC using the equivalent serial KC

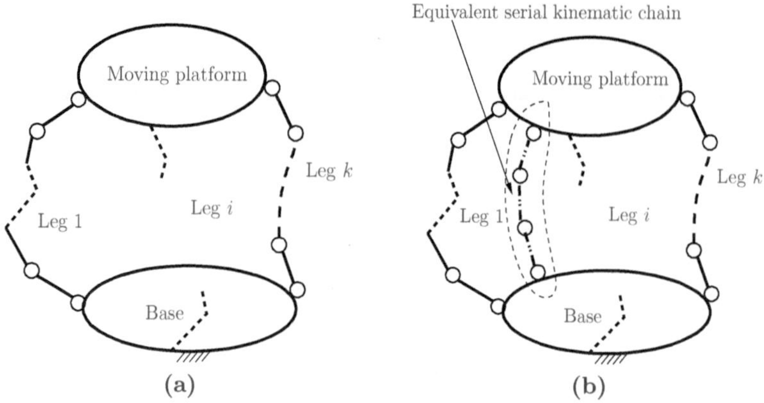

Fig. B.1. k-legged PKC of an m-legged PKC: (**a**) The original one and (**b**) The one with the equivalent KC added

[Fig. B.1b] without affecting the instantaneous mobility of the k-legged PKC. The instantaneous mobility of the $(k+1)$-legged PKC obtained above will be $(\mathcal{F}_{[k]} + R_e)$, where $(\mathcal{F}_{[k]}$ is the mobility of the k-legged PKC, and R_e denotes the number of independent parameters to determine the configuration of the equivalent serial KC with the moving platform fixed. Just as in the above examples, the equivalent serial KC is generally selected in such a way that $R_e = 0$.

It is evident that a sufficient condition for the twist systems of the equivalent serial KC and the k-legged parallel KC to be equal to each other under any change of configuration of the equivalent serial KC and the k-legged PKC is that the equivalent serial KC and each leg of the k-legged PKC form a $(\mathcal{C}_{[k]}+R_e+R^i)$-DOF (degree-of-freedom) single-loop KC of full-cycle mobility. Here, $\mathcal{C}_{[k]}$ and R^i denote, respectively, the order of the twist system of the moving platform and the number of independent parameters to determine the configuration of the leg i with the moving platform fixed. The equivalent serial KC of a k-legged PKC satisfying the above condition is called a *full-cycle equivalent serial KC.*

B.1.2 A Sufficient Condition for a PM to Have Full-Cycle Mobility

Using the concept of equivalent serial KC, a *sufficient condition* for a PKC to have full-cycle mobility can be proposed as follows:

A PM has full-cycle mobility if (a) it has a full-cycle equivalent serial KC or (b) it is composed of more than one PKC having a full-cycle equivalent serial KC and all the full-cycle equivalent serial KCs form a non-overconstrained KC.

Since multi-DOF single-loop non-overconstrained KCs are easy to identify and the types of multi-DOF single-loop overconstrained KCs have been identified in Chap. 3 in details, the mobility analysis of PMs can be performed based on the above sufficient condition.

B.2 Procedure for the Mobility Analysis

The mobility analysis of a PKC includes the instantaneous mobility analysis and the full-cycle mobility inspection.

B.2.1 Instantaneous Mobility Analysis

The *instantaneous mobility analysis* of PKCs has been dealt with in Sect. 2.2.3 and [69]. As in Sect. 2.2.3, m denotes the number of legs in a PM, while c and $\mathcal{F}$ denote, respectively, the order of the wrench system $\mathcal{W}$ and mobility (or DOF) of the PKC. Similarly, c^i and f^i denote the order of the wrench system $\mathcal{W}^i$ and DOF of leg i. R^i denotes the redundant DOF of leg i. In the mobility analysis of PMs, the following three indices should be computed:

The order $\mathcal{C}$ of the twist system of the PKC (also called the *connectivity* of the moving platform) is (see (2.18))

$$\mathcal{C} = 6 - c. \tag{B.1}$$

The *mobility* (or the *degree of freedom*) $\mathcal{F}$ of the PKC is (see (2.21))

$$\mathcal{F} = \mathcal{C} + R = 6 - c + \sum_{i=1}^{m} R^i. \tag{B.2}$$

where

$$R^i = f^i - (6 - c^i) = f^i - 6 + c^i \tag{B.3}$$

The *number of overconstraints* (also *passive constraints* or *redundant constraints*) if $\Delta > 0$ is (see (2.22))

$$\Delta = \sum_{i=1}^{m} c^i - c \tag{B.4}$$

Consider the 3-$(\text{PRRR})_A$ PKC shown in Fig. B.2a. The mobility analysis of this mechanism has been discussed in [72, 106]. In this PKC, all the axes of the R joints within a same leg are parallel. The axis of a P joint is not perpendicular to the axes of the R joints within the same leg. The axes of the R joints on the moving platform are not all parallel. The wrench system of each leg is a 2-ζ_∞-system, which is composed of all the ζ_∞ whose axes are perpendicular to the axes of all the R joints within a same leg. The wrench system of the PKC is the 3-ζ_∞-system. We have $c^i = 2$, $c = 3$, $R^i = 0$. Using (B.1), (B.2) and (B.4), we obtain

$$\mathcal{C} = 6 - c = 3$$

and

$$\mathcal{F} = \mathcal{C} + \sum_{i=1}^{3} R^i = 3.$$

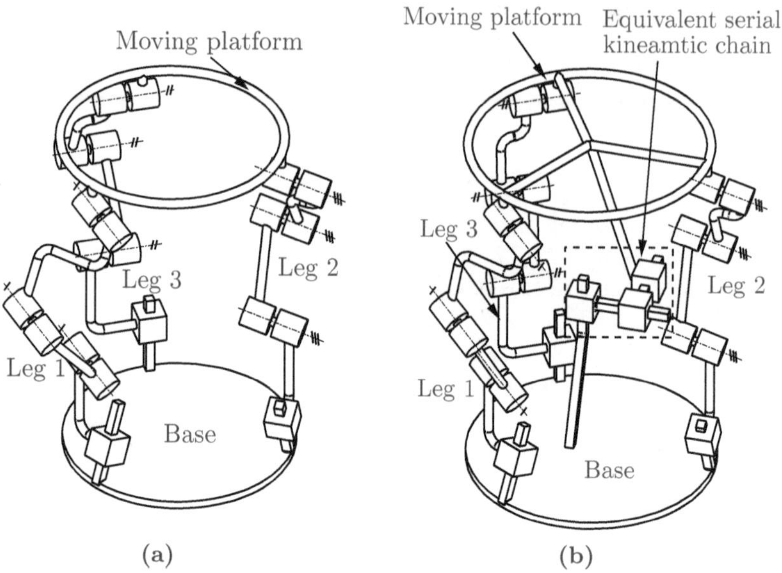

Fig. B.2. Mobility analysis of the 3-$(PRRR)_A$ PKC: (**a**) The original KC and (**b**) The KC with an equivalent serial KC added

The number of overconstraints of this 3-legged PKC is

$$\Delta = \sum_{i=1}^{3} c^i - c = 6 - 3 = 3.$$

B.2.2 Full-Cycle Mobility Inspection

The mobility of a PM obtained using (B.2) is usually instantaneous. In this section, we will discuss how to determine whether the PM has *full-cycle mobility* or not.

Based on the sufficient condition for a PM to have full-cycle mobility, a procedure for the full-cycle mobility inspection of a PKC is proposed as follows.

Step 1. Check if $\Delta = 0$ for a PKC. If yes, then the PKC has full-cycle mobility and inspection ends. Otherwise, go to the next step.

Step 2. Identify and remove inactive joints.
The purpose of this step is to simplify the full-cycle mobility inspection. An inactive joint is a joint in an PKC which always loses its DOF. When an inactive joint is removed from a mechanism, the relative motion within the PKC is unchanged. Inactive joints can be identified using the methods proposed in the literature [72, 79].

Step 3. Check if the PKC obtained in Step 2 has a full-cycle equivalent serial KC. If yes, the PKC has full-cycle mobility and inspection ends. Otherwise, go to the next step.

Step 4. For $j = 2$ to $(m-1)$, check if the PKC can be decomposed into j m_i-legged PKCs which all have full-cycle equivalent serial KCs and whose full-cycle equivalent serial KCs form a non-overconstrained KC. Here, $\sum_{i=1}^{j} m_i = m$. If yes for any one j, the PKC has full-cycle mobility and inspection ends. Otherwise, go to the next step.

Step 5. The PKC does not have full-cycle mobility[1] and inspection ends.

We have found that the instantaneous mobility of the 3-$(\text{PRRR})_A$ PKC shown in Fig. B.2a is 3. Now, let us discuss the full-cycle mobility inspection. Following the above procedure, we have

Step 1. Since $\Delta = 3 \neq 0$, go to the next step.

Step 2. There are no inactive joints in this PKC.

Step 3. It can be found that the PPP equivalent serial KC [Fig. B.2b] and each $(\text{PRRR})_A$ leg form a 3-DOF single-loop KC. Thus, the PPP equivalent serial KC of the 3-$(\text{PRRR})_A$ PKC is its full-cycle equivalent serial KC. Then we conclude that the PKC has full-cycle mobility.

B.3 Examples

The method for the mobility analysis proposed above will now be illustrated using the following five examples.

Example B.1. Consider the $(\text{PRRR})_A$R-2-$(\text{PRRR})_A$ PKC shown in Fig. B.3. In this PKC, all the axes of the R joints within a same "()" are parallel. The axis of a P joint is not perpendicular to the axes of the R joints within the same "()".

Instantaneous mobility analysis

The wrench system of the $(\text{PRRR})_A$R leg is a 1-ζ_∞-system, in which the base wrench is a ζ_∞ whose axis is perpendicular to the axes of all the R joints within the same leg. The wrench system of each $(\text{PRRR})_A$ leg is a 2-ζ_∞-system, which is composed of all the ζ_∞ whose axes are perpendicular to the axes of all the R joints within a same leg. The wrench system of the PKC is the 3-ζ_∞-system. We have $c^1 = 1$, $c^2 = c^3 = 2$, $c = 3$, $R^i = 0$. Using (B.1), (B.2) and (B.4), we obtain

$$\mathcal{C} = 6 - c = 3$$

and

$$\mathcal{F} = \mathcal{C} + \sum_{i=1}^{3} R^i = 3.$$

The number of overconstraints of this 3-legged PKC is

$$\Delta = \sum_{i=1}^{3} c^i - c = 1 + 2 + 2 - 3 = 2.$$

[1] Although there may exist paradoxical PKCs, they are excluded from the scope of this book.

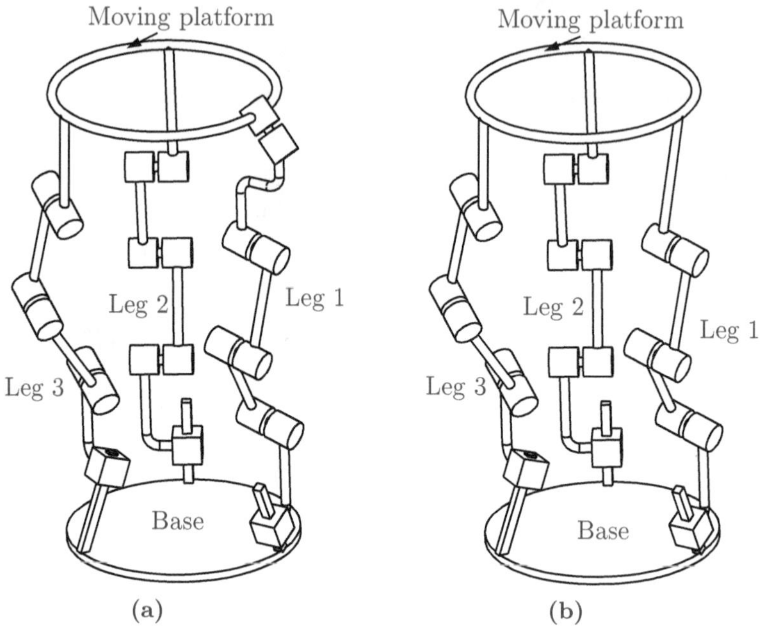

Fig. B.3. Mobility analysis of the $(PRRR)_A$R-2-$(PRRR)_A$ PKC: (**a**) The original KC and (**b**) The KC with inactive joints removed

Full-cycle mobility inspection

Following the procedure for the full-cycle mobility inspection, we have

Step 1. Since $\Delta = 2 \neq 0$, go to the next step.

Step 2. For this PKC, the R joint in the $(PRRR)_A$R leg which is located on the moving platform is inactive [72, 79]. Removing the inactive joint, one obtains the 3-$(PRRR)_A$ PKC shown in Fig. B.3b.

Step 3. As we have found above, the 3-$(PRRR)_A$ PKC has a full-cycle equivalent serial KC. The full-cycle equivalent serial KC can be represented by a PPP serial KC. Thus, the PKC has full-cycle mobility.

Example B.2. Consider the 3-$\check{R}\check{R}\check{R}\hat{R}\hat{R}$ PKC shown in Fig. B.4. In this PKC, the axes of all the $\check{R}$ joints pass through one common point, while the axes of the $\hat{R}$ joints pass through another common point. The mobility analysis of this PKC can be performed as follows.

Instantaneous mobility analysis

The wrench system of each leg is a 1-ζ_0-system in which the base wrench can be a ζ_0 whose axis passes through the above two common points. The wrench system of the PKC is still a 1-ζ_0-system. We have $c^i = 1$, $c = 1$, $R^i = 0$. Using (B.1), (B.2) and (B.4), we obtain

$$\mathcal{C} = 6 - 1 = 5$$

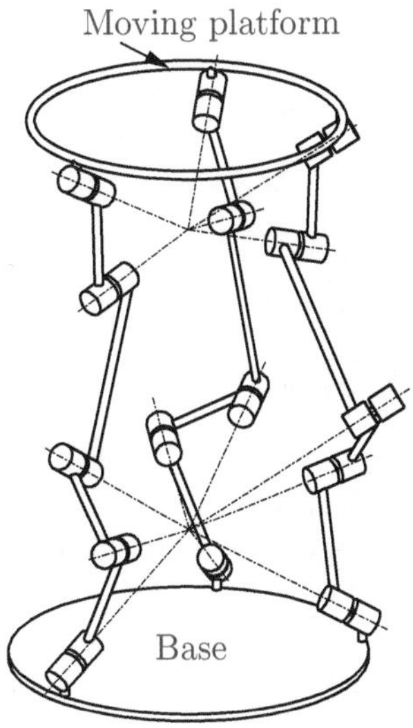

Fig. B.4. Mobility analysis of the 3-ŘŘŘR̂R̂ PKC

and

$$\mathcal{F} = \mathcal{C} + \sum_{i=1}^{3} R^i = 5.$$

The number of overconstraints of this 3-legged PKC is

$$\Delta = \sum_{i=1}^{3} c^i - c = 3 - 1 = 2.$$

Full-cycle mobility inspection
Following the procedure for the full-cycle mobility inspection, we have

Step 1. Since $\Delta = 2 \neq 0$, go to the next step.
Step 2. For this PKC, there are no inactive joints.
Step 3. It can be found that the PKC has a full-cycle equivalent serial KC. The full-cycle equivalent serial KC can be represented by leg 1. Thus, the PKC has full-cycle mobility.

Example B.3. Consider the 4-R̋R̋R̀R̀R̋ PKC shown in Fig. B.5a[78]. In this PKC, the axes of all the R̋ joints are parallel, while the axes of the R̀ joints within the same leg are parallel. The mobility analysis of this PKC can be performed as follows.

Instantaneous mobility analysis
The wrench system of each leg is a 1-$\boldsymbol{\zeta}_\infty$-system in which the base wrench can be a $\boldsymbol{\zeta}_\infty$ whose axis is perpendicular to the axes of all the R joints within the

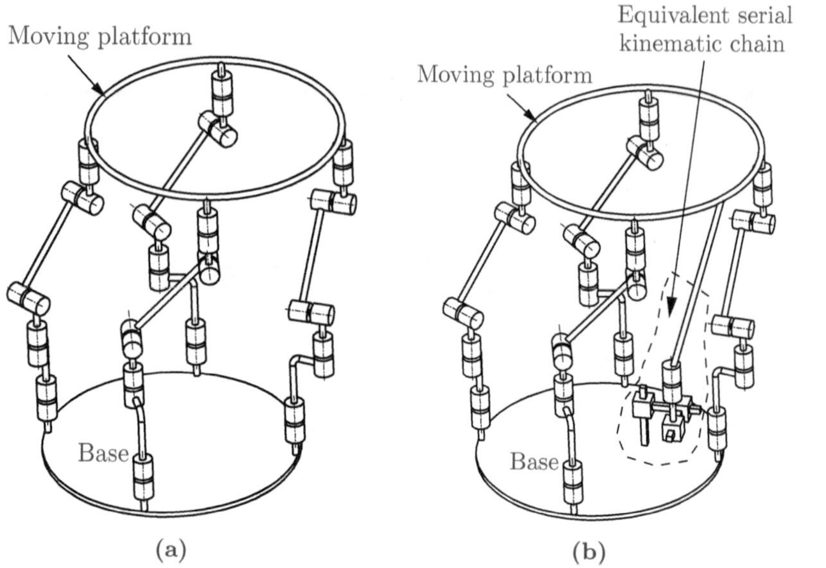

Fig. B.5. Mobility analysis of the 4-R̋R̋R̀R̀R̋ PKC: (**a**) The original KC and (**b**) The KC with an equivalent serial KC added

leg. The wrench system of the PKC is a 2-ζ_∞-system, which is composed of all the $\boldsymbol{\zeta}_\infty$ whose axes are perpendicular to the axes of the R̋ joints. We have $c^i = 1$, $c = 2$, $R^i = 0$. Using (B.1), (B.2) and (B.4), we obtain

$$\mathcal{C} = 6 - 2 = 4$$

and

$$\mathcal{F} = \mathcal{C} + \sum_{i=1}^{4} R^i = 4.$$

The number of overconstraints of this 4-legged PKC is

$$\Delta = \sum_{i=1}^{4} c^i - c = 4 - 2 = 2.$$

Full-cycle mobility inspection

Following the procedure for the full-cycle mobility inspection, we have

Step 1. Since $\Delta = 2 \neq 0$, go to the next step.

Step 2. For this PKC, there are no inactive joints.

Step 3. It can be found that the PKC has a full-cycle equivalent serial KC. The full-cycle equivalent serial KC is a PPPR serial KC in which the axis of the R joint is parallel to the axes of the R̋ joints [Fig. B.5b]. Thus, the PKC has full-cycle mobility.

Example B.4. Consider the 2-R̄R̄R̄R̈-R̋R̀R̀R̀R̋ PKC shown in Fig. B.6a[77]. In this PKC, the axes of all the R̄ joints are parallel, the axes of the R̈ joints are coaxial, the axes of the R̋ joints are parallel to the axes of the R̈ joints, while the axes of the R̀ joints within a same leg are parallel. The mobility analysis of this PKC can be performed as follows.

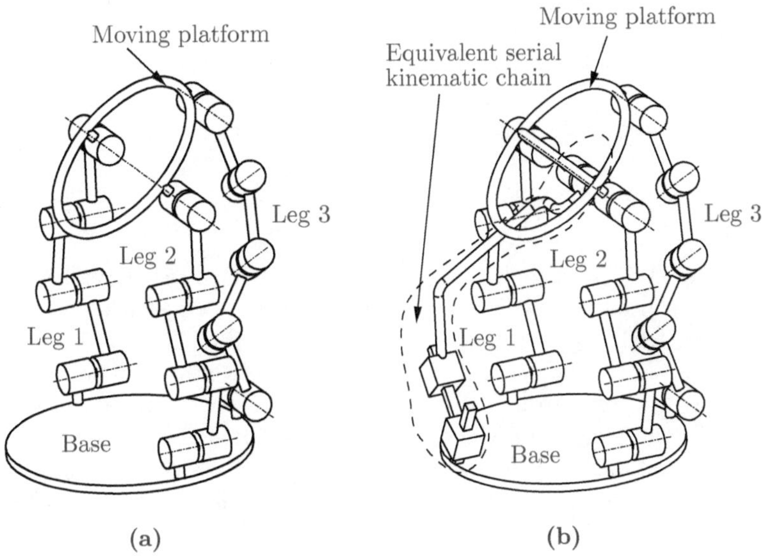

Fig. B.6. Mobility analysis of the 2-R̄R̄R̄R̈-R̋R̀R̀R̀R̋ PKC: (**a**) The original KC and (**b**) The KC with an equivalent serial KC added

Instantaneous mobility analysis

The wrench system of each R̄R̄R̄R̈ leg is a 1-ζ_0-1-ζ_∞-system, whose base can be represented by a ζ_∞ whose axis is perpendicular to the axes of all the R joints and a ζ_0 whose axis intersects the axes of the R̈ joint and is parallel to the axes of the R̄ joints. The wrench system of the R̋R̀R̀R̀R̋ leg is a 1-ζ_∞-system in which the base wrench can be a ζ_∞ whose axis is perpendicular to the axes of all the R joints within the leg. The wrench system of the PKC is a 1-ζ_0-2-ζ_∞-system, which is composed of all the ζ_∞ whose axes are perpendicular to the axes of the R̈ joints and all the ζ_0 whose axes are parallel to the axes of the R̄ joints and intersect the axes of the R̈ joints. We have $c^1 = c^2 = 2$, $c^3 = 1$, $c = 3$, $R^i = 0$. Using (B.1), (B.2) and (B.4), we obtain

$$\mathcal{C} = 6 - 3 = 3$$

and

$$\mathcal{F} = \mathcal{C} + \sum_{i=1}^{3} R^i = 3.$$

The number of overconstraints of this 3-legged PKC is

$$\Delta = \sum_{i=1}^{3} c^i - c = 2 + 2 + 1 - 3 = 2.$$

Full-cycle mobility inspection

Following the procedure for the full-cycle mobility inspection, we have

Step 1. Since $\Delta = 2 \neq 0$, go to the next step.
Step 2. For this PKC, there are no inactive joints.
Step 3. It can be found that the PKC has a full-cycle equivalent serial KC. The full-cycle equivalent serial KC is a PPR serial KC [Fig. B.6b] in which the axis of the R joint is coaxial with the axes of the R̈ joints located on the moving platform and the directions of the P joints are all perpendicular to the axes of the R̄ joints. Thus, the PKC has full-cycle mobility.

Example B.5. Consider the 4-legged 2-ŘŘŘR̂R̂_2-R̋R̋R̋R̂R̂ PKC shown in Fig. B.7a. The 4-legged PKC is composed of two leg-groups: the leg-group of the two ŘŘŘR̂R̂ legs and the leg-group of the two R̋R̋R̋R̂R̂ legs. The geometric constraints on the location of the joints that are indicated by the notation of the joints need only to be satisfied by the joints within the same leg-group. In this PKC, the axes of all the Ř joints in the two ŘŘŘR̂R̂ legs pass though a common point, the axes of all the R̂ joints in the two ŘŘŘR̂R̂ legs pass though a second common point, the axes of all the R̂ joints in the two R̋R̋R̋R̂R̂ legs pass through a third common point, the axes of all the R̋ joints are parallel. The mobility analysis of this PKC can be performed as follows.

Instantaneous mobility analysis

The wrench system of each ŘŘŘR̂R̂ leg is a 1-ζ_0-system in which the base wrench can be a ζ_0 whose axis passes through the common point of the axes of all the Ř joints and the common point of the axes of all the R̂ joints. The wrench system of each R̋R̋R̋R̂R̂ leg is a 1-ζ_0-system in which the base wrench can be a ζ_0 whose axis passes through the common point of the axes of all the R̂ joints and is parallel to the axes of the R̋ joints. The wrench system of the PKC is a 2-ζ_0-system, which is a linear combination of the above two 1-ζ_0-systems. We have $c^i = 1$, $c = 2$, $R^i = 0$. Using (B.1), (B.2) and (B.4), we obtain

$$\mathcal{C} = 6 - 2 = 4$$

and

$$\mathcal{F} = \mathcal{C} + \sum_{i=1}^{4} R^i = 4.$$

The number of overconstraints of this 4-legged PKC is

$$\Delta = \sum_{i=1}^{4} c^i - c = 4 - 2 = 2.$$

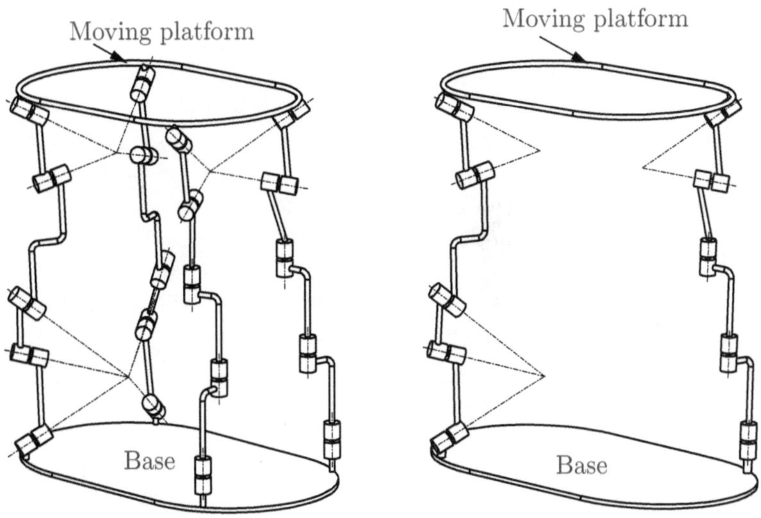

(a) (b)

Fig. B.7. Mobility analysis of the 2-ŘŘŘR̂R̂_2-R̋R̋R̋R̂R̂ PKC: (**a**) The original KC and (**b**) The KC composed of two equivalent serial KCs

Full-cycle mobility inspection

Following the procedure for the full-cycle mobility inspection, we have

Step 1. Since $\Delta = 2 \neq 0$, go to the next step.

Step 2. For this PKC, there are no inactive joints.

Step 3. The PKC does not have a full-cycle equivalent serial KC. Go to the next step.

Step 4. It is found that the PKC composed of legs 1 and 2 has a full-cycle equivalent serial KC, which can be represented by leg 1, and that the PKC composed of legs 3 and 4 has a full-cycle equivalent serial KCs, which can be represented by leg 3. In addition, legs 1 and 3 form a non-overconstrained KC. Thus, the PKC has full-cycle mobility.

The above examples show that the mobility analysis of PMs has been well solved using the proposed method. It is noted that the mobility analysis of PMs such as the mechanisms in Example 1 can also be dealt with using the approaches proposed in [22, 23, 36, 47, 49, 69, 72, 91, 106, 107, 121, 130]. However, the mobility of some mechanisms such as the mechanisms in Examples 4 and 5 cannot be well solved using the previous approaches in their current state.

It is also noted that there exist PMs with only instantaneous mobility and mixed instantaneous and full-cycle mobility [74]. Such PMs deserve further investigations.

B.4 Summary

This appendix has presented a systematic approach for the mobility analysis of PMs. Using the proposed approach, the mobility analysis is performed in two steps. The first step is the instantaneous mobility analysis, and the second step is the full-cycle mobility inspection. The first step can be performed based on screw theory. The second step has been solved using the concept of equivalent serial KC and the types of multi-DOF single-loop overconstrained KC.

The results introduced in this paper will facilitate the understanding and the application of PMs. The extension of the proposed approach to more complicated multi-loop mechanisms deserves further investigation.

C. Method Based on the Displacement Group Theory

In this appendix, we briefly introduce the method based on the displacement group theory[3, 34, 42, 44, 47, 81, 105, 112] for the type synthesis of PMs. This method was first proposed in [42, 47, 112] for the type synthesis of PPP= parallel mechanisms and later extended to the type synthesis of other classes of PMs such as S= PMs [64], PPPR= PMs [3] and PPS= PMs [41]. Using this method, PMs with finite motion are generated directly. There is no need to deal with the instantaneous motion or instantaneous constraint of the PMs. Recently, a method was proposed based on the displacement group theory for the type synthesis of linear PPP= PM [81]. A linear PPP= PM is a PPP= PM for which the forward displacement analysis can be performed by solving a set of linear equations. However, there are currently certain limitations associated with the method based on the displacement group theory.

C.1 Displacement Groups and Their Generators

The set of all possible displacements has the structure of a group and is thus called the *displacement group* $\mathcal{D}$[3, 34, 42]. The displacement group $\mathcal{D}$ has 12 displacement subgroups including $\mathcal{D}$ itself (Table C.1). To facilitate the description of the displacement subgroups — especially the intersection of displacement subgroups — notations similar to the ones proposed in [3] are adopted. Displacement subgroups as well as the axes of R and H joints are denoted with calligraphic fonts. Vectors are denoted with lower-case boldfaces. Any mechanical system, which produces a displacement subgroup is called its *mechanical generator*. Table C.1 lists the mechanical generators of these displacement subgroups, which are composed of P, R and H joints. As pointed out in [47], planar parallelograms can be used in the generators of displacement subgroups $\mathcal{T}(\mathbf{n})$, $\mathcal{T}$, $\mathcal{F}(\mathbf{n})$, $\mathcal{Y}(\mathbf{u}, p)$, $\mathcal{X}(\mathbf{u})$ and $\mathcal{D}$. Spatial parallelograms [3, 35] can also be used in the generators of displacement subgroups $\mathcal{T}$, $\mathcal{X}(\mathbf{u})$ and $\mathcal{D}$ [3].

In the literature, different notations are proposed to represent displacement subgroups. In order to facilitate the review of the literatures on the displacement group theory, different notations of displacement subgroups are summarized in Table C.2.

Table C.1. Displacement subgroups and their generators

Dimension	Subgroup	Description	Generator
0	$\mathcal{I}$	No displacement	-
1	$\mathcal{P}(\mathbf{u})$	Translations in the direction $\mathbf{u}$.	P where the direction of the P joint is parallel to $\mathbf{u}$.
	$\mathcal{R}(\mathcal{A})$	Rotations about axis $\mathcal{A}$.	R where the axis of the R joint is $\mathcal{A}$.
	$\mathcal{H}(\mathcal{A}, p)$	Rotations ϕ about $\mathcal{A}$ and translations s along the same direction that are related by the pitch $s = p\phi$.	H which has axis $\mathcal{A}$ and pitch p.
2	$\mathcal{T}(\mathbf{n})$	Translations in the directions of two distinct unit vectors $\mathbf{u}$ and $\mathbf{v}$ that are perpendicular to $\mathbf{n}$.	PP in which the directions of P joints are perpendicular to $\mathbf{n}$.
	$\mathcal{C}(\mathcal{A})$	Independent rotations about $\mathcal{A}$ and translations along the same direction.	PR, RP, PH, HP, RH, HR or HH in which the axis of each R or H joint has axis $\mathcal{A}$ and the direction of the P joint is parallel to $\mathcal{A}$.
3	$\mathcal{T}$	Translations in the directions of three distinct unit vectors.	PPP
	$\mathcal{F}(\mathbf{n})$	Two independent translations in the directions of two distinct unit vectors $\mathbf{u}$ and $\mathbf{v}$ that are perpendicular to $\mathbf{n}$ as well as one rotation about an axis parallel to $\mathbf{n}$.	PPR, PRP, RPP, PRR, RPR, RRP, or RRR in which the axes of the R joints are parallel to $\mathbf{n}$, and the directions of the P joints are perpendicular to $\mathbf{n}$.
	$\mathcal{S}(\mathcal{O})$	Rotations about point $\mathcal{O}$.	RRR in which the axes of the R joints pass through $\mathcal{O}$.
	$\mathcal{Y}(\mathbf{u}, p)$	Rotations ϕ about an axis parallel to $\mathbf{u}$ and translations s along the same direction that are related by the pitch $s = p\phi$ as well as two independent translations in the directions of two distinct unit vectors normal to $\mathbf{u}$.	PPH, PHP, HPP, PHH, HPH, HHP, and HHH in which each H joint has a pitch of p and an axis parallel to $\mathbf{u}$, and the direction of each P joint is perpendicular to $\mathbf{u}$.

continued on next page

CONTINUED FROM PREVIOUS PAGE

Dimension	Subgroup	Description	Generator
4	$\mathcal{X}(\mathbf{u})$	Rotations ϕ about an axis parallel to $\mathbf{u}$ as well as translations in the directions of three distinct unit vectors.	PPPX, PPXP, PXPP, XPPP, PPXX, PXXX, XPXX, XXPX, XXXP, and XXXX in which each X (R of H) joint has an axis parallel to **u**.
6	$\mathcal{D}$	Three independent translations and three independent rotations.	omitted

Table C.2. Notations of displacement subgroups

Dimension	Notation in [44]	Notation in [34]	Notation in [3]	Notation in this book
0	$\{E\}$	-	$\mathcal{I}$	$\mathcal{I}$
1	$\{T(\mathbf{u})\}$	P	$\mathcal{P}(\mathbf{e})$	$\mathcal{P}(\mathbf{u})$
	$\{R(N,\mathbf{u})\}$	R	$\mathcal{R}(\mathcal{A})$	$\mathcal{R}(\mathcal{A})$
	$\{H(N,\mathbf{u},p)\}$	H	$\mathcal{H}(\mathcal{A},p)$	$\mathcal{H}(\mathcal{A},p)$
2	$\{T(P)\}$	P2	$T_2(\mathbf{u},\mathbf{v})$	$\mathcal{T}(\mathbf{n})$
	$\{C(N,\mathbf{u})\}$	C	$\mathcal{C}(\mathcal{A})$	$\mathcal{C}(\mathcal{A})$
3	$\{T\}$	P3	$\mathcal{T}_3$	$\mathcal{T}$
	$\{G(P)\}$	F	$\mathcal{F}(\mathbf{u},\mathbf{v})$	$\mathcal{F}(\mathbf{n})$
	$\{S(N)\}$	S	$\mathcal{S}(\mathcal{O})$	$\mathcal{S}(\mathcal{O})$
	$\{Y(\mathbf{w},p)\}$	HP2	$\mathcal{Y}(\mathbf{e},p)$	$\mathcal{Y}(\mathbf{u},p)$
4	$\{X(\mathbf{w})\}$	RP3	$\mathcal{X}(\mathbf{e})$	$\mathcal{X}(\mathbf{u})$
6	$\{D\}$	D	$\mathcal{D}$	$\mathcal{D}$

C.2 Operations on Displacement Subgroups

There are two types of operations on displacement subgroups: the intersection and the product. Let $\mathcal{G}_1$, $\mathcal{G}_2$, $\ldots$, $\mathcal{G}_n$ be n displacement subgroups, the intersection and product of these n displacement subgroups are denoted by $\mathcal{G}_1 \cap \mathcal{G}_2 \cap \ldots \cap \mathcal{G}_n$ and $\mathcal{G}_1 \bullet \mathcal{G}_2 \bullet \ldots \bullet \mathcal{G}_n$. Both operations are associative. The intersection is also commutative. We have

$(\mathcal{G}_1 \cap \mathcal{G}_2) \cap \mathcal{G}_3 = \mathcal{G}_1 \cap (\mathcal{G}_2 \cap \mathcal{G}_3)$

$\mathcal{G}_1 \cap \mathcal{G}_2 = \mathcal{G}_2 \cap \mathcal{G}_1$

$(\mathcal{G}_1 \bullet \mathcal{G}_2) \bullet \mathcal{G}_3 = \mathcal{G}_1 \bullet (\mathcal{G}_2 \bullet \mathcal{G}_3)$.

In Table C.3, we list the results obtained when intersecting two displacement subgroups [34]. These results are closely related to the type synthesis of PMs. For brevity, the product of two displacement subgroups, which was addressed in [3, 42], will not be discussed in this book.

Table C.3. Non-identity intersections of displacement subgroups

Subgroup 1	Subgroup 2	Intersection	Condition
$\mathcal{P}(\mathbf{u}_1)$	$\mathcal{P}(\mathbf{u}_2)$	$\mathcal{P}(\mathbf{u}_1)$	$\mathbf{u}_1 = \mathbf{u}_2$
	$\mathcal{T}(\mathbf{n}_2)$	$\mathcal{P}(\mathbf{u}_1)$	$\mathbf{u}_1 \perp \mathbf{n}_2$
	$\mathcal{C}(\mathcal{A}_2)$	$\mathcal{P}(\mathbf{u}_1)$	$\mathbf{u}_1 \parallel \mathcal{A}_2$
	$\mathcal{T}$	$\mathcal{P}(\mathbf{u}_1)$	
	$\mathcal{F}(\mathbf{n}_2)$	$\mathcal{P}(\mathbf{u}_1)$	$\mathbf{u}_1 \perp \mathbf{n}_2$
	$\mathcal{Y}(\mathbf{u}_2, p_2)$	$\mathcal{P}(\mathbf{u}_1)$	$\mathbf{u}_1 \perp \mathbf{u}_2$
	$\mathcal{X}(\mathbf{u}_2)$	$\mathcal{P}(\mathbf{u}_1)$	
	$\mathcal{D}$	$\mathcal{P}(\mathbf{u}_1)$	
$\mathcal{R}(\mathcal{A}_1)$	$\mathcal{R}(\mathcal{A}_2)$	$\mathcal{R}(\mathcal{A}_1)$	$\mathcal{A}_1 = \mathcal{A}_2$
	$\mathcal{C}(\mathcal{A}_2)$	$\mathcal{R}(\mathcal{A}_1)$	$\mathcal{A}_1 = \mathcal{A}_2$
	$\mathcal{F}(\mathbf{n}_2)$	$\mathcal{R}(\mathcal{A}_1)$	$\mathcal{A}_1 \parallel \mathbf{n}_2$
	$\mathcal{S}(\mathcal{O}_2)$	$\mathcal{R}(\mathcal{A}_1)$	$\mathcal{A}_1$ passes through $\mathcal{O}_2$
	$\mathcal{X}(\mathbf{u}_2)$	$\mathcal{R}(\mathcal{A}_1)$	$\mathcal{A}_1 \parallel \mathbf{u}_2$
	$\mathcal{D}$	$\mathcal{R}(\mathcal{A}_1)$	
$\mathcal{H}(\mathcal{A}_1, p_1)$	$\mathcal{H}(\mathcal{A}_2, p_2)$	$\mathcal{H}(\mathcal{A}_1, p_1)$	$\mathcal{A}_1 = \mathcal{A}_2$ and $p_1 = p_2$
	$\mathcal{C}(\mathcal{A}_2)$	$\mathcal{H}(\mathcal{A}_1, p_1)$	$\mathcal{A}_1 = \mathcal{A}_2$
	$\mathcal{Y}(\mathbf{u}_2, p_2)$	$\mathcal{H}(\mathcal{A}_1, p_1)$	$\mathcal{A}_1 \parallel \mathbf{u}_2$ and $p_1 = p_2$
	$\mathcal{X}(\mathbf{u}_2)$	$\mathcal{H}(\mathcal{A}_1, p_1)$	$\mathcal{A}_1 \parallel \mathbf{u}_2$
	$\mathcal{D}$	$\mathcal{H}(\mathcal{A}_1, p_1)$	
$\mathcal{T}(\mathbf{n}_1)$	$\mathcal{T}(\mathbf{n}_2)$	$\mathcal{T}(\mathbf{n}_1)$	$\mathbf{n}_1 = \mathbf{n}_2$
		$\mathcal{P}(\mathbf{u})$ where $\mathbf{u} \perp \mathbf{n}_1$ and $\mathbf{u} \perp \mathbf{n}_2$	$\mathbf{n}_1 \neq \mathbf{n}_2$
	$\mathcal{C}(\mathcal{A}_2)$	$\mathcal{P}(\mathbf{u})$ where $\mathbf{u} \parallel \mathcal{A}_2$	$\mathbf{n}_1 \perp \mathcal{A}_2$
	$\mathcal{T}$	$\mathcal{T}(\mathbf{n}_1)$	
	$\mathcal{F}(\mathbf{n}_2)$	$\mathcal{T}(\mathbf{n}_1)$	$\mathbf{n}_1 = \mathbf{n}_2$
		$\mathcal{P}(\mathbf{u})$ where $\mathbf{u} \perp \mathbf{n}_1$ and $\mathbf{u} \perp \mathbf{n}_2$	$\mathbf{n}_1 \neq \mathbf{n}_2$
	$\mathcal{Y}(\mathbf{u}_2, p_2)$	$\mathcal{T}(\mathbf{n}_1)$	$\mathbf{n}_1 = \mathbf{u}_2$
		$\mathcal{P}(\mathbf{u})$ where $\mathbf{u} \perp \mathbf{n}_1$ and $\mathbf{u} \perp \mathbf{u}_2$	$\mathbf{n}_1 \neq \mathbf{u}_2$
	$\mathcal{X}(\mathbf{u}_2)$	$\mathcal{T}(\mathbf{n}_1)$	
	$\mathcal{D}$	$\mathcal{T}(\mathbf{n}_1)$	
$\mathcal{C}(\mathcal{A}_1)$	$\mathcal{C}(\mathcal{A}_2)$	$\mathcal{C}(\mathcal{A}_1)$	$\mathcal{A}_1 = \mathcal{A}_2$
		$\mathcal{P}(\mathbf{u})$ where $\mathbf{u} \parallel \mathcal{A}_1$	$\mathcal{A}_1 \parallel \mathcal{A}_2$
	$\mathcal{T}$	$\mathcal{P}(\mathbf{u})$ where $\mathbf{u} \parallel \mathcal{A}_1$	
	$\mathcal{F}(\mathbf{n}_2)$	$\mathcal{P}(\mathbf{u})$ where $\mathbf{u} \parallel \mathcal{A}_1$	$\mathcal{A}_1 \perp \mathbf{n}_2$
		$\mathcal{R}(\mathcal{A}_1)$	$\mathcal{A}_1 \perp \mathbf{n}_2$
	$\mathcal{S}(\mathcal{O}_2)$	$\mathcal{R}(\mathcal{A}_1)$	$\mathcal{A}_1$ passes through $\mathcal{O}_2$
	$\mathcal{Y}(\mathbf{u}_2, p_2)$	$\mathcal{P}(\mathbf{u})$ where $\mathbf{u} \parallel \mathcal{A}_1$	$\mathcal{A}_1 \perp \mathbf{u}_2$
		$\mathcal{H}(\mathcal{A}_1, p_2)$	$\mathcal{A}_1 \parallel \mathbf{u}_2$

continued on next page

CONTINUED FROM PREVIOUS PAGE

Subgroup 1	Subgroup 2	Intersection	Condition
		$\mathcal{P}(\mathbf{u})$ where $\mathbf{u} \parallel \mathcal{A}_1$	$\mathcal{A}_1 \perp \mathbf{u}_2$
	$\mathcal{X}(\mathbf{u}_2)$	$\mathcal{C}(\mathcal{A}_1)$	$\mathcal{A}_1 \parallel \mathbf{u}_2$
		$\mathcal{P}(\mathbf{u})$ where $\mathbf{u} \parallel \mathcal{A}_1$	$\mathcal{A}_1 \nparallel \mathbf{u}_2$
	$\mathcal{D}$	$\mathcal{C}(\mathcal{A}_1)$	
$\mathcal{T}$	$\mathcal{T}$	$\mathcal{T}$	
	$\mathcal{F}(\mathbf{n}_2)$	$\mathcal{T}(\mathbf{n}_2)$	
	$\mathcal{Y}(\mathbf{u}_2, p_2)$	$\mathcal{T}(\mathbf{u}_2)$	
	$\mathcal{X}(\mathbf{u}_2)$	$\mathcal{T}$	
	$\mathcal{D}$	$\mathcal{T}$	
$\mathcal{F}(\mathbf{n}_1)$	$\mathcal{F}(\mathbf{n}_2)$	$\mathcal{F}(\mathbf{n}_1)$	$\mathbf{n}_1 = \mathbf{n}_2$
		$\mathcal{P}(\mathbf{u})$ where $\mathbf{u} \perp \mathbf{n}_1$ and $\mathbf{u} \perp \mathbf{n}_2$	$\mathbf{n}_1 \neq \mathbf{n}_2$
	$\mathcal{S}(\mathcal{O}_2)$	$\mathcal{R}(\mathcal{A})$ where $\mathcal{A}$ passes through $\mathcal{O}_2$ and is parallel to $\mathbf{n}_1$	
	$\mathcal{Y}(\mathbf{u}_2, p_2)$	$\mathcal{T}(\mathbf{n}_1)$	$\mathbf{n}_1 = \mathbf{u}_2$
		$\mathcal{P}(\mathbf{u})$ where $\mathbf{u} \perp \mathbf{n}_1$ and $\mathbf{u} \perp \mathbf{u}_2$	$\mathbf{n}_1 \neq \mathbf{u}_2$
	$\mathcal{X}(\mathbf{u}_2)$	$\mathcal{F}(\mathbf{n}_1)$	$\mathbf{n}_1 = \mathbf{u}_2$
		$\mathcal{T}(\mathbf{n}_1)$	$\mathbf{n}_1 \neq \mathbf{u}_2$
	$\mathcal{D}$	$\mathcal{F}(\mathbf{n}_1)$	
$\mathcal{S}(\mathcal{O}_1)$	$\mathcal{S}(\mathcal{O}_2)$	$\mathcal{S}(\mathcal{O}_1)$	$\mathcal{O}_1 = \mathcal{O}_2$
		$\mathcal{R}(\mathcal{A})$ where $\mathcal{A}$ passes through $\mathcal{O}_1$ and $\mathcal{O}_2$	$\mathcal{O}_1 \neq \mathcal{O}_2$
	$\mathcal{X}(\mathbf{u}_2)$	$\mathcal{R}(\mathcal{A})$ where $\mathcal{A}$ passes through $\mathcal{O}_1$ and is parallel to $\mathbf{u}_2$	
	$\mathcal{D}$	$\mathcal{S}(\mathcal{O}_1)$	
$\mathcal{Y}(\mathbf{u}_1, p_1)$	$\mathcal{Y}(\mathbf{u}_2, p_2)$	$\mathcal{Y}(\mathbf{u}_1, p_1)$	$\mathbf{u}_1 = \mathbf{u}_2$ and $p_1 = p_2$
		$\mathcal{T}(\mathbf{u}_1)$	$\mathbf{u}_1 = \mathbf{u}_2$ and $p_1 \neq p_2$
		$\mathcal{P}(\mathbf{u})$ where $\mathbf{u} \perp \mathbf{u}_1$ and $\mathbf{u} \perp \mathbf{u}_2$	$\mathbf{u}_1 \neq \mathbf{u}_2$
	$\mathcal{X}(\mathbf{u}_2)$	$\mathcal{Y}(\mathbf{u}_1, p_1)$	$\mathbf{u}_1 = \mathbf{u}_2$
		$\mathcal{T}(\mathbf{u}_1)$	$\mathbf{u}_1 \neq \mathbf{u}_2$
	$\mathcal{D}$	$\mathcal{Y}(\mathbf{u}_1, p_1)$	
$\mathcal{X}(\mathbf{u}_1)$	$\mathcal{X}(\mathbf{u}_2)$	$\mathcal{X}(\mathbf{u}_1)$	$\mathbf{u}_1 = \mathbf{u}_2$
		$\mathcal{T}$	$\mathbf{u}_1 \neq \mathbf{u}_2$
	$\mathcal{D}$	$\mathcal{X}(\mathbf{u}_1)$	
$\mathcal{D}$	$\mathcal{D}$	$\mathcal{D}$	

C.3 Kinematic Bond

A *kinematic bond* (mechanical bond or liaison) $\mathcal{L}(i, j)$ between two rigid bodies i and j is the set of allowed relative displacements between two rigid bodies [3, 42]. Generally, a kinematic bond can be represented by a displacement subgroup or a product of displacement subgroups. Any mechanical system, which produces the kinematic bond is called its *mechanical generator.*

Similarly to displacement subgroups, the operations on kinematic bonds also include the intersection and the product.

C.4 Steps for the Type Synthesis of Parallel Kinematic Chains

In an m-legged PM (Fig. 1.1), the motion pattern of the moving platform, which is described here using a kinematic bond $\mathcal{L}$, is the intersection of the kinematic bonds corresponding to its legs $\mathcal{L}_i$ $(i = 1, 2, \ldots, m)$, i.e.,

$$\mathcal{L} = \mathcal{L}_1 \cap \mathcal{L}_2 \cap \ldots \cap \mathcal{L}_m \tag{C.1}$$

To perform the type synthesis of m-legged PMs with a given motion pattern, one can first determine the set of m kinematic bonds of legs whose intersection is the specified kinematic bond [(C.1)]. Then, for each kinematic bond, select a generator and obtain an $m-$legged PKC.

Let us take the type synthesis of 3-legged PPP= PKCs as an example to illustrate the above procedure. The required motion pattern of a PPP= PKC is the $\mathcal{T}$ displacement subgroup. For brevity, we limit ourselves to kinematic bonds of legs which are all single displacement subgroups.

From Table C.3, we can find the sets of two displacement subgroups whose intersections are $\mathcal{T}$:

$$\mathcal{T} = \mathcal{T} \cap \mathcal{T} \tag{C.2}$$

$$\mathcal{T} = \mathcal{T} \cap \mathcal{X}(\mathbf{u}) \tag{C.3}$$

$$\mathcal{T} = \mathcal{T} \cap \mathcal{D} \tag{C.4}$$

$$\mathcal{T} = \mathcal{X}(\mathbf{u}_1) \cap \mathcal{X}(\mathbf{u}_2) \tag{C.5}$$

where $\mathbf{u}_1 \neq \mathbf{u}_2$

To find the sets of three displacement subgroups whose intersections are $\mathcal{T}$, we need to further replace one displacement subgroup in each of the above sets of two displacement subgroups with its corresponding set of two displacement subgroups using (C.2)–(C.5) or the following equations, which are also listed in Table C.3:

$$\mathcal{X}(\mathbf{u}_1) = \mathcal{X}(\mathbf{u}_1) \cap \mathcal{X}(\mathbf{u}_2) \tag{C.6}$$

where $\mathbf{u}_1 = \mathbf{u}_2$

$$\mathcal{X}(\mathbf{u}) = \mathcal{X}(\mathbf{u}) \cap \mathcal{D} \tag{C.7}$$

$$\mathcal{D} = \mathcal{D} \cap \mathcal{D} \tag{C.8}$$

Substituting (C.2)–(C.8) into (C.2)–(C.5), we have

$$\mathcal{T} = \mathcal{T} \cap \mathcal{T} \cap \mathcal{T} \tag{C.9}$$

$$\mathcal{T} = \mathcal{T} \cap \mathcal{T} \cap \mathcal{X}(\mathbf{u}) \tag{C.10}$$

$$\mathcal{T} = \mathcal{T} \cap \mathcal{T} \cap \mathcal{D} \tag{C.11}$$

$$\mathcal{T} = \mathcal{T} \cap \mathcal{X}(\mathbf{u}_1) \cap \mathcal{X}(\mathbf{u}_2) \tag{C.12}$$

$$\mathcal{T} = \mathcal{T} \cap \mathcal{X}(\mathbf{u}) \cap \mathcal{D} \tag{C.13}$$

$$\mathcal{T} = \mathcal{T} \cap \mathcal{D} \cap \mathcal{D} \tag{C.14}$$

$$\mathcal{T} = \mathcal{X}(\mathbf{u}_1) \cap \mathcal{X}(\mathbf{u}_2) \cap \mathcal{X}(\mathbf{u}_3) \tag{C.15}$$

where not all $\mathbf{u}_1$, $\mathbf{u}_2$ and $\mathbf{u}_3$ are parallel.

$$\mathcal{T} = \mathcal{X}(\mathbf{u}_1) \cap \mathcal{X}(\mathbf{u}_2) \cap \mathcal{D} \tag{C.16}$$

where $\mathbf{u}_1 \neq \mathbf{u}_2$.

From the above equations, we obtain the following nine sets of 3 displacement subgroups whose intersection is the $\mathcal{T}$ displacement subgroup: $\mathcal{T}$-$\mathcal{T}$-$\mathcal{T}$, $\mathcal{T}$-$\mathcal{T}$-$\mathcal{X}(\mathbf{u})$, $\mathcal{T}$-$\mathcal{T}$-$\mathcal{D}$, $\mathcal{T}$-$\mathcal{X}(\mathbf{u}_1)$-$\mathcal{X}(\mathbf{u}_2)$, $\mathcal{T}$-$\mathcal{X}(\mathbf{u})$-$\mathcal{D}$, $\mathcal{T}$-$\mathcal{D}$-$\mathcal{D}$, $\mathcal{X}(\mathbf{u}_1)$-$\mathcal{X}(\mathbf{u}_2)$-$\mathcal{X}(\mathbf{u}_3)$ (not all $\mathbf{u}_1$, $\mathbf{u}_2$ and $\mathbf{u}_3$ are parallel), and $\mathcal{X}(\mathbf{u}_1)$-$\mathcal{X}(\mathbf{u}_2)$-$\mathcal{D}$ ($\mathbf{u}_1 \neq \mathbf{u}_2$).

For each of the above set of three displacement subgroups, we can replace each displacement subgroup with its generators (see Table C.1) and obtain a number of PPP= PKCs.

For example, for the $\mathcal{X}(\mathbf{u}_1)$-$\mathcal{X}(\mathbf{u}_2)$-$\mathcal{X}(\mathbf{u}_3)$ set of displacement subgroups, we obtain the following PPP= PKCs:

3-PPPX, 3-PPXP, 3-PXPP, 3-XPPP, 3-PPXX, 3-PXXX, 3-XPXX, 3-XXPX, 3-XXXP, and 3-XXXX.

By selecting a set of three actuated joints for each PPP= PKCs, a large number of PPP= PMs can be obtained. The PPP= PMs include the Y-star robot and H-robot [112] based on the 3-<u>R</u>HPR PPP= PM, the Tripteron [66, 72] based on the 3-<u>P</u>RRR PPP= PM, the Orthoglide [102] based on the 3-<u>P</u>RPR PPP= PM, and the 3-<u>P</u>PP PPP= PM [35] based on the 3-<u>P</u>PP PPP= PM. In the Y-star robot, H-robot [112] and the Orthoglide [102], each unactuated P joint is replaced with a planar parallelogram. In the PPP= PM [35], the two unactuated P joints in each leg are replaced by a spatial parallelogram.

The above PMs belong to the following eight classes of the 16 classes of 3-DOF PPP= PMs proposed in Chap. 6 (see Table 5.2): 3-3-3, 3-3-2, 3-3-0, 3-2-2, 3-2-0, 3-0-0, 2-2-0 and 3-0-0.

Recently, the type synthesis of PPP= PKCs involving legs with a kinematic bond which is a product of two displacement subgroups $\mathcal{F}(\mathbf{u})$ has been presented in [90]. The PPP= PMs proposed [90] belongs to the class 1-1-1 proposed in Chap. 6.

Comparing the method based on displacement group theory with the virtual chain approach, it is found that the types of PPP= PKCs obtained using the former approach are a subset of the types obtained using the latter approach. This is due to that it is very difficult to find all the kinematic bonds of legs that include the specified motion pattern (kinematic bond) of the moving platform as well as the generators of these kinematic bonds of legs. For example, the PŔR̀R̀Ŕ leg (Table 6.1) for PPP= PMs has not been obtained using the method based on the displacement group theory. This issue is currently the focus of the research on the type synthesis based on the displacement group approach [41, 64, 90, 92] and still needs further investigation.

References

[1] S.K. Agrawal. Rate kinematics of in-parallel manipulators systems. In *Proceedings of the 1990 IEEE International Conference on Robotics and Automation*, pages 104–109, Cincinnati, USA, 1990.

[2] S.K. Agrawal. Study of an in-parallel mechanism using reciprocal screws. In *Proceedings of the Ninth World Congress on the Theory of Machines and Mechanisms*, pages 405–408, Prague, 1991.

[3] J. Angeles. The qualitative synthesis of parallel manipulators. In C.M. Gosselin and I. Ebert-Uphoff, editors, *Proceedings of the Workshop on Fundamental Issues and Future Research Directions for Parallel Mechanisms and Manipulators*, pages 160–169, Quebec, October 3–4 2002.

[4] W.T. Appleberry. Anti-rotation positioning mechanism. United States Patent, 1992.

[5] L. Birglen, C.M. Gosselin, N. Pouliot, B. Monsarrat, and Laliberté T. Shade, a new 3-DOF haptic device. *IEEE Transactions on Robotics and Automation*, 18(2):166–175, 2002.

[6] I. Bonev. Delta parallel robot — the story of success. http://www.parallemic. org/Reviews/Review002.html, 2002.

[7] I. Bonev. What is going on with parallel robots? http://www.robotics.org/public/articles/archivedetails.cfm?id=798, 2002.

[8] I. Bonev. Bibliography on parallel mechanisms. http://www.robotics.org, 2002.

[9] I. Bonev and C.M. Gosselin. Advantages of the modified Euler angles in the design and control of PKMs. In *Proceedings of the 3rd Chemnitz parallel kinematics seminar*, pages 171–188, Chemnitz, Germany, 2002.

[10] T. Brogårdh. PKM research - important issues, as seen from a product development perspective at abb. In C.M. Gosselin and I. Ebert-Uphoff, editors, *Proceedings of the Workshop on Fundamental Issues and Future Research Directions for Parallel Mechanisms and Manipulators*, pages 68–82, Quebec, October 3–4 2002.

[11] B. Brumson. Parallel kinematic robots. http://www.robotics.org/public/articles /archivedetails.cfm?id=797, 2002.

[12] H. Bruyninckx. Unified kinematics for serial, parallel and mobile robots. In J. Lenarcic and M. Husty, editors, *Advances in Robotic Kinematics: Analysis and Control*, pages 343–352. Kluwer Academic Publishers, 1998.

[13] M. Carricato and V. Parenti-Castelli. A family of 3-DOF translational parallel manipulators. In *Proceedings of the 2001 ASME Design Engineering Technical Conferences & Computers and Information in Engineering Conference*, pages DETC2001/DAC–21035, Pittsburgh, PA, 2001.

[14] M. Carricato and V. Parenti-Castelli. Position analysis of a new family of 3-DOF translational parallel robot manipulators. In *Proceedings of the 2001 ASME Design Engineering Technical Conferences & Computers and Information in Engineering Conference*, pages DETC2001/DAC–21036, Pittsburgh, PA, 2001.

[15] V.K. Chan and I. Ebert-Uphoff. Constraint singularities of parallel mechanisms. In *Proceedings of the 2001 IEEE International Conference on Robotics and Automation*, volume 2, pages 1313–1320, Seoul, Korea, 2001.

[16] R. Clavel. Device for the movement and positioning of an element in space. United States Patent, 1990.

[17] O. Company, F. Marquet, and F. Pierrot. A new high-speed 4-dof parallel robot synthesis and modeling issues. *IEEE Transactions on Robotics and Automation*, 19(3):411–420, 2003.

[18] D.J. Cox and D. Tesar. The dynamic model of a three-degree-of-freedom parallel robotic shoulder module. In *Proceedings of the Fourth International Conference on Advanced Robotics*, Columbus, 1989.

[19] W.M. Craver. Structural analysis and design of a three-degree-of-freedom robotic shoulder module. Master's thesis, The University of Texas at Austin, 1989.

[20] J.S. Dai and J. Rees Jones. A linear algebraic procedure in obtaining reciprocal screw systems. *Journal of Robotic systems*, 20(7):401–412, 2003.

[21] J.S. Dai and J. Rees Jones. Interrelationship between screw systems and corresponding reciprocal systems and applications. *Mechanism and Machine Theory*, 36(5):633–651, 2001.

[22] J.S. Dai, Z. Huang, and H. Lipkin. Screw system analysis of parallel mechanisms and applications to constraint and mobility study. In *Proceedings of the 2004 ASME Design Engineering Technical Conferences & Computers and Information in Engineering Conference*, pages DETC2004/MECH–57604, Salt Lake City, USA, 2004.

[23] J.S. Dai, Z. Huang, and H. Lipkin. Mobility of overconstrained parallel mechanisms. *ASME Journal of Mechanical Design*, 128(1):220–229, 2006.

[24] R. Di Gregorio. A new parallel wrist using only revolute pairs: the 3-RUU wrist. *Robotica*, 19(3):305–309, 2001.

[25] R. Di Gregorio. Kinematics of a new spherical parallel manipulator with three equal legs: the urc wrist. *Journal of Robotic Systems*, 18(5):213–219, 2001.

[26] R. Di Gregorio. Translational parallel manipulators: New proposals. *Journal of Robotic Systems*, 16(12):595–603, 2002.

[27] R. Di Gregorio and V. Parenti-Castelli. Mobility analysis of the 3-UPU parallel mechanism assembled for a pure translational motion. In *Proceedings of the 1999 IEEE/ASME International Conference on Advanced Intelligent Mechatronics*, pages 520–525, 1999.

[28] R. Di Gregorio and V. Parenti-Castelli. A translational 3-DOF parallel manipulator. In J. Lenarčič and M. Husty, editors, *Advances in Robot Kinematics: Analysis and Control*, pages 49–58, Boston, 1998. Kluwer Academic Publishers.

[29] C.F. Earl and J. Rooney. Some kinematic structures for robot manipulator design. *Journal of Robotic Systems*, 18(5):213–219, 1983.

[30] A.G. Erdman. *Modern Kinematics: Developments in the Last Forty Years (Chapter 3)*. John Wiley & Sons Inc, 1993.

[31] Y. Fang and L.-W. Tsai. Structure synthesis of a class of 4-DOF and 5-DOF parallel manipulators with identical limb structures. *The International Journal of Robotics Research*, 21(9):799–810, 2002.

[32] P. Fanghella, C. Galletti, and E. Giannotti. Parallel robots that change their group of motion. In Lenarčič J. and Roth B., editors, *Advances in Robot Kinematics — Mechanisms and Motion*, pages 49–56, The Netherlands, 2006. Springer.

[33] A. Frisoli, D. Checcacci, F. Salsedo, and M. Bergamasco. Synthesis by screw algebra of translating in-parallel actuated mechanisms. In J. Lenarčič and M.M. Stanišič, editors, *Advances in Robot Kinematics*, pages 433–440. Kluwer Academic Publishers, 2000.

[34] C. Galletti. Mobility analysis of single-loop kinematic chains: an algorithmic approach based on displacement groups. *Mechanism and Machine Theory*, 29(8):1187–1204, 1994.

[35] F. Gao, W. Li, X. Zhao, Z. Jin, and H. Zhao. New kinematic structures for 2-, 3-, 4-, and 5-DOF parallel manipulator designs. *Mechanism and Machine Theory*, 37(11):1395–1411, 2002.

[36] G. Gogu. Mobility criterion and overconstraints of parallel manipulators. In *Proceedings of CK 2005, International Workshop on Computational Kinematics*, pages paper 20–CK2005, Cassino, Italy, 2005.

[37] C.M. Gosselin and J. Angeles. The optimum kinematic design of a spherical three-degree-of-freedom parallel manipulator. *ASME Journal of Mechanisms, Transmission and Automation in Design*, 111(2):202–207, 1989.

[38] C.M. Gosselin and J.-F. Hamel. The agile eye: a high-performance three-degree-of-freedom camera-orienting device. In *Proceeding of the 1996 IEEE International Conference on Robotics and Automation*, pages 781–786, San Diego, 1996.

[39] C.M. Gosselin and X. Kong. Cartesian parallel manipulators. United States Patent, 2004.

[40] C.M. Gosselin, M. Tale Masouleh, V. Duchaine, P.-L. Richard, S. Foucault, and X. Kong. Parallel mechanisms of the multipteron family: Kinematic architectures and benchmarking. In *Proceedings of the 2007 IEEE International Conference on Robotics and Automation*, page (in press), Roma, Italy, 2007.

[41] J. M. Hervé. The planar-spherical kinematic bond: implementation in parallel mechanisms. http://www.parallemic.org/ Reviews/Review013.html, 2003.

[42] J.M. Hervé. The lie group of rigid body displacements, a fundamental tool for mechanisms design. *Mechanism and Machine Theory*, 34(5):719–730, 1999.

[43] J.M. Hervé. Parallel mechanisms with pseudo-planar motion generators. In J. Lenarčič and C. Galletti, editors, *On Advances in Robot Kinematics*, pages 431–440, Boston, 2004. Kluwer Academic Publishers.

[44] J.M. Hervé. Analyse structurelle des mécanismes par groupe des déplacements. *Mechanism and Machine Theory*, 13(4):437–450, 1978.

[45] J.M. Hervé. Design of parallel manipulators via the displacement group. In *Proceedings of the 9th World Congress on the Theory of Machines and Mechanisms*, pages 2079–2082, Milan, Italy, 1995.

[46] J.M. Hervé and M. Karouia. The novel 3-RUU wrist with no idle pair. In C.M. Gosselin and I. Ebert-Uphoff, editors, *Proceedings of the Workshop on Fundamental Issues and Future Research Directions for Parallel Mechanisms and Manipulators*, pages 284–285, Quebec, October 3–4 2002.

[47] J.M. Hervé and F. Sparacino. Structural synthesis of parallel robots generating spatial translation. In *Proceedings of the fifth International Conference on Advanced Robotics*, pages 808–813, Pisa, Italy, 1991.

[48] L.L. Howell. *Compliant Mechanisms.* John Wiley & Sons Inc., Boca Raton, 2001.

[49] Z. Huang and Q.J. Ge. A simple method for mobility analysis using reciprocal screws. In *Proceedings of the 2006 ASME Design Engineering Technical Conferences & Computers and Information in Engineering Conference*, pages DETC200699677, Philadelphia, USA, 2006.

[50] Z. Huang and X. Kong. Kinematic analysis on the spatial parallel mechanisms with redundant degree of freedom. *Chinese Journal of Mechanical Engineering*, 31(3):44–50, 1995.

[51] Z. Huang and Q.C. Li. Two novel symmetrical 5-DOF parallel mechanisms. *Journal of Yanshan University*, 25(4):283–286, 2001.

[55] Z. Huang and Q.C. Li. General methodology for the type synthesis of lower-mobility symmetrical parallel manipulators and several novel manipulators. *The International Journal of Robotics Research*, 21(2):131–145, 2002.

[53] Z. Huang and Q.C. Li. On the type synthesis of lower-mobility parallel manipulators. In C.M. Gosselin and I. Ebert-Uphoff, editors, *Proceedings of the Workshop on Fundamental Issues and Future Research Directions for Parallel Mechanisms and Manipulators*, pages 272–283, Quebec, October 3–4 2002.

[54] Z. Huang and Q.C. Li. Construction and kinematic properties of 3-UPU parallel mechanisms. In *Proceedings of the 2002 ASME Design Engineering Technical Conferences & Computers and Information in Engineering Conference*, pages DETC2002/MECH–34321, Montreal, Canada, 2002.

[55] Z. Huang and Q.C. Li. General methodology for the type synthesis of lower-mobility symmetrical parallel manipulators and several novel manipulators. *The International Journal of Robotics Research*, 21(2):131–145, 2002.

[56] Z. Huang and T.S. Zhao. Theory and application of selecting components of spatial mechanisms. *Chinese Journal of Mechanical Engineering (in Chinese)*, 36(10):81–85, 2000.

[57] K.H. Hunt. *Kinematic Geometry of Mechanisms.* Cambridge University Press, Cambridge, 1990.

[58] K.H. Hunt. Constant-velocity shaft couplings: a general theory. *ASME Journal of Engineering for Industry*, 95, 1973.

[59] K.H. Hunt. Structural kinematics of inparallel-actuated robot-arms. *ASME Journal of Mechanical Design*, 105(4):705–712, 1983.

[60] K.A. Jensen, Lusk C.P., and Howell L.L. An XYZ micromanipulator with three translational degrees of freedom. *Robotica*, 24(3):305–314, 2006.

[61] Q. Jin and T.-L. Yang. Position analysis for a class of novel 3-DOF translational parallel robot mechanisms. In *Proceedings of the 2001 ASME Design Engineering Technical Conferences & Computers and Information in Engineering Conference*, pages DETC2001/DAC21151, Pittsburgh, PA, 2001.

[62] Q. Jin, T.-L. Yang, A.-X. Liu, H.-P. Shen, and F.-H. Yao. Structure synthesis of a class of 5-DOF parallel robot mechanisms based on single-opened-chain units. In *Proceedings of the 2001 ASME Design Engineering Technical Conferences & Computers and Information in Engineering Conference*, pages DETC2001/DAC–21153, Pittsburgh, PA, 2001.

[63] Y. Jin, I.-M. Chen, and G. Yang. Kinematic design of a 6-DOF parallel manipulator with decoupled translation and rotation. *IEEE Transactions on Robotics*, 22(3):545–551, 2006.

[64] K. Karouia and J.M. Hervé. A three-DOF tripod for generating spherical rotation. In J. Lenarčič and M.M. Stanišič, editors, *Advances in Robot Kinematics*, pages 395–402. Kluwer Academic Publishers, 2000.

[65] D. Kim and W. K. Chung. Kinematic condition analysis of three-DOF pure translational parallel manipulators. *ASME Journal of Mechanical Design*, 125(3):323–331, 2003.

[66] H.S. Kim and L.-W. Tsai. Evaluation of a cartesian parallel manipulator. In Lenarčič J. and Thomas F., editors, *Advances in Robot Kinematics – Theory and Applications*, pages 21–28, Boston, 2002. Kluwer Academic Publishers.

[67] X. Kong. Detection of input interference in spatial linkages. *Journal of Mechanical Transmission (in Chinese)*, 23(4):23–25, 1999.

[68] X. Kong. Detection of type-II passive joints in spatial kinematic chains. *Mechanical Science and Technology (in Chinese)*, 19(1):73–74, 1999.

[69] X. Kong. *Type synthesis and kinematics of general and analytic parallel mechanisms*. PhD thesis, Université Laval, 2003.

[70] X. Kong and C.M. Gosselin. Generation of parallel manipulators with three translational degrees of freedom using screw theory. In *Proceedings of the CCToMM Symposium on Mechanisms, Machines and Mechatronics*, pages paper M3–01–012, Montreal, Canada, 2001.

[71] X. Kong and C.M. Gosselin. Type synthesis of linear translational parallel manipulators. In Lenarčič J. and Thomas F., editors, *Advances in Robot Kinematics – Theory and Applications*, pages 411–420, Boston, 2002. Kluwer Academic Publishers.

[72] X. Kong and C.M. Gosselin. Kinematics and singularity analysis of a novel type of 3-C̲RR 3-DOF translational parallel manipulator. *The International Journal of Robotics Research*, 21(9):791–798, 2002.

[73] X. Kong and C.M. Gosselin. A class of 3-DOF translational parallel manipulators with linear input-output equations. In Gosselin C.M. and Ebert-Uphoff I., editors, *Proceedings of the Workshop on Fundamental Issues and Future Research Directions for Parallel Mechanisms and Manipulators*, pages 25–32, Quebec, October 3–4 2002.

[74] X. Kong and C.M. Gosselin. Mobility analysis of parallel mechanisms based on screw theory and the concept of equivalent serial kinematic chain. In *Proceedings of the 2005 ASME Design Engineering Technical Conferences and Computers and Information in Engineering Conference*, pages DETC2005/MECH–85337, Long Beach, USA, September 24–28 2005.

[75] X. Kong and C.M. Gosselin. Type synthesis of 3-DOF spherical parallel manipulators based on screw theory. *ASME Journal of Mechanical Design*, 126(1):101–108, 2004.

[76] X. Kong and C.M. Gosselin. Type synthesis of 4-DOF SP-equivalent parallel manipulators: A virtual-chain approach. *Mechanism and Machine Theory*, 41(11):1306–1319, 2006.

[77] X. Kong and C.M. Gosselin. Type synthesis of 3-DOF PPR-equivalent parallel manipulators based on screw theory. *ASME Journal of Mechanical Design*, 127(11):1113–1121, 2005.

[78] X. Kong and C.M. Gosselin. Type synthesis of 3T1R 4-DOF parallel manipulators based on screw theory. *IEEE Transactions on Robotics and Automation*, 20(2):181–190, 2004.

[79] X. Kong and C.M. Gosselin. Type synthesis of 3-DOF translational parallel manipulators based on screw theory. *ASME Journal of Mechanical Design*, 126(1):83–92, 2004.

[80] X. Kong and C.M. Gosselin. Type synthesis of 3-DOF translational parallel manipulators based on screw theory and a virtual joint. In *Proceedings of the 15th CISM-IFToMM Symposium on Robot Design, Dynamics and Control*, pages paper rom04–06, Saint-Hubert, Montreal, Canada, 2004.

[81] X. Kong and C.M. Gosselin. Type synthesis of 3-DOF linear translational parallel manipulators. In *Proceedings of the 12th World Congress in Mechanism and Machine Science*, page (in press), Besançon, France, 2007.

[82] X. Kong and C.M. Gosselin. Type synthesis of 3-DOF UP-equivalent parallel manipulators using a virtual chain approach. In Lenarčič J. and Roth B., editors, *Advances in Robot Kinematics – Mechanisms and Motion*, pages 123–132, The Netherlands, 2006. Springer.

[83] X. Kong and C.M. Gosselin. Generation and forward displacement analysis of RPR-PR-RPR analytic planar parallel manipulators. *ASME Journal of Mechanical Design*, 124(2):294–300, 2002.

[84] X. Kong and C.M. Gosselin. Type synthesis of three-degree-of-freedom spherical parallel manipulators. *The International Journal of Robotics Research*, 23(2):237–245, 2003.

[85] X. Kong and T.-L. Yang. Detection of improper general spatial kinematic chains involving deferent types of joints. In *Proceedings of the 1998 ASME Design Engineering Technical Conferences, DE-Vol. 70*, pages 323–328, 1994.

[86] X. Kong, C.M. Gosselin, and P.-L. Richard. Type synthesis of parallel mechanisms with multiple operation modes. In *Proceedings of the 2006 ASME Design Engineering Technical Conferences & Computers and Information in Engineering Conference*, pages DETC2006–99628, Philadelphia, USA, 2006.

[87] Y. Koseki, T. Tanikawa, Koyachi N., and Arai T. Kinematic analysis of translational 3-dof micro-parallel mechanism using matrix method. In *IEEE International Conference on Intelligent Robots and Systems (IROS)*, pages 786–792, Takamatsu, Japan, 2000.

[88] V. Kumar, K.J. Waldron, G. Chrikjian, and H. Lipkin. Applications of screw system system theory and lie theory to spatial kinematics: A tutorial. Presented at the 2000 ASME Design Engineering Technical Conferences & Computers and Information in Engineering Conference, 2000.

[89] N. Labontiu. *Compliant Mechanisms: Design of Flexure Hinges.* CRC Press, Boca Raton, 2003.

[90] C.-C. Li and J.M. Herve. Translational parallel manipulators with doubly planar limbs. *Mechanism and Machine Theory*, 41(4):433–455, 2006.

[91] Q.C. Li and Z. Huang. Mobility analysis of lower-mobility parallel manipulators based on screw theory. In *Proceedings of the 2003 IEEE International Conference on Robotics & Automation*, pages 1179–1183, Taipei, Taiwan, 2003.

[92] Q.C. Li, Z. Huang, and J.M. Herve. Type synthesis of 3R2T 5-DOF parallel mechanisms using the lie group of displacements. *IEEE Transactions on Robotics and Automation*, 20(2):173–180, 2004.

[93] Y. Li and Q. Xu. Design and analysis of a new 3-DOF compliant parallel positioning platform for nanomanipulation. In *Proceedings of 5th IEEE Conference on Nanotechnology*, pages 93–98, Nagoya, Japan, 2005.

[94] X.-J. Liu and J. Wang. Some new parallel mechanisms containing the planar four-bar parallelogram. *The International Journal of Robotics Research*, 22(9):717–732, 2003.

[95] X.J. Liu, J. Wang, F. Gao, and Wang L.-P. On the analysis of a new spatial three-degrees-of-freedom parallel manipulator. *IEEE Transactions on Robotics and Automation*, 17(6):959–968, 2001.

[96] Y. Luo, T. Yang, and A. Seireg. Structural types synthesis of multi-loop spatial kinematic chains with general variable constraints. In *Proceedings of the 1998 ASME Design Engineering Technical Conferences*, pages DETC98/MECH–5895, Atlanta, USA, 1998.

[97] H. Makino and N. Furuya. SCARA robot and its family. In *Proceedings of the International Conference on Assembly Automation*, pages 433–444, Boeblingen, Germany, 1982.

[98] J.-P. Merlet. Parallel robots: Open problems. http://www-sop.inria.fr/coprin/equipe/merlet/Problemes/isrr99-html.html, 2002.

[99] J.P. Merlet. *Parallel Robots.* Kluwer Academic Publishers, Dordrecht, 2002.

[100] J.P. Merlet. An initiative for the kinematics study of parallel manipulators. In C.M. Gosselin and I. Ebert-Uphoff, editors, *Proceedings of the Workshop on Fundamental Issues and Future Research Directions for Parallel Mechanisms and Manipulators*, pages 2–9, Quebec, October 3–4 2002.

[101] P.A. Millman and J.E. Colgate. Design of a four degree-of-freedom force reflecting manipulandum with a specified force/torque workspace. In *Proceedings of the 1991 IEEE International Conference on Robotics and Automation*, volume 2, pages 1488–1493, Sacramento, USA, 1991.

[102] A. Pashkevich, Wenger P., and D. Chablat. Design strategies for the geometric synthesis of orthoglide-type mechanisms. *Mechanism and Machine Theory*, 40(8):907–930, 2004.

[103] H.H. Pham, H.C. Yeh, and I-M. Chen. Micromanipulation system design based on selective actuation mechanisms. *The International Journal of Robotics Research*, 25(1):171–186, 2006.

[104] P.-L. Richard, C.M. Gosselin, and X. Kong. Kinematic analysis and prototyping of a partially decoupled 4-DOF 3T1R parallel manipulator. In *Proceedings of the 2006 ASME Design Engineering Technical Conferences & Computers and Information in Engineering Conference*, pages DETC2006–99570, Philadelphia, USA, 2006.

[105] J.M. Rico and B. Ravani. On mobility analysis of linkages using group theory. *ASME Journal of Mechanical Design*, 125(1):70–80, 2003.

[106] J.M. Rico, L.D. Aguilera, J. Gallardo, R. Rodriguez, H. Prozco, and J.M. Barrera. A more general mobility criterion for parallel platforms. In *Proceedings of the 2004 ASME Design Engineering Technical Conferences & Computers and Information in Engineering Conference*, pages DETC2004/MECH–57064, Salt Lake City, USA, 2004.

[107] J.M. Rico, L. D. Aguilera, J. Gallardo, R. Rodriguez, H. Orozco, and J.M. Barrera. A more general mobility criterion for parallel platforms. *ASME Journal of Mechanical Design*, 128(1):207-219, 2006.

[108] J.M. Rico, J.J. Cervantes-Schez, A. Tadeo-Chez, G.I. Pez-Soto, and J. Rocha-Chavarr. A comprehensive theory of kinematic synthesis of parallel platforms. In *Proceedings of the 2006 ASME Design Engineering Technical Conferences & Computers and Information in Engineering Conference*, pages DETC2006–99070, Philadelphia, USA, 2006.

[109] L. Rolland. About parallel robots. http://www.loria.fr/~rolland/apropos_eng. html, 2002.

[110] L.H. Rolland. The manta and the kanuk: Novel 4-DOF parallel mechanisms for industrial handling. In *Proceedings of 1999 International Mechanical Engineering Congress and Exposition*, pages 831–844, Nashville, USA, 1999.

[111] S.T. Smith. *Flexures: Elements of Elastic Mechanisms.* Gordon and Breach Science Publishers, Singapore, 2000.

[112] F. Sparacino and J.M. Hervé. Synthesis of parallel manipulators using Lie-groups: Y-star and H-robot. In *Proceedings of the 1993 IEEE/Tsukuba International Workshop on Advanced Robotics*, pages 75–80, Tsukuba, Japan, 1993.

[113] R. Taylor and D. Stoianovici. Medical robotics in computer-integrated surgery. *IEEE Transactions on Robotics and Automation*, 19(5):765–781, 2003.

[114] L.-W. Tsai. The enumeration of a class of three-DOF parallel manipulators. In *Proceedings of the 10th World Congress on the Theory of Machines and Mechanisms*, pages 1121–1126, Oulu, Finland, 1999.

[115] L.-W. Tsai. *Robot analysis: The Mechanics of Serial and Parallel Manipulators.* John Wiley & Sons Inc, 1999.

[116] P. Vischer and R. Clavel. Argos: A novel 3-DOF parallel wrist mechanism. *The International Journal of Robotics Research*, 19(1):5–11, 2000.

[117] S. Vorndran. Low-inertia parallel-kinematics systems for submicron alignment and handling. http://www.parallemic.org/Reviews/Review012.html, 2002.

[118] K.J. Waldron. Hybrid overconstrained linkages. *Journal of Mechanisms*, 3:73–78, 1967.

[119] J. Wang and C.M. Gosselin. Kinematic analysis and singularity representation of spatial five-degree-of-freedom parallel mechanisms. *Journal of Robotic systems*, 14(12):851–869, 1998.

[120] Y. Wang and C.M. Gosselin. On the design of a 3-PRRR spatial parallel compliant mechanism. In *Proceedings of the 2004 ASME Design Engineering Technical Conferences and Computers and Information in Engineering Conference*, pages DETC2004–57472, Salt Lake City, USA, 2004.

[121] T. Yang, Q. Jin, A. Liu, H. Shen, and Y. Luo. Structural synthesis and classification of the 3-DOF translational parallel robot mechanisms based on the units of single-opened-chain. *Chinese Journal of Mechanical Engineering*, 38(9):31–36, 2002.

[122] T.-L. Yang. Structural analysis and number synthesis of spatial mechanisms. In *Proceedings of the 6th World Conference on the Theory of Machines and Mechanism*, pages 280–283, New Dehli, 1983.

[123] T.-L. Yang. *Topology Structure Design of Robot Mechanisms (in Chinese)*. China Machine Press, Beijing, 2004.

[124] T.-L. Yang and D.-J. Sun. A general formula of degree-of-freedom for parallel manipulators and its application. In *Proceedings of the 2006 ASME Design Engineering Technical Conferences & Computers and Information in Engineering Conference*, pages DETC200699129, 2006.

[125] T.-L. Yang and H.-F. Yao. Topological characteristics and automatic generation of structural synthesis of spatial mechanisms: Part II – automatic generation of structural types of kinematic chains. In *Proceedings of the 1992 ASME Design Technical Conference*, pages 187–190, Scottsdale, AZ, USA, 1992.

[126] T.-L. Yang, Q. Jin, A.-X. Liu, F.-H. Yao, and Y. Luo. Structure synthesis of 4-DOF (3-translation and 1-rotation) parallel robot mechanisms based on the units of single-opened-chain. In *Proceedings of the 2001 ASME Design Engineering Technical Conferences & Computers and Information in Engineering Conference*, pages DETC2001/DAC–21152, Pittsburgh, PA, 2001.

[127] T.L. Yang. Kinematic structural analysis and spatial single-loop chains. In *Proceedings of the 1986 ASME Design Engineering Technical Conferences*, pages paper 86–DET–181, Columbus, USA, 1986.

[128] B.-Y. Yi, G.B. Chung, H.Y. Na, W.K. Kim, and I.H. Suh. Design and experiment of a 3-DOF parallel micromechanism utilizing flexure hinges. *IEEE Transactions on Robotics and Automation*, 19(4):604–612, 2003.

[129] J. Yoon and J. Ryu. A new family of hybrid 4-DOF parallel mechanisms with two platforms and its application to a footpad device. *Journal of Robotic Systems*, 22(5):287–298, 2005.

[130] J.-S. Zhao, K. Zhou, and Z.-J. Feng. A theory of degrees of freedom for mechanisms. *Mechanism and Machine Theory*, 39(5):621–643, 2004.

[131] T.S. Zhao and Z. Huang. A novel three DOF translational platform mechanism and its kinematics. In *Proceedings of the 2000 ASME Design Engineering Technical Conferences & Computers and Information in Engineering Conference*, pages DETC2000/MECH–14101, Baltimore, USA, 2000.

[132] T.S. Zhao, J.S. Dai, and Z. Huang. Geometric synthesis of spatial parallel manipulators with fewer than six degrees of freedom. *Journal of Mechanical Engineering Science, Proc, IMechE*, 216:1175–1186, 2002.

[133] D. Zlatanov, I. Bonev, and C.M. Gosselin. Constraint singularities of parallel mechanisms. In *Proceedings of the 2002 IEEE International Conference on Robotics and Automation*, volume 1, pages 496–502, Washington DC, USA, 2002.

[134] D. Zlatonov and C.M. Gosselin. A new parallel architecture with four degrees of freedom. In *Proceedings of the 2nd Workshop on Computational Kinematics*, pages 57–66, Seoul, Korea, 2001.

[135] M. Zoppi, D. Zlatanov, and C.M. Gosselin. Analytical kinematics models and special geometries of a class of 4-DOF parallel mechanisms. *IEEE Transactions on Robotics*, 21(6):1046–1055, 2005.

Index

Springer Tracts in Advanced Robotics

Edited by B. Siciliano, O. Khatib and F. Groen

Vol. 33: Kong, X.; Gosselin, C.
Type Synthesis of Parallel
Mechanisms
272 p. 2007 [978-3-540-71989-2]

Vol. 32: Milutinović, D.; Lima, P.
Cells and Robots: Modeling and Control of
Large-Size Agent Populations
130 p. 2007 [978-3-540-71981-6]

Vol. 31: Ferre, M.; Buss, M.; Aracil, R.;
Melchiorri, C.; Balaguer C. (Eds.)
Advances in Telerobotics
500 p. 2007 [978-3-540-71363-0]

Vol. 30: Brugali, D. (Ed.)
Software Engineering for Experimental Robotics
490 p. 2007 [978-3-540-68949-2]

Vol. 29: Secchi, C.; Stramigioli, S.; Fantuzzi, C.
Control of Interactive Robotic Interfaces – A
Port-Hamiltonian Approach
225 p. 2007 [978-3-540-49712-7]

Vol. 28: Thrun, S.; Brooks, R.; Durrant-Whyte, H.
(Eds.)
Robotics Research – Results of the 12th
International Symposium ISRR
602 p. 2007 [978-3-540-48110-2]

Vol. 27: Montemerlo, M.; Thrun, S.
FastSLAM – A Scalable Method for the
Simultaneous Localization and Mapping
Problem in Robotics
120 p. 2007 [978-3-540-46399-3]

Vol. 26: Taylor, G.; Kleeman, L.
Visual Perception and Robotic Manipulation – 3D
Object Recognition, Tracking and Hand-Eye
Coordination
218 p. 2007 [978-3-540-33454-5]

Vol. 25: Corke, P.; Sukkarieh, S. (Eds.)
Field and Service Robotics – Results of the 5th
International Conference
580 p. 2006 [978-3-540-33452-1]

Vol. 24: Yuta, S.; Asama, H.; Thrun, S.;
Prassler, E.; Tsubouchi, T. (Eds.)
Field and Service Robotics – Recent Advances in
Research and Applications
550 p. 2006 [978-3-540-32801-8]

Vol. 23: Andrade-Cetto, J,; Sanfeliu, A.
Environment Learning for Indoor Mobile Robots
– A Stochastic State Estimation Approach
to Simultaneous Localization and Map Building
130 p. 2006 [978-3-540-32795-0]

Vol. 22: Christensen, H.I. (Ed.)
European Robotics Symposium 2006
209 p. 2006 [978-3-540-32688-5]

Vol. 21: Ang Jr., H.; Khatib, O. (Eds.)
Experimental Robotics IX – The 9th International
Symposium on Experimental Robotics
618 p. 2006 [978-3-540-28816-9]

Vol. 20: Xu, Y.; Ou, Y.
Control of Single Wheel Robots
188 p. 2005 [978-3-540-28184-9]

Vol. 19: Lefebvre, T.; Bruyninckx, H.;
De Schutter, J. Nonlinear Kalman Filtering
for Force-Controlled Robot Tasks
280 p. 2005 [978-3-540-28023-1]

Vol. 18: Barbagli, F.; Prattichizzo, D.;
Salisbury, K. (Eds.)
Multi-point Interaction with Real
and Virtual Objects
281 p. 2005 [978-3-540-26036-3]

Vol. 17: Erdmann, M.; Hsu, D.; Overmars, M.;
van der Stappen, F.A (Eds.)
Algorithmic Foundations of Robotics VI
472 p. 2005 [978-3-540-25728-8]

Vol. 16: Cuesta, F.; Ollero, A.
Intelligent Mobile Robot Navigation
224 p. 2005 [978-3-540-23956-7]

Vol. 15: Dario, P.; Chatila R. (Eds.)
Robotics Research – The Eleventh
International Symposium
595 p. 2005 [978-3-540-23214-8]

Vol. 14: Prassler, E.; Lawitzky, G.; Stopp, A.;
Grunwald, G.; Hägele, M.; Dillmann, R.;
Iossifidis. I. (Eds.)
Advances in Human-Robot Interaction
414 p. 2005 [978-3-540-23211-7]

Vol. 13: Chung, W.
Nonholonomic Manipulators
115 p. 2004 [978-3-540-22108-1]

Vol. 12: Iagnemma K.; Dubowsky, S.
Mobile Robots in Rough Terrain –
Estimation, Motion Planning, and Control
with Application to Planetary Rovers
123 p. 2004 [978-3-540-21968-2]

Vol. 11: Kim, J.-H.; Kim, D.-H.; Kim, Y.-J.;
Seow, K.-T.
Soccer Robotics
353 p. 2004 [978-3-540-21859-3]

Vol. 10: Siciliano, B.; De Luca, A.; Melchiorri, C.;
Casalino, G. (Eds.)
Advances in Control of Articulated and
Mobile Robots
259 p. 2004 [978-3-540-20783-2]

Vol. 9: Yamane, K.
Simulating and Generating Motions
of Human Figures
176 p. 2004 [978-3-540-20317-9]

Vol. 8: Baeten, J.; De Schutter, J.
Integrated Visual Servoing and Force
Control – The Task Frame Approach
198 p. 2004 [978-3-540-40475-0]

Vol. 7: Boissonnat, J.-D.; Burdick, J.;
Goldberg, K.; Hutchinson, S. (Eds.)
Algorithmic Foundations of Robotics V
577 p. 2004 [978-3-540-40476-7]

Vol. 6: Jarvis, R.A.; Zelinsky, A. (Eds.)
Robotics Research – The Tenth
International Symposium
580 p. 2003 [978-3-540-00550-6]

Vol. 5: Siciliano, B.; Dario, P. (Eds.)
Experimental Robotics VIII – Proceedings
of the 8th International Symposium ISER02
685 p. 2003 [978-3-540-00305-2]

Vol. 4: Bicchi, A.; Christensen, H.I.;
Prattichizzo, D. (Eds.)
Control Problems in Robotics
296 p. 2003 [978-3-540-00251-2]

Vol. 3: Natale, C.
Interaction Control of Robot Manipulators –
Six-degrees-of-freedom Tasks
120 p. 2003 [978-3-540-00159-1]

Vol. 2: Antonelli, G.
Underwater Robots – Motion and Force Control
of Vehicle-Manipulator Systems
268 p. 2006 [978-3-540-31752-4]

Vol. 1: Caccavale, F.; Villani, L. (Eds.)
Fault Diagnosis and Fault Tolerance for
Mechatronic Systems – Recent Advances
191 p. 2003 [978-3-540-44159-5]

Printing: Mercedes-Druck, Berlin
Binding: Stein + Lehmann, Berlin